The Behavior of Communicating

The Behavior

HARVARD UNIVERSITY PRESS

CAMBRIDGE, MASSACHUSETTS

AND LONDON, ENGLAND   1977

# of Communicating

An Ethological Approach

W. JOHN SMITH

Library of Congress Cataloging in Publication Data

Smith, William John.
     The behavior of communicating.

     Includes bibliographical references and index.
     1. Animal communication. I. Title.
QL776.S64   591.5′9   76-43021
ISBN 0-674-06465-8

To Ernst Mayr

Acknowledgments

Fourteen months spent in research at the Smithsonian Tropical Research Institute during 1970 and 1971 gave me an opportunity to think about the conceptual structure required for a comprehensive approach to the ethological study of communication, and to consider exploring this in a book. While the concern grew from the directions my research has taken since 1959, work on the book did not begin until the autumn of 1971. The impetus then was an advanced undergraduate course in animal communication that I started to teach at the University of Pennsylvania. This teaching has since provided a yearly need to review the diverse explorations under way on the behavior of communicating, to investigate the relations between different approaches, and to devise a framework within which the aims of research on different aspects of the topic could be integrated.

Many people have helped me learn to study behavior. I am particularly grateful to Ernst Mayr and Martin Moynihan, who first encouraged and guided my evolutionary and ethological interests; to my parents, who earlier encouraged a love of nature and study; and to those who taught and advised me at all stages, especially Bill Baldwin, Peter Millman, Bert Nesbitt, George Wallace, Ed Wilson, Don Griffin, Robert MacArthur, and Paul Rozin. Special thanks are offered to those who have read and commented on all or large parts of drafts of the manuscript: David Ainley, Malte Andersson, Colin Beer, Hugh Dingle, Erving Goffman, Ilan Golani, Don Griffin, Robert Hinde, Michi Ishida, Adam Kendon, Amy Kramer, Ernst Mayr, Ed Miller, Martin Moynihan, Michael Simpson, and Ed Wilson. Similar gratitude is due those who advised on more limited sections: Norm Adler, Mike Baker, Marc Bekoff, Penny Bernstein, Dick Estes, Frank Gill, Henry and Lila Gleitman, Jack Hailman, Gail Haslett, Bill Labov, Peter Marler, Gene Morton, Roger Payne, Stan

Rand, Bob Ricklefs, Burt Rosner, Steve Rothstein, Paul Rozin, Neal Smith, Sharon Smith, Sue Smith, Chuck Snowdon, Margie Sonntag, Bob Storer, William A. Wilson, Jr., Gail Zivin, and Naida Zucker.

Figure 1.2 was devised and drawn by Robin Yeaton; the other figures were prepared by Laurie Sand and John and Bette Wollsey, with financial assistance from the Institute of Neurological Sciences. The authors and publishers who have permitted figures to be reproduced are appreciatively thanked.

Institutions have been important throughout in providing colleagues, resources, and bases for my work. My main appointments are to the University of Pennsylvania, where I am a member of the Departments of Biology and Psychology and interact frequently with colleagues and students in other departments, institutes, graduate and professional schools; and to the Smithsonian Tropical Research Institute, where I have additional colleagues and do much of my field work. I also have formal affiliations with the Philadelphia Zoological Society and the Academy of Natural Sciences of Philadelphia, and began the research as a graduate and then a postdoctoral student at the Biological Laboratories and the Museum of Comparative Zoology of Harvard University. Many other institutions with which I have no formal ties have been extremely helpful over the years and have been acknowledged in my published reports of research. Part of the time spent in writing was aided by the National Science Foundation, which has been the major source of support for my research. The Wenner-Gren Foundation for Anthropological Research and Tom Sebeok deserve special thanks for bringing me together on several occasions with behavioral scientists from diverse disciplines and thereby greatly expanding my perspective.

I could not have managed the task of preparing this book without the continual understanding and support of Sue and Ben, who have been exceedingly patient.

Contents

The Behavior of Communicating

1

Introduction

In the silence of a night, a wolf stands on a ridge and howls. Immediately, a wolf in an adjacent valley turns and trots toward it. In the darkness of a hive, a honey bee pauses between foraging excursions to dance an intricate pattern on the surface of a comb. Touching the first with its antennae a second forager follows briefly, then flies alone from the hive to the same region where the first had collected from flowers rich in nectar. A cornered cat waves its bushed-out tail and snarls; its opponent hesitates, then pulls back from it. A bird sings; its mate flies to join it. A waking baby cries; its mother comes and lifts it from its crib. A person speaks; another listens and waits to speak in turn.

In each of these events one participant has communicated with another: it makes information available to be shared. Where the participants are apart or in the darkness this information comes through the use of special cries or dances. Even when they can see one another, special movements, postures, and sounds are important to their communicating, although the participants also provide each other with relevant information in other ways. For instance, the cat's opponent can see that the animal with which it is fighting is indeed a cat, and can also see that it is cornered and crouching, perhaps ready to leap. What it cannot see as readily is what the cornered cat is likely to do if it leaps—it may not know whether the other cat is prepared to defend itself or attempt to escape. The waving, bushed-out tail and the snarls tell it: the cornered individual is still forgoing the offensive, but will attack if pushed further. Information that was essentially private, about the actions the cornered cat is preparing to undertake, is thus made public.

The information received in each event is useful to its recipient. It enables the recipient to select a course of action that is appropriate in dealing with the other individual's behavior or probable behavior. In

fact, both individuals gain from sharing information: the wolves are able to reform their pack after separating during a hunt, both bees contribute to their hive's winter stores by gathering nectar, each cat avoids being injured by the other, the birds reaffirm their pair bond, the baby's needs are met and the mother further aids the carrier of her genes, and the two people have an orderly conversation, perhaps with a fruitful exchange of ideas.

If animals can gain by sharing information, we should not be surprised that they have specializations that facilitate the process by providing access to information which might otherwise be less readily available. Communicating is an important feature of their lives, because each individual must continuously base its behavior on available information (*information* in fact can be defined as that which permits choices to be made). The more informed it is, the more likely to be pertinent are its choices. The more informed its associates are, at any rate within certain spheres, the more likely is their behavior to be useful to it.

This, much oversimplified, is an ethologist's view of the behavior of communicating. It is behavior that enables the sharing of information between interacting individuals as they respond to each other. The aim of this book is to develop, from this viewpoint, an analysis of communicating: the kinds of behavioral specializations that are involved and the kinds of information they make available (that is, information about what); the other kinds of sources of information that are used in addition to these special acts; the ways recipients of the information respond; and the functions of these responses. The material is discussed within a theoretical framework designed for analyzing communicating as a component of interactional behavior. Other analytic frameworks are considered and their contributions to an understanding of communicating are reviewed, along with some nonbiological perspectives. The book represents an attempt to synthesize the contributions of each field and approach, and to map promising directions for further research.

The Naturalistic, Evolutionary Perspective of Ethology

Ethology is a field of evolutionary biology. Of the various biological disciplines that are involved with behavior, ethology has been most concerned with the study of communicating. But what is distinctive about ethology? What sets it apart from, say, comparative physiology and other related fields? What point of view can ethology offer to the study of a kind of behavior that is also of fundamental interest to psychologists, sociolinguists, anthropologists, and others?

The name *ethology* is based upon the Greek *ethos,* which has a variety of connotations depending on the form of its initial letter. It can

refer to a character, characteristic disposition, or habit that is a distinguishing feature of some individual or group, and it was in essentially this sense that it initially entered the biological domain. This was in 1859, when Isidore Geoffroy Saint-Hilaire chose ethology to refer to the study of animals as living beings in their natural environments (see Jaynes 1969), a field now divided between the concern of ecologists with relationships and ethologists with behavior. Ethology remains characterized by its naturalistic emphasis; in one of the most recent attempts to distinguish it from other behavioral sciences, Beer (1975a) has remarked that "the best ethological work . . . has a quality that is emergent from a combination of profound curiosity about, refined perception of, and exquisite feeling for the patterns of behaviour shown by different kinds of animals in nature."

Ethology was not among the predominating fields of behavioral science at first. Cuvier's controlled, laboratory approach was ascendant—the approach that came to characterize behavior as studied experimentally in comparative psychology. Following Cuvier, few biologists employed ethological procedures to study behavior. Yet these few did well. The field survived and developed, and was dramatically acclaimed in 1973 with the award of Nobel prizes to Lorenz, von Frisch, and Tinbergen. Through these men and a handful of predecessors such as Darwin, Julian Huxley, and Heinroth, ethology has become the field in which the study of behavior is founded on observation of animals as they deal with natural or nearly natural environments.

To Lorenz (1950), naturalistic observation is the "inviolate" requirement that distinguishes ethology from all other behavioral sciences. It is inviolate because the opportunities and exigencies of the environment of each species lead to the evolution of behavior that is adaptive in the Darwinian sense. These opportunities provide the forces of natural selection that favor animals with propensities for appropriate behavior for survival and breeding in natural circumstances. Individuals who have survived and bred more successfully than have those whose behavior left them less able, or who have somehow helped their close relatives to do this, have contributed disproportionately to the genetics of subsequent generations. (In evolutionary terms, the former have been more "fit"; see Mayr 1963 for the fundamental concepts and terms of modern evolutionary theory, and W. D. Hamilton 1964 for the further concept of "inclusive fitness" that covers effects on relatives.) As a result, the members of a species have come to share typical behavioral responses to recurrent situations. However, if an animal is removed from the circumstances in which it evolved, its typical behavior may no longer be appropriate. The animal, in fact, will likely select only a portion of its activities and try to solve its new problems as best it can with these

limited tools. Only within the kind of environment in which it evolved can the full range of a species' behavior be expected, and its adaptiveness revealed.

As many sources of variation as possible are customarily eliminated or held nearly constant in laboratory experiments, which makes it easier to test preconceived postulates about single phenomena. The laboratory environment is usually extremely impoverished: an empty white box, for instance, is an "open field" to a comparative psychologist. To a rodent or other animal it must be a severely barren environment in which it is constrained by walls, exposed, and vulnerable; unlike the richer and less orderly environments of nature, there is no place it can run to seek cover. Controlling the variability of test situations may even remove environmental properties whose very existence is not expected, but which are essential to the animals' natural responsiveness. Loss of such properties, even the loss of variation itself, may yield results that are irrelevant, misleading, or uninterpretable.

Nonetheless, controlled experiments can be very valuable to ethologists. For instance, they are used in identifying specific features of a situation that may affect its outcome, and in indicating how these features may contribute. Fortunately, experiments can often be designed without savage improverishment of everything natural and can mimic events that do occur in nature. Such experiments deviate minimally from conditions that have figured in the evolution of the system being studied.

Having a naturalist's perspective helps an ethologist to examine questions that can be safely isolated for testing, especially after enough is known about the behavior of a species to identify problems sufficiently significant to merit the narrow focus necessary in experimental manipulation. Yet experimental design remains a problem and requires a flexible approach. For instance, Colin Beer (1973) tells of two experiments in which he tried to elicit evidence from young laughing gulls *Larus atricilla* that they could recognize the voices of their individual parents and also that they could respond appropriately to the species' alarm call. Because young laughing gulls hide alone in the grass when not with their parents, he felt that they could reasonably be tested in a spacious box within a laboratory building. He brought naturally raised gull chicks from the marshes into his nearby laboratory, let them settle down in the box, and played sample tape recordings of what appeared to be the relevant calls of wild adults to them. Although he was correct in suspecting that the young were capable of both individual recognition and alarm responses, the chicks refused to confirm either. Thwarted, he resorted to playing a long and unedited field recording made in the gull colony. Both responses appeared, dramatically. He had missed seeing individual recognition of voice because he had not been playing the

appropriate vocalization, and he had missed eliciting fleeing and hiding because these responses are given only to the clamor of the colony's alarm chorus, not to isolated alarm calls. As he remarked: "my initial assumptions had led me to design an experiment that was too removed from nature to retain the phenomenon I was seeking to study." His naturalist's perspective, however, led him to believe that the responses could be elicited and enabled him to find the appropriate design.

Because even truly natural circumstances vary, they can produce test situations having many of the characteristics of experimental manipulation. In natural events that provide tests, most apparently crucial features differ only slightly from other events, while the feature under study changes in a single but important way. The experienced observer seeks recurrent natural variations, in either behavior or in the circumstances in which responses recur, that can be compared and used to test hypotheses. Tinbergen (1958, 1959a) has called this the use of "natural experiments" (see chapter 10 for examples). In fact, ethologists are not alone among behavioral scientists in exploiting this technique. In linguistics and psychiatry using natural events is known as the use of "decisional situations" (Condon and Ogston 1967). Some psychologists have also recognized the advantages of combining "representative design" of research, in which the probabilistic nature of environmental circumstances is retained or accurately reflected, and "systematic design" in which much more experimental control is imposed (Petrinovich 1973).

Ethologists have an advantage over scientists who study only human behavior: they can very often test hypotheses by comparing the behavior of different species. The use of interspecies comparisons is part of ethology's heritage as a branch of evolutionary biology: they were employed in comparative physiology, comparative anatomy, and systematics before ethology began, and have continued to be useful. The species chosen for comparison are usually alike in many respects, but differ in ways that have influenced the nature of their adaptations. For instance, one of two closely related, monogamous, nonflocking species may live in a more open habitat than does the other. Its members are thus more frequently or continuously visible to their mates than in the other species, and also more exposed to predators that hunt by sight. Both these features should affect their display behavior in predictable ways. Effectively, each such comparison makes use of experiments in which ecology established the constants and variables, phylogeny the relationships, and natural selection the impetus for divergence, convergence, parallelism, or other changes. The biologist can see the results of experiments that have proceeded over large numbers of generations. The process, of course, can only be inferred.

As evolutionary biologists, ethologists view most naturally occurring

behavior as *adaptive*. That is, they contend that the behavior with which animals respond to their environments functions to sustain and protect the individuals, and to ensure their ability to contribute to future generations. The statement "birds migrate to warm climates in order to escape the low temperatures and food shortages of winter" is teleological because it assigns a goal to the behavior of migrating (Mayr 1974a). Mayr shows that the statement is also legitimate in science, objective, and useful in indicating causal relationships. His treatment greatly clarifies the virtues inherent in teleological language, as well as the hoary problems associated with it. He further argues that because "the occurrence of goal-directed processes is perhaps the most characteristic feature of the world of living organisms" we should accept teleological descriptions of these processes. He suggests the term "teleonomic" (as distinct from the automatic "teleomatic" processes that act on inanimate objects), and defines a teleonomic process or behavior pattern as "one which owes its goal-directedness to the operation of a program." A "program," in turn, is "coded or prearranged information that controls a process (or behavior) leading it toward a given end." The programs with which ethologists have been concerned are products of natural selection; that is, animals with behavioral propensities that lead to appropriate goals have reproduced more successfully than those without. Evolution produces both "closed" programs that supply all the necessary information in an animal's genotype, and "open" programs that can incorporate additional information as an individual gains experience.

Teleonomic statements, implying that behavior is goal-directed and functional, are made throughout this book—and throughout the ethological and other evolutionary literature. Readers from other disciplines who may harbor residual suspicions that opportunities for causal explanation are being circumvented (when in fact they are being clarified by suggesting why particular patterns have evolved) would do well to read Mayr's perceptive analysis of teleology.

Display Behavior: An Ethological Concept Central to the Study of Communicating

Extensive observation is the backbone of ethological research, and the concepts of ethology have been developed in attempts to account for the regular patterns of behavior that emerge from prolonged watching. Some patterns are remarkably regular: ethologists realized quite soon that major blocks of behavior recur in nearly unchanged form from event to event. These blocks quickly assumed central positions as a distinctive class of activities in the conceptual framework of ethology.

Such behavior is characterized by acts that, once elicited, tend to be

carried out in nearly invariant sequences that have been called "fixed action patterns." In fact, the word *fixed* is an overstatement, but the phenomenon of behavior organized into relatively stereotyped packages is nonetheless striking. Stereotyped behavioral patterns have been found to be crucial in the organization of many kinds of activities, ranging from methods of searching for food, to tactics for responding to predators, to the repetitive sequences of twists, turns, pullings, and pushings that birds use in building their species-specific nests. The adaptiveness of many such patterns is obvious in the results they consistently produce.

Yet the most conspicuous of all fixed action patterns often produce no obvious results, at least by acting directly on things in the environments of the individuals who perform them. These unusually conspicuous, even bizarre, stereotyped patterns are not actions in which food is physically uncovered, prey seized or torn asunder, or opponents or predators pushed away or beaten back. Instead they occur as animals interact, as they court, dispute over territories, or defer to or dominate one another, for example. The performing individuals are usually not dealing with other participants by physical force when they do these stereotyped acts; if one is performed and the performer's opponent withdraws it is not because he was pushed away—he elects to leave under his own steam.

Ethologists realized from such observations that these acts must achieve their ends indirectly by functioning as stimuli capable of influencing the behavior of other individuals. That is, these acts are specialized as signals. They are among the most consistently useful sources of information that are available to individuals participating in interactions.

These signal acts are called "displays," a term originating in Huxley's work (1914) on the postures, "dances," and other signaling movements of great crested grebes (see figure 1.1). Following Huxley, the term was initially applied to visible acts, but it was clear that the class of signaling behavior comprises many other kinds of specializations: displays can also be audible, tactile, chemical releases, or even electrical discharge patterns. Moynihan (1956, 1960) finally provided the first general definition of displays as any behavior specially adapted "in physical form or frequency to subserve social signal functions."

Ethologists have found displays to be very widespread. Darwin saw them in dogs, humans, and many other species; Craig in pigeons; Heinroth and Lorenz in ducks and geese; Huxley in grebes, loons, and shorebirds; von Frisch in honeybees; and Tinbergen in fish and gulls. Some species of at least arthropods, cephalopods, and vertebrates have elaborate repertoires of displays, usually with several acts suited to be received by each of two more sensory modalities.

We are casually familiar with many display acts: songs of birds,

Figure 1.1. Courtship of the great crested grebe *Podiceps cristatus*. These are some of the displays seen by Huxley in his pioneering studies of the social behavior of great crested grebes. He found the displays shown here to be performed in a sequence in which a female would approach a male, calling. The male would finally look toward her and then, when he dived, she would instantly adopt the striking "cat attitude" pose shown in *A*, turning from side to side. The male would emerge about a meter from her, appearing to grow vertically out of the water until standing erect in the "ghostly penguin" pose shown in *B*, and revolving until at his fullest height he faced the female. He would then sink slowly onto the surface as the female put down her wings and raised her neck, and the two would face and shake their heads as shown in *C*. (From Huxley 1914.)

howls of dogs and wolves, droning of cicadas, the back-arched posturing and erection of hairs of a Halloween cat, or the urinating of a dog on the same series of fire hydrants, poles, and trees in a daily circuit of its neighborhood. Far more display acts escape our casual, speech-oriented attention. Yet our human preoccupation with language should not so narrow our view that we fail to appreciate the specialized behavior patterns with which other species communicate. Nor should their acts appear foreign to us. Many of our so-called nonverbal modes of communicating are fundamentally similar in form and origin and are very important in supplementing, enriching, and in some circumstances overshadowing or replacing our complex linguistic communication. In fact, although our speech behavior is so intricately specialized and productive as to be in a class by itself, specializations of facial expression, bodily movement, and vocalization undoubtedly preceded it in the course of

human evolution, have coexisted with it throughout its evolution, and by accomplishing a part of the social task of sharing information must have influenced the evolutionary paths open to it. We know relatively little about human communicative behavior other than speech, but crying, blushing, smiling, and shaking one's fist are all specialized acts in common human use. In many important respects all are similar to bird calls, the color changes of fish, the facial movements of monkeys and apes, and the leg-waving of salticid spiders. Even though much remains to be learned about how the displays of nonhuman species contribute to communication, the techniques developed to study them help us to understand better how we ourselves communicate.

Ethologists investigate displays within the evolutionary perspective. The rationale with which they account for the adaptiveness of display behavior as a class determines their general expectations of the characteristics of display behavior. This rationale is basic to the distinctiveness of ethology among the behavioral sciences, and must be understood if ethology's contributions are to be evaluated. It can be summarized briefly.

Animals need to be informed. Obtaining information is crucial to their ability to respond actively to their surroundings. In the great sweep of evolutionary time, natural selection has repeatedly had the effect of enlarging the capacities of animals to tap more sources of information and different kinds of sources. Individual animals are among the most pertinent sources of information available to each other. For example, if an individual can be seen to be in any way strange, this may indicate that it is a source of danger. Or sudden indications that a customary associate is startled, alert, and perhaps ready to flee may correlate with the coming of a predator that as yet only it has detected. Selection favors individuals able to obtain and use such information from each other.

This is only part of the story, however. Individuals can benefit not just from obtaining information from each other, but also from making it available to each other. There are evolutionary pressures upon animals to become more useful as sources of information (W. John Smith 1969a): it is adaptive to provide information that may lead a recipient to behave appropriately to the communicator's needs. The information might, for instance, reduce the ambiguity in a developing social interaction, perhaps calming a recipient and making it more receptive to a mating approach. Or, in events in which immediate responses are crucial, such information might lead the recipient to escape from danger it had not perceived and in surviving remain available to provide a similar service for the communicator in the future. It might lead the recipient to escape from the communicator, or at least to threaten rather than to attack it, in either case forestalling or avoiding a fight in which either in-

dividual might be injured. The range of recipient behavior that is appropriate to a communicator's needs in different circumstances is enormous.

The range of responses that can evolve has one important limitation, at least in the majority of intraspecific communication: evolution favors responses that are appropriate from the standpoint of the recipient as well. This constraint is also in effect for some kinds of interspecific interactions. In other cases, information may be concealed or provided in displays that are misleading, restricting the advantage to only one individual—to an individual attempting to control access to limited resources in intraspecific competition, or attempting to take another's life in interactions between predators and their prey, for instance. This kind of evolutionary product is unstable, a new advantage for one side creating a selection pressure for counteradaptation by the other.

An evolutionary perspective sometimes leads to a preoccupation with phenomena that are under relatively strict genetic control. Such a preoccupation has influenced the ways in which display behavior has traditionally been viewed by ethologists. For instance, displays have been described as behavior specialized for informative functions through a process of genetic evolution called "ritualization." However, ethologists studying displays are usually concerned less with developing evolutionary theory than with gaining a comprehensive understanding of the behavior, and they readily accept the effects of learning on the development and performance of displays. This understanding does not require them to reject evolutionary models because the propensity to learn is also a product of natural selection. Instead, ethologists must forego the narrower preoccupation with extreme genetic control and broaden their basic concepts to encompass effects in which individual experience plays a role. Thus a concept like ritualization that deals only with the effects of genetic evolution must be supplemented by the admission of processes of specialization based on individual experience. In this book the term "conventionalization" is adopted as a name for the latter, and conventionalization and ritualization are grouped as parallel processes of "formalization" (see chapter 11, in which the sources and evolution of displays are considered).

More serious consequences for the study of communicating have resulted from an ethological preoccupation with individual animals. This is consistent with evolutionary concerns, because natural selection works on the reproductive success of individuals, and displays can be viewed as adaptive properties of individuals (including their "inclusive fitness") comparable to webbed feet or opposable thumbs. Yet displays are adaptive not just for the individuals who perform them, but also for those who respond—and, in fact, adaptive only to the extent that responses are elicited. An event in which communication occurs through the agency of displays involves the behavior of at least two individuals, a

communicator and one or more recipients, usually to their mutual advantage. Consideration of mutual advantages violates no evolutionary principles, even though it complicates their application.

In analyzing displays as individual properties, ethologists have usually concentrated on the individuals who display and the "causes" and functions of their displays. Functions have usually not been studied but inferred; for instance, birds' songs are inferred from the circumstances in which they are used to attract mates and repel competitors (see chapter 10). Causes have been analyzed in terms of the internal states and the stimuli that lead an individual to perform a display, which emphasizes the mechanisms controlling the behavior of an individual. Displays have thus been studied in much the same way as have fixed action patterns of kinds not fundamentally involved in interacting.

An interaction, however, is only partly under the control of any individual participant. Even if one individual dominates an interaction, the others must agree to submit; they thus play their parts in determining its course and its outcome. In the behavioral mechanisms by which an interaction is mutually controlled, individuals are thus dependent on one another. These mechanisms cannot be studied by observing individual animals alone.

Knowing an animal's physiological state and the stimuli to which it was responding when it displayed is not the same as knowing what its display contributed to an interaction. To know what advantages came to that animal after it performed the display (it obtained a mate, or it established a boundary with a territorial opponent) is to know something of the functions that can follow, but not how these functions were achieved: that is, not how the interaction was managed to produce such outcomes. What is missing in the traditional emphasis on individuals is the mechanics of their interactions—the interactions to which displays contribute, and in which social functions are achieved. Yet few interactions are so simple that one participant displays and the other produces a definitive response, certainly not most interactions by which bonds or other persisting social relationships are established.

Departing from the tradition emphasizing the mechanisms controlling the behavior of individuals, an interactional perspective is adopted throughout this book. This perspective is fully compatible with that focused on the control of individual behavior (as discussed in chapter 8) and no less ethological in its naturalistic, evolutionary biases.

An Interactional Perspective

The behavior of communicating involves sharing. It differs from eating, sleeping, migrating, or avoiding a predator, all of which can be done alone. The performance of a display by a solitary animal is wasted

energy if no other individual responds. Only when there is response (even a response that is not immediately evident) by one individual to another does display behavior become functional in communicating. And it has consequences for both the communicator and the responding recipient. The behavior of each affects the other; they interact.

Interaction is the arena for communication, the sphere within which the behavior of communicating serves those who engage in it. Therefore, however else communication is analyzed, it must be considered from an interactional perspective that seeks to explain the contributions of the behavior of communicating to the task of interacting. That a great deal remains to be learned about the general characteristics of interactions is not a crippling disadvantage. It is sufficient to say that interactions characteristically follow orderly, roughly repeatable patterns to which each participant makes its behavioral contributions.

Interactions are important to all animals. No animal of any kind leads an entirely solitary existence. Whether infrequently or continuously, animals must interact with one another to breed, to be raised or to raise their young, to find food, to avoid predators, to limit competition. With such a broad spectrum of functions, interactional behavior must take many forms. And yet chaotic interactions are not often seen in natural circumstances: at least with members of their own species, animals usually employ orderly procedures to deal quickly with most difficulties. Individuals who meet can compete for such abstract resources as territories and, without actually fighting, test each other and allocate the spoils. Others meet without competition but test one another and, as a result of their testing, form bonds and remain together to raise families. In some species, most individuals spend their lives as members of large organizations that are made stable and advantageous to all because many kinds of patterns of interacting are kept within acceptable limits. The diversity and kinds of manageable interactions may be very great indeed, particularly in persistent societies such as the elaborately differentiated troops of many species of primates or the flocks of some species of birds.

Interactions are joint products. If they are to be orderly, they must be managed jointly. The course or flow of an interaction must be mutually guided or sustained, making each participant dependent on the others, and requiring each to cooperate and adjust its actions to those of the other participants. However, each does not yield any more control than it must. The participants need not be equal or receive equal benefits to make interacting profitable for each. But each must deal with the others in such a way that it does remain worthwhile for them to participate. This requirement limits what each can do, fosters interdependence, and leads to the use of orderly procedures for interacting.

To be able to act jointly in managing their interactions, each participant must be able to anticipate the others' behavior. Usually considerable information is available that each can use in selecting the tactics it will employ. Each participant usually knows something of the circumstances within which they meet, and each knows the others and some of their characteristic behavior patterns, or knows them to be strangers and hence relatively unpredictable. Each has certain "social identities" (Goodenough 1965) based on such things as age, sex, social relationship, and activities that it assumes with respect to the others, for example, adult, parent, bringing food and immature, offspring, seeking to be fed; or adult, male, neighbor versus adult, male, neighbor facing off across their territorial border. Such conventions as territorial borders are informative; so are the statuses participants may assume relative to one another when not separated by spatial boundaries. The ability of animals to use histories and conventions as sources of information gives continuity and structure to their relationships with each other and provides a framework for interacting that permits them to anticipate at least the general limits and directions of each others' behavior.

Such frameworks are loose, however, and often do not provide sufficient information to permit interactions to run smoothly. The moves of each participant can differ in countless ways as circumstances change moment by moment. Adherence to the kinds of behavior typical of their mutually understood relationships does not by itself enable each to predict the other's moves in detail. Nor is the behavior that is determined by adherence to conventionalized relationships and habits what we commonly mean by communicating, however informative it may be.

When we speak of communicating, we commonly imply the use of specialized signals. They are not strictly necessary; we could define communication as any sharing of information from any source, as is regularly done in biology and some other sciences. (It is less often defined so broadly in human-oriented sciences where many authors—for example, Goffman 1969 and Ekman and Friesen 1969a—restrict the term to the "intentional" provision of information; see McKay 1972 and Hinde 1972:86–90 for a discussion of different uses, alternative terms, and problems.) Nonetheless, the narrowness implied in the common usage can be fruitful. To *begin* a study of communicating by analyzing the special signals does make sense. They have been honed for the task of being informative, and each signal used is relevant to the circumstances in which it appears—though many other potential sources of information may not be. Specialized signals make available information about their users that might otherwise remain private—for example, information about the kinds of acts in which the signalers may engage.

Because they deal with information about the behavior that can be expected, specialized signals often provide sources for the information of most immediate pertinence to interacting animals, and contribute crucially to the orderliness of their interactions. They enable an individual participant to predict another's moves more accurately, and thus to select its responses and adjust its tactics appropriately. They also enable it to make its forthcoming responses predictable to the other participant, in turn giving the latter more lead time to select its response.

Having taken an interactional perspective within this book, I am concerned with the social communication that occurs between living animals. All kinds of sources for shared information are discussed, and I contend that they are necessary to the process of communicating. The focus, however, is primarily upon formalized behavioral sources of information: displays and some related phenomena. Thus "communication" is accepted in something very like the sense suggested by common use of the word. Where more precise terms are necessary they should be found in other words; "communication" is too widely and loosely used to be now endowed with uncommon precision.

Analyzing Communication within an Interactional Perspective

To learn how the performance of display behavior contributes to the orderliness and fruitfulness of interactions, we must understand the information that it shares, the characteristics of displays that make them vehicles for this information, the problems recipients have in obtaining the information, the effects it has on their behavior, and how these effects in turn affect the animals who display. We also need to evaluate the importance of other sources of information in determining recipients' responses to displays in different circumstances. In other words, we must dissect and describe the process of communicating so that we may study its parts and the ways they are interrelated.

To understand communicating we must first map out what happens in an event when communication occurs. The process as a whole is complex, but its essential structure is fairly evident. It requires three essential units: a communicator who provides a signal and one or more individuals who receive that signal (W. John Smith 1963, 1965).

Other things must also be considered, but how this skeletal description is expanded depends on which features are being analyzed. In one typical elaboration, the communicator is viewed as a source of information with transmitting devices that send signal units from a shared code over a channel to the receiving devices of a destination for the information. This description permits mapping of the effects and relationships of a larger number of components that are involved in an act of communi-

cating. It is useful, for instance, in investigations of the adaptiveness of particular physical characteristics of a signal for which the characteristics of the transmitting and receiving devices and the channel must be understood. Yet even this description is still very abstract and of limited use. Real events are much more complex, especially because many more sources of information than one signal have a significant effect on a recipient's behavior. Expanding the minimal description to accommodate these other sources of information is vital to understanding the process of cummunicating.

The communicator's contribution to the event is not adequately described just by saying that it performed a particular signal. Its behavior has other components, some of which must be evident to a recipient of the signal: it occurs at a particular place or a particular time, for instance. The communicator may be seen to be standing or heard to be moving. It may be inescapably obvious that the communicator is fighting or eating. The communicator may be performing more than one signal, as in figure 1.2. All its acts, formalized or not, are informative. Its shape, size, silhouette, and markings are informative. And apart from the communicator, the setting of the event provides recipients with many other sources of information. Recipients also bring to the event sources of information that are characteristic of their own physiological states, memories, or genetically predetermined predispositions.

Not all the available and pertinent sources of information are effective in any one event. Some are not attended: the recipient who responds in figure 1.2 did not initially see the communicator's pose, hand, waving tail, or erect head hair, and missed any of those that ceased before it turned to face the direction of the communicator's vocalization. "Noise" from at least two audible sources interfered with its reception of the signal "Hi." Of course, this is noise only as defined with respect to that signal; the bird and the sleeping animal may each be a relevant source of information for the recipient, perhaps even more important than the signal that has caught our attention as observers of the event.

Information must actually be shared between communicator and recipient and have some effect if communication is to occur. Signals go unattended in some events, and the process of communicating is not completed. Or signals may be addressed at large—"to whom it may concern," as it were—and no potential recipient may be within sensory range. Yet, even if an apparently attentive recipient is present, it can be quite impractical for an observer to get an indication of a response. The only effect of the signal may be to cause the shared information to be stored for reference in the indefinite future, yet this is a response, and may affect future behavior of the recipient. We do see overt responses

Figure 1.2. Communication. The participants in this interaction are generalized animals, each perched at the end of a series of small figures representing stages they have gone through as developing individuals, gaining experience, and sitting in a tree that represents their phylogeny—the information with which they are endowed genetically by their evolutionary histories.

The nearer recipient in the upper diagram hears the communicator's signal "Hi!" while looking elsewhere and receiving diverse visual, audible, and

tactile stimuli contextually to the signal. Some of these stimuli—such as the bird's song and the snoring of the third individual of their species—are effectively "noise" interfering with the signal's reception.

Of the communicator's various signals, only the vocalization is initially effective. The communicator's posture, direct gaze, proffered hand, waving tail, and erect head hair are unseen. So, too, are other informative features of that individual such as its height, age-correlated characteristics of appearance, and individual peculiarities. The recipient might find several of these sources of information to be pertinent, although which would depend in part on the circumstances of this interaction.

In the lower diagram the recipient of the first "Hi!" is responding, having sorted through the information from various sources and selected an appropriate way to behave. In uttering "Hi!" it now becomes the communicator of the moment, and its neighbor the recipient. In this event not only its vocalization but also its gaze, upright tail, and erect head hair are being attended.

The individual at the base of the tree is also a recipient of the vocalizations, but one lacking information from most other sources (having been asleep). In the lower diagram this individual has responded by awakening, pricking up an ear, and glancing about with one eye open, but is apparently confused and is performing little display behavior other than a slight smile. (Drawn by Robin Yeaton.)

when signals are performed, however, albeit not in every event. In fact, we often see different responses in different events, even when the same kinds of signals are performed by the communicators. These diverse responses lead to the possibility that each signal has numerous functions. To understand its contribution to the tasks of interacting, we must study it in many different events.

The questions raised by an interactional perspective lead our attention to different features of a communication event, shifting from communicator, to signal, to response, to other sources of information, and so on. For example, a question such as What information is made available by a signal? usually leads to a focus on the communicator. In principle, the information content of a signal could be ascertained by studying either the communicator or a recipient of the signal because the behavior of both can reflect the information. However, our description of the structure of a communication event suggests that recipients can differ from one another in their predispositions to respond, and in the additional sources of information available to each in different events. The pertinence of a given type of signal, and thus the ways in which recipients respond to it, can thus vary. Communicators are a less heterogeneous group. Each communicator performing a signal is in circumstances that render the signaling behavior appropriate, and thus in

circumstances in which the signal's information content is pertinent to it. On the other hand, a question such as What functions are engendered by the performance of a given signal? leads us to focus initially on recipient individuals, and then on communicators as well. Functions accrue only when signals elicit responses, and the responses have consequences for both parties.

The different questions that must be asked lead us to develop a range of analytic procedures, each appropriate for emphasizing different features of the process of communicating. Developing these procedures, and clarifying the relationships among their goals, requires a unifying theoretical framework—a conceptual model—of the relationships among the components of communication. Although ethology has not produced a framework of this sort, the field of semiotics (the study of signs) has. Its theory, particularly the analytic distinctions it incorporates, are useful to ethologists.

Semiotic theory employs three major distinctions or levels of abstraction. Each level is devoted to the analysis of particular components and kinds of relationships within the process of communicating (C. W. Morris 1946). The first, syntactic level of analysis is concerned just with signals, the basic tools of communicating. The signals are analyzed in abstraction from the events in which they occur, and treated only as physical entities. For instance, their forms are described and compared, as well as the relative capacity of each for carrying information. When feasible, the latter operation yields a quantitative measure of the amount of information contained in a signal, but without studying which *kinds* of information are represented. Syntactics, in short, is the level of analysis at which we describe the Cheshire cat's grin without the Cheshire cat. This is as curious as Alice thought it to be; in real life the grin is not separable from the cat. But communication as a process is not the subject matter of this level. Rather, the utility of syntactic procedures is that they abstract signal units from communication and enable their characteristics to be examined without the distractions afforded by the variable behavioral matrix in which they are used. Among the main syntactic concerns developed in this book are a review of the diversity of forms of displays (chapter 2) and a reexamination of the display concept, seeking to determine what constitutes a signal unit and to ask what other formalized units should be recognized (chapters 13 and 14).

The second "semantic" level (figure 1.2) is concerned with the kinds of information made available by a display or other signal. What, for instance, does a grin tell you? Any signal is useful only because it has a consistent relationship with something else: with entities and their properties (behavioral and otherwise) that are called its "referent," it can be thought of as providing information about that referent. Semantic anal-

yses attempt to describe the kinds of information that are, so to speak, contained in the signal. Each kind of information content is often called a "message" of the signal (W. John Smith 1965, 1968, following Cherry 1966 and earlier editions). The different messages now known to be widespread—that is, to be carried by the displays of diverse species— are discussed in chapters 4, 5, and 6.

The semantic level is concerned only with what the messages are, not with the uses that are made of them. Thus it, like the syntactic level, does not deal with communicating as such, but with a component abstracted from the process; that component is the information content of the signals which are studied at the preceding level. To say that a signal has an information content, that it makes information available, does not imply that the functions of performing the signal lie simply in being "informative." The information made available by signaling may function to bluff, coerce, appease, or accomplish many things, but these are pragmatic, not semantic, issues.

The third, "pragmatic" level of semiotics is the most inclusive, embracing the signals and their messages studied at the previous levels, and investigating their use by participants who are actually communicating (figure 1.2). In a lucid discussion of human communication, Cherry (1966:244) described the subject matter of this complex level as comprising the "specific circumstances and environments" in which individuals signal, and "all questions of value or usefulness of messages, all questions of sign recognition and interpretation, and all other aspects which we could regard as psychological in character . . . the concepts of meaning *to* specific people." *Meaning* is a difficult term. As he stipulatively defined it, meaning has to do with the responses made to signals by their recipients—a response is an indication of the significance attached to a given signal at a given time—as well as the responses that communicators intend to elicit by providing the signal. (Ethologists studying nonhuman animals seldom have been concerned with the expectations a communicator might have about the effect of performing a display.)

At this level of analysis the most accessible component of a communication event is an overt response by a recipient. Ethological research dealing with the responses made after displays are performed in different events, and with the functions accruing to these responses, is discussed in chapter 10. However, a recipient's responses in most events in which displays are performed are determined in considerable part by the sources of information that are available contextually. Without contextual sources of information displays would literally be meaningless, because the information provided by a display can rarely be sufficient to enable a recipient animal to choose an appropriate

response. Thus the matrix of unspecialized acts, specialized and unspecialized forms of interacting, and nonbehavioral characteristics of circumstances within which displays are employed includes essential sources of information (chapter 9).

The evolutionary development of the characteristics that adapt display behavior to the task of being informative are pragmatic issues, because natural selection acts on the abilities of displays to elicit responses. Evolutionary questions have been the subject of much ethological research, and are dealt with in chapters 11 and 12.

Isolating components of the process of communicating at the three analytic levels of semiotics reduces the number of complex relationships that must be considered simultaneously in research. This makes it easier to develop procedures suited to specific, narrowly defined relationships. The artificial divisions and limitations this reduction imposes, however, can obscure pervasive relationships that are of fundamental concern. For instance, because function depends on responses and is studied only at the pragmatic level, the evolutionary adaptiveness of signaling behavior is not assessed at the syntactic and semantic levels. Yet neither the form of the signaling behavior nor the kinds of information it makes available can be fully understood except as adaptive specializations, shaped by the requirements of communicating to the advantage of the communicator, and usually of recipients as well. If we are not to lose sight of the basic evolutionary orientation underlying the ability of ethology to contribute to the study of communicating, we must keep in mind that every syntactic or semantic feature has pragmatic-level implications and is molded and constrained by the need to be adaptive. When working at the more abstract levels of analysis we may miss important features unless adaptive issues force us to consider why signals and messages should have their apparent characteristics. Evolutionary theory should not guide initial description, but it can help us discover ways in which that description is inadequate.

The three levels of semiotic theory have seldom been used by ethologists studying display behavior, although they have contributed to the development of analytic procedures in linguistic, logical, and engineering studies (see review by Cherry 1966) and in the psychological study of language use (for example, George Miller 1964). A pragmatic level scheme of C.W. Morris was adopted by Marler (initially in 1961a) as a means of classifying displays. The linguist T. A. Sebeok (1962, 1965) was among the first to introduce semiotic theory to biology, reviewing and interpreting the ethological literature from a semiotic perspective. Sebeok even introduced the term "zoosemiotics" to cover the combined fields of ethology and semiotics (see Sebeok 1970, in which he traces the history of his term), recognizing the conceptual enrichment each

field offers the other. My use of the analytic divisions of semiotics in ethological research began primarily with an attempt to systematize a study of the kinds of information carried by displays (W. John Smith 1963, 1965, 1966; and chapters 4 to 6 herein), and to understand the implications of these (semantic) findings for considerations of the formal characteristics of displays (syntactics) and the ways in which displays function (pragmatics), as explained in chapter 7. This work began with studies of various species of birds and has been extended to mammals, including humans. Its main procedures, and the kinds of results obtained from the study of the display repertoire of each species, are illustrated by an extensive example developed in chapter 3.

Plan and Focus

The book is organized in a sequence that largely parallels the semiotic divisions for analyzing signals (syntactics, chapter 2), referents (semantics, chapters 4, 5, and 6), and usage (pragmatics, chapters 10, 11, and 12). The interpolated chapters serve special purposes. Chapter 3 is an example of a case study, largely syntactic and semantic, of the display repertoire of a single species. Chapter 7 offers a model to account for semantic generalizations developed in the preceding chapters, incorporating syntactic and pragmatic issues. This approach differs from but is compatible with the emphases of traditional ethological approaches. Chapter 8 compares the existing ethological approaches, and, briefly, approaches originating outside biology. Chapter 9 describes other sources of information than display behavior available to the participants in interactions, making possible the consideration of responses, functions, and evolutionary topics in the next three chapters. Chapter 13 reevaluates and redefines the display concept, and chapter 14 considers the evidence for classes of formalized behavior in addition to displays; chapter 15 is a brief recapitulation of major points.

The focus of this book is on the behavior of communicating as an adaptive component of interactional behavior: the information made available, the responses, the evolutionary modifications that enhance sharing information, and its effectiveness. Topics are treated in detail to the extent that they bear on this focus. Thus evolutionary and interactional theory are not reviewed in any detail because these areas simply provide general perspectives for the focus. General semiotic theory is not addressed; rather, its basic analytic divisions are simply borrowed and found useful. And some traditional ethological approaches to display behavior are reviewed only briefly because they are oriented to issues peripheral to the topic of communicating. For example, the well-worn instinct-learning controversy has often been concerned with dis-

play behavior, but research on it is designed primarily to determine how particular behavior patterns originate and develop in individual repertoires, not how they contribute to communicating. Similarly, research on the physiological (motivational or "drive") states of individuals who participate in interactions is largely peripheral because it is concerned with the mechanisms controlling the behavior of each individual, rather than those controlling the processes of interacting. These topics are not central issues in the study of communicating, although they are important to the field of ethology as a whole. The book is not intended to offer an annotated listing of all the studies bearing on communicating, however indirectly, but rather a critical coverage of the ethological findings, postulates, and theories that are most relevant within an interactional perspective.

2

The Diversity of Displays

The historically strong interest of ethologists in stereotyped behavior, and especially in "displays," stereotyped acts that are specialized to make information available, was indicated in the introductory chapter. This chapter surveys the kinds of acts that ethologists have recognized as displays.

Enough display behavior has been described to make clear that the diversity with which we must cope as we study communication comparatively among animals is very great indeed. Without being a book in itself, a survey of that diversity can only be an annotated outline, selectively skimming from the detailed knowledge now available. (A much more detailed survey is provided in a forthcoming book edited by T. A. Sebeok [in press], although no reviews can now encompass the entire literature on displays. And ethologists have really just begun to describe displays, the vast majority of which must remain unknown in a world containing perhaps a million species of animals [Mayr 1946]).

Even though they provide too vast a subject to be covered in detail, these specialized acts provide some general knowledge for a useful first step in studying the behavior of communicating. Displays are the basic behavioral tools available to communicators, and many are readily apparent as animals communicate. This is not to say that we fully understand the nature of displays, but it is not the task of this chapter to examine the implications and adequacy of the display concept. The traditional definition and the criteria that have been implicit in practice are accepted here, because this practice has produced virtually all of what is now known about display behavior. The concept must be reevaluated and refined, which will become evident in succeeding chapters, but such redefinition points to research yet to come (see chapters 13 and 14).

Surveying the Diversity

The number of ways in which displays can differ from one another is enormous. Any naturalist who has tried to learn just the songs and calls of common birds can testify to this, even though these are only a portion of all bird displays and only a minute fraction of all animal displays. Birds also have formalized postures and movements, and in many species patterns of touching and nonvocal kinds of sounds. Other kinds of animals bring other sensory systems into the process of communicating, employing olfaction, taste, vibrational receptors that are tuned to other than audible or ultrasonic frequencies, and even special electric current sensors.

Ethologists have scarcely begun to catalog this diversity. They started by describing those behavior patterns directly accessible to human sensory organs. As the name suggests, the displays first described were primarily visible movements and postures, but they also included many sounds, vocal and otherwise. Equipment that extends the human senses has become increasingly available, facilitating the task of recognition and cataloging. The analysis of movements, for instance, has been greatly assisted by videotaping or cine-filming behavior, which enables sequences to be examined in the brief samples represented by successive frames, and to be replayed to reveal details that would otherwise be missed. Research on audible displays lagged until portable magnetic tape recorders became available and until equipment was devised for the analysis of frequency/time patterns of human voice. The latter (commercially known as the Sona-graph [Kay Elemetrics Company]) has also been useful in analyzing these physical properties of nonspeech sounds. Along with oscilloscope traces of amplitude/time envelopes, it permits increased resolution of the properties of sounds and enables us to work with kinds of differences that our ears do not easily appreciate. The Sona-graph also permits more objective description than just relying on individually idiosyncratic human perceptions. The calls of birds or frogs now can be described so that different workers can be sure they are working with physically comparable material, not just with what sound like similar displays. Finally, our understanding of the sources of information available to sensory modalities with which we cannot cope directly or in which we are especially limited is beginning to be enhanced by instruments that detect the chemicals made available in some displaying, the electric fields generated by some aquatic animals, and ultrasonic cries. Although such advances hold great promise, visible and audible displaying, in that order, remain the best described.

Figure 2.1. Posture of a cat, terrified by a dog. The cat stands at its full height, arches its back, erects much of its bodily hair (note especially that on the tail), lifts the base of its tail and throws the dangling tip of the tail to one side, flattens back its ears, and exposes its teeth. (From Darwin 1872.)

Visible Displays

Highly specialized postures have been described for many species of animals. Darwin's illustration of a cat terrified by a dog is an excellent example: its back and tail are highly arched, its legs fully extended, its neck withdrawn and ears pulled back, and its fur is standing on end (figure 2.1). Darwin (1872) also pictured a hostile dog in a very different pose: head and ears thrust forward, tail held high, and hair bristling, primarily on the back of its neck. The postures of cats and dogs are not uniquely dramatic. Figure 2.2 shows examples of elaborate poses from three species of birds, the crouched display postures of Adelie penguins in Antarctica and of the tyrant flycatcher *Myiozetetes similis* in the Panamanian lowlands, and the wing-raising display of an upland Andean tyrant *Muscisaxicola albilora*. All these birds were sketched as they threatened territorial opponents, but although their striking poses and feather rufflings have much in common, there are also marked differences among the species.

Figure 2.2. Display poses of birds of three species threatening territorial opponents. *A.* Two Adelie penguins facing one another in crouch displays. Note their flexed legs, horizontal bodies, and withdrawn necks. The gaping mouths and bulging downturned eyes are separable displays here combined with the crouch, as are the erected crests (seen as a bump on each bird where feathers are raised behind the eye, and made more conspicuous by the flattening of the forehead feathers). (From Ainley 1974a.) *B.* A *Myiozetetes similis* flycatcher facing another *M. similis* with which it has just had a grappling fight. It is crouched down with its tail slightly cocked. Its throat, flanks, and back are ruffled, but not its crown, chest, and rump feathers. *C.* A *Muscisaxicola albilora* flycatcher facing a rival in a pause between attacks and chases. Its crown, chest, and flanks are ruffled, but its throat, back, and rump are not. Its tail is fanned. It is standing upright and partly raising one wing. (From W. John Smith 1971a.)

The pioneering studies of displays by Julian Huxley (1914) on the great crested grebe must surely have been prompted in part by the utterly bizarre performances that birds of this species execute on the water: their "cat attitude" (see figure 1.1), "penguin dance," and "ghostly penguin attitude." Just as surely, the strange, elaborate movements of courting ducks (figure 2.3) attracted Heinroth (1911) and later his student Lorenz. The many highly ritualized acts of gulls (see figures 8.1, 11.1, and 12.2) provided the initial focus for the now classical display studies of Tinbergen and his students. Other workers

Figure 2.3. Sequences of displays by a courting mallard drake. Based on cine film records made by Konrad Lorenz, these drawings show sequences of display postures and movements identified by the following numbers: (1) bill-shake; (2) head-flick; (3) tail-shake; (4) grunt-whistle; (5) head-up—tail-up; (6) turn toward female; (7) nod-swimming; and (8) turning the back of the head. (From Lorenz 1958. Copyright © 1958 by Scientific American, Inc. All rights reserved.)

have recorded the grimacing of baboons, the head-bobbing of lizards, the graceful movements, posturings, and odd color changes of fish, and the mating flights of butterflies.

These examples are not meant to imply that all visible displays are blatant or bizarre. Much visible displaying is less spectacular, but many striking displays have commanded a disproportionate share of the attention of ethologists. The less spectacular are often described, but perhaps also often missed. As experienced and excellent an observer as Tinbergen admits (1959b:20) that only after he and his students had seen a very striking head flagging display in the black-headed (figure 12.2) and kittiwake gulls did he recognize that a less striking form is also used by the herring gull which he had studied for years; most of us have had similar experiences.

The diversity of visible display in vertebrate animals alone precludes any simple cataloging. Displaying individuals prance, stretch, or cower; they wave, freeze, and leap; raise or lower fur, feathers, fins, gill covers, spines, or dewlaps; inflate air sacs, change the colors of their body surfaces, and even manipulate objects. They act singly, in synchronized simultaneous or alternating performances with mates, and sometimes in large groups.

Many invertebrate species with adequate vision also have diverse postures, movements, and color changes. For instance, males of hunting spiders perform stereotyped leg-waving movements as they approach females for mating (figure 2.4A and B). Fiddler crabs wave their large claws when defending their burrows and attracting passing females (figure 2.4C, D, E, and F; see also figure 9.4); these claws account for about a third of a fiddler crab's weight, and their great size seems to have evolved entirely for use in signaling (Crane 1975). Various butterflies and dragonflies maneuver elaborately during courtship flights (Bastock 1967), and some bees perform stereotyped maneuvers to "lead" their hive mates in the direction of food sources (Lindauer 1967). Alarmed or mating squids very rapidly alter their color patterns (Moynihan 1975).

Light is produced and emitted by many kinds of animals (and even plants) and can make visible signaling possible even when darkness prevails. Little is known about the uses to which this light is put, although a considerable literature deals with the structure of light-emitting organs and the chemistry of bioluminescence. In at least fireflies and other insects, however, communicative functions have been established for some bioluminescent emissions. Courting fireflies identify themselves to one another with flashes that are species specific in duration, brightness, and spectral composition, organized within a distinctive flashing schedule, and given off by luminescent organs distinctive in size

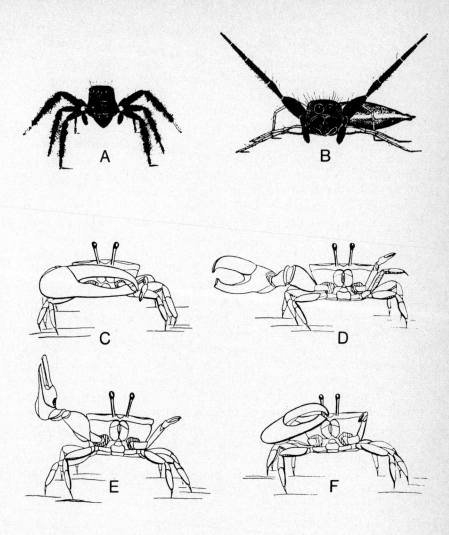

Figure 2.4. Visible displays of spiders and crabs. Upper: Courting display poses of two species of salticid spiders: A: *Corythalia fulgipedia,* and B: *Ashtabula furcillata.* (From Crane 1949.) Lower: A fiddler crab (*Uca lactea* of Fiji Island) displays with a Lateral Wave of his large cheliped, a claw whose size seems to have been developed entirely for signalling functions. The cheliped starts from its flexed position (C) and is unflexed outward (D), then raised (E), and finally returned (F) to the starting point. (From Crane 1957. Copyright © 1957 by the New York Zoological Society.)

and shape (see review by Lloyd 1971). A firefly can add further distinctiveness by maneuvering in the air while flashing.

Of the several flashing patterns an individual firefly may have, however, one may not be formalized as a signal. It instead is used to provide illumination that aids the firefly in locating suitable places to alight (Lloyd 1968). Bioluminescence may not be specialized for intraspecific communication in other cases, but instead may function to mislead predators (see chapter 13). For instance, the occasional luminescence of ocean waves or of ships' wakes as tiny marine creatures are stirred up may involve such tactics as those used by the copepod *Metridia lucens:* it flashes immediately upon being jolted and then, having excreted its reacting chemicals into the water, leaves its light behind and darts off into the darkness (David and Connover 1961).

Headlightlike organs filled with luminescent bacteria are found just below the eyes of the flashlight fish *Photoblepharon palpebratus.* A fish can turn off the continuously emitted light from these organs simply by raising black lids to cover them (Morin et al. 1975). Fish of this species swim singly or in groups along Red Sea coral reefs on moonless nights. When undisturbed, they leave their lights on most of the time, covering them only in very brief flashes a few times per minute. The light is thought to be useful to the fish in feeding, both attracting their phototactic crustacean prey and helping them to see it. The blinking can also be combined with a swimming pattern that may help a flashlight fish to avoid predators: when swimming over parts of a coral reef that offer few refuges from predators, a fish will abruptly change direction and increase its swimming speed when it blinks. In addition to its apparent applications in prey capture and predator avoidance behavior, the bioluminescence of the flashlight fish is also modifiable for intraspecific signaling. For instance, in a territorial pair the female will respond to a territorial intruder by swimming back and forth, then turning off her lights and swimming to the intruder, showing the lights again on arriving next to the other fish. In all observed cases, the intruder has then departed.

The versatility with which a flashlight fish appears to use its bioluminescence exceeds that known for any other animal, although research on the functions of the bioluminescence of marine organisms is just beginning. Thus far, most functions proposed for light emission by fish have been in the categories of luring prey, for example, in angler fish, see Wickler (1968), Hulet and Musil (1968), or predator evasion. For example, in daytime the pony fish emits light from its ventral surface, thereby camouflaging its silhouette (Hastings 1971). Young and Roper (1976) have shown that captive squid (*Abraliopsis* sp.) will luminesce when overhead lights are turned on and will extinguish their photo-

phores when the lights are turned off; the ease with which observers could see the silhouettes of the squid from below diminished as the photophores began to glow, and squids that matched the intensity of the overhead lights disappeared from view. This "photophore countershading" may explain why many fish, squid, and shrimp have ventral luminescence (Clarke 1963), although in other cases the restriction of lighting to the undersides of fish may relate to the larger numbers of predators that would be in a position to see dorsal lighting (McAllister 1967), and the lights themselves probably have other functions, including intraspecific signaling.

Audible Displays

Both audible and visible displays are so enormously diverse as to defy cataloging. The majority of audible displays have yet to be described even casually. The most familiar might be loosely classified as "vocal": sounds produced by expelling air from the lungs, first past some sort of vibrating mechanism and then one or more resonating chambers or tubes. These vocalizations include the sounds of human language and of laughter, crying, sobbing, and screaming.

Man is the animal whose communication is most dependent on complexly patterned vocal utterances, but his nearest rivals are more commonly found among birds than among other primates. The many species of monkeys and apes, as in other groups of vertebrate animals, differ considerably in the extent to which their vocal display repertoires are elaborated. Only arboreal species appear to have loud "singing" performances, and thus far only the South American titi monkey *Callicebus moloch* has been shown to have an unusually elaborate, very frequently used vocal repertoire (Moynihan 1966a). The relatively terrestrial primates may have evolved more visible than audible display elaborations, although this must be very carefully verified with wild populations. Itani (1963), for instance, has found that although captive Japanese macaques produce few vocalizations their wild counterparts employ an "extremely rich" vocal repertoire. Certainly, however, many terrestrial primates perform visible displays far more frequently than they vocalize. For example, the gorilla, perhaps man's closest living relative, is a strangely quiet creature. Fossey (1972) found fewer than four vocalizations per hour in 2,255 hours of quantified observations, and the great majority of those were uttered only by silverback males.

Not just primates, but most mammals and many other kinds of animals have vocal displays. Dolphins whistle and squeal. Bulls, whether cattle, bison, or Steller's sea lions, roar. Many rodents (Sales 1972; Brooks and Banks 1973; Sales and Pye 1974) and perhaps other

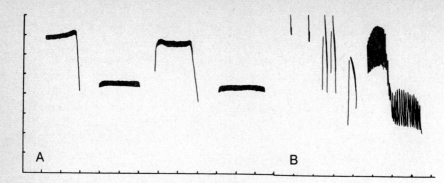

Figure 2.5. Vocal displays of the Carolina chickadee. The song (*A*), a clear, high-pitched, tonal sound, was recorded from a male moving about in his territory. The nonsong vocalization (*B*), a set of brief, quickly stuttered notes followed by a rasping sound that breaks abruptly from high to lower pitch, was recorded from a male near a territorial border where he had been skirmishing with a neighbor during the preceding several hours; similar calls are also uttered in skirmishes. (From S. T. Smith 1972.)

The figure is based on tracings of sonagrams and illustrates how audible displays are often analyzed by portraying the distributions of the sound's louder frequencies (measured in kiloHertz, or thousands of cycles per second) in time (measured in tenths of seconds). Comparable frequency/time distributions are shown by printed musical notation, although the standard convention employed for music cannot represent all the characteristics of these vocal displays. Musical notation, however, permits experienced readers to imagine the sounds it represents much more readily than do sonagrams, which are best used for making visual and abstract comparisons.

mammals can communicate when the individuals are very close together with vocalizations that are, to human ears, ultrasonic. Other groups of animals that make significant use of ultrasonic signaling include such insects as bush crickets (Tettigoniidae), other orthopterans, and forest-dwelling cicadas, but their ultrasounds are not vocal in origin; our current understanding of ultrasonic communication is reviewed by Sales and Pye (1974). Many bird species are highly vocal (figure 2.5), uttering both elaborate "songs" and other vocalizations that ornithologists traditionally relegate to secondary status as "calls." This secondary status is not justified, because most bird species conduct the main portion of their face-to-face interactions with calls and not song. Frogs also "sing" and call, and even tortoises have vocal courtships (see Blair 1968), although most lizards are largely or wholly silent.

Vocalizations, although usually the first kinds of audible displays of which we think, are by no means the only products of specialized sound-producing behavior. Others include the chest-beating of gorillas and,

more rarely, of chimpanzees (van Lawick–Goodall 1968). Chest-beating has an analogue in the behavior of the North American ruffed grouse, which "drums" a rolling beat by compressing air in front of its chest with sweeping wing strokes. Other species like the spruce grouse produce a loud clap by striking the wings together above the back while in a flight display (MacDonald 1968).

Numerous kinds of animals from beetles to birds generate noise in display behavior by striking objects or substrates. Gorillas, chimpanzees, humans, and other primates will display by slapping on tree trunks and branches, or against the ground. Branches are shaken and jumped on by various displaying primates, including some New World monkeys. Among nonprimate mammals, beavers and seals slap the water surface when alarmed, and rabbits and some rodents drum on the ground with their feet. Among birds, most woodpeckers have a loud display in which they drum with their bills on dead limbs. Kilham (1972) has reported on a Panamanian woodpecker that has developed several forms of drumming and is less dependent on vocalizations than is a related and sympatric species. Woodpeckers and chimpanzees will often go to favorite branches and trees to produce their percussive displays, because a good drum is necessary to obtain optimal resonance.

Many animals cannot properly "vocalize." Some fishes can produce sounds by the somewhat comparable means of expelling air from their air bladders, which they constrict with specialized muscles, but other fish use quite different mechanisms, such as scraping parts of their gill structures together. Scraping is abundantly employed by arthropods and is usually called "stridulation": grasshoppers rub their hind legs against their wing covers, and different species of insects have specialized filelike structures on various articulated body parts that can be rubbed together to produce sounds. Cicadas actually snap a drumlike membrane, the tymbal. Fruit flies produce species-specific sounds by vibrating their wings (Waldron 1964; Ewing and Bennet-Clark 1968; Bennet-Clark and Ewing 1969). In reviewing the acoustical communication of arthropods, Alexander (1967) found diversity reflected in "innumerable" onomatopoetic terms such as click, rasp, lisp, rattle, tick, beep, chirp, pulse, trill, and buzz, as well as in a great many terms for the sound-producing movements that employ different devices: wingstroke, tymbal pop, toothstroke, head-jerk, wing-snap, abdomen-waggle, leg-stroke, stridulation, or crepitation. Yet even vertebrates that employ vocal sound production often have sound-making anatomical specializations of other body parts: tail feathers, for instance, are modified in snipe to yield a winnowing sound during flight display, and wing feathers in manakins to produce snapping and ripping noises in flight.

Some "sound" is actually produced as display behavior by animals

lacking earlike transducers. For instance, many social insects such as honeybees in their dances (Wenner 1959; Esch 1961) or ants that have been buried by cave-ins of their nests (Wilson 1971a) produce sounds that are audible to us and yet apparently not audible to them. Even leeches (Annelida) may court by tapping a duet on the leaves of their substrate (Frings and Frings 1968). Most such species in effect receive these displays tactually, as substrate-borne vibrations. Some substrates are especially suitable for transmitting vibrations: in the mating of various spiders a male approaching a female signals by strumming her web in a pattern different from that produced by entrapped prey; female web-building spiders are too shortsighted to be trusted not to confuse males with food (see review by Frings and Frings 1968). The surface of water is a substrate for a few kinds of animals, and at least one species of water strider (insects in the family Gerridae) courts by propagating special wave patterns (Wilcox 1972).

Tactile Displays

Formalized patterns of touching behavior are common and widely distributed among mammals, and a wide variety have been described. The nose, lips, teeth, tongue, neck, hands, arms, flanks, tail, and genitalia may be involved in making contact. Noses are often pressed against noses, or a nose may be gently grasped by a mouth; in at least tenrecs and shrews there may be gentle biting or mouth-to-mouth grasping (Eisenberg and Gould 1966), a form of tactile behavior also seen in many fish species. Mammalian arms can embrace, and hands can touch or grasp. Nonhuman primates have elaborately developed embracing and touching displays (see, for example, Weber 1973, accounts in DeVore 1965, and figure 11.2 in this book, in which the palm-up visible display of the young gorilla invites a tactile display of palmslapping or handholding); humans employ touch extensively in intimate behavior (D. Morris 1971). Mammals may throw their hips against other individuals' bodies (in shrews, E. Gould 1969; in young captive gorillas, personal observation), or rub their heads along others' bodies or necks. A neck like a giraffe's is ideal for tactile "necking sessions" (Dagg 1970). Torsos or flanks can be rubbed alongside other bodies, as they are in the contact sunning of black-tailed prairie dogs (W. John Smith et al. 1973 and in prep.), or can be pressed hard together as in threatening encounters of arctic ground squirrels (Carl 1971; Watton and Keenleyside 1974). From two to four resting titi monkeys may entwine their tails, like the stems of a bouquet (Moynihan 1966a; and figure 2.6). Female rats require tactile stimulation from sufficient and correctly timed intromissions to induce the hormonal conditions prerequisite to

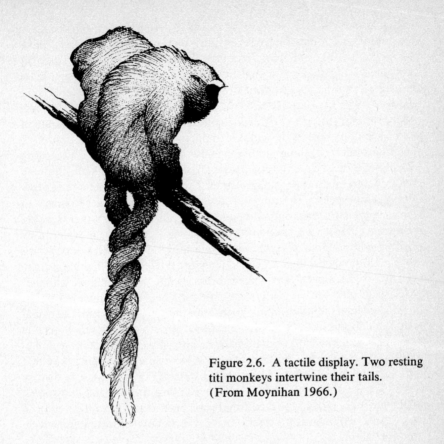

Figure 2.6. A tactile display. Two resting titi monkeys intertwine their tails. (From Moynihan 1966.)

ovum implantation and to facilitate sperm transport through the cervix (Adler 1969; Wilson, Adler, and le Boeuf 1965).

In many primate species the most commonly seen tactile displaying is during allogrooming, in which one or more individuals groom each other, or groom some other individual (see chapter 13). The avian analogue of allogrooming is allopreening, in which one individual nibbles and cleans the feathers of another.

Although relatively few species of birds allopreen (Sparks 1967) and birds are usually thought to engage in little tactile displaying, several different kinds have been described. These often involve mutual bill-touching: mated storks "clatter" their bills together in greeting (for example, Kahl 1971), and other procedures are used by gannets (Armstrong 1965), albatrosses (Fisher 1972), gulls (Howell, Araya, and Millie 1974), herons (Meyerreicks 1960), woodpeckers (partially reviewed by Kilham 1972), and even some songbirds (for example,

various finches, Hinde 1955; Coutlee 1968). Some of this bill-touching may relate to allopreening of the head region (for murres, see Tuck 1960) but much resembles instead a "courtship feeding" display in which one individual gives food to another, often its mate. When pairs are well established the bill-touching of courting is replaced by courtship feeding in some species. Other kinds of avian tactile displays seem less widely distributed: drakes tread on the backs of ducks when mounted for copulation, and male killdeer plovers also tread while mounted, sometimes after pushing bodily against the female before mounting (Phillips 1972). Mated *Agapornis* parrots may gently nip a partner's toes in getting it to sidle along a branch to a more suitable site (Dilger 1960). One species of grebe (*Rollandia rolland*) has a ceremony in which mates face, then one dives and comes up under the other, striking it breast to breast with enough force to knock it clear of the water; they repeat the "bumping" several times, each bird playing both roles (Storer 1967). In the pied-billed grebe *Podilymbus podiceps* the female usually rubs her head against her mate's breast while he copulates with her (McAllister and Storer 1963); the male rubs his head against her breast during reverse mounting early in the mating season (Storer, personal communication). N. G. Smith (1966a) reported that when female Kumlien's gulls solicit copulation from unresponsive mates they sometimes use head-tossing, customarily a visible display, in such a way that the movements begin to prod the male's breast. Perhaps a tactile display in the earliest stages of evolution from a visible display, this movement might not have been seen had Smith not been experimentally reducing the males' willingness to mate (his experiments are summarized in chapter 9).

Tactile displays are common in fish, at least in that their lateral-line organs are in part effectively pressure sensors: as fish align while displaying they must get much information about each other's fin and tail movements from sensing water turbulence. Direct contact is also involved in displaying, as when a male three-spined stickleback nudges a female with his snout before she begins to lay eggs (Tinbergen 1953a).

Tactile communication is apparently universal in colonial insects, although it is not usually complexly developed (Wilson 1971:233). However, in honeybee communication it is fundamental to the famous dance performance by which a dancer indicates characteristics of a flight that will lead to some desired site (von Frisch 1950, 1967, 1974; Lindauer 1961, 1967, 1971; Esch 1967; and see figure 6.4). In the usual case in which the site provides a source of food or water, she dances within the hive on the vertical surface of a comb. The direction of the flight, measures correlating with the distance (if the source is sufficiently distant), and worth of the source are encoded by features of

"waggle runs" linking successive circling patterns of the dancer into a rough figure eight (see chapter 6, in which the information provided by this performance is discussed). The geometry of the dance pattern and the duration of a 250 Hz tone produced by the dancer's vibrating wings must be perceived by bees who may be recruited to seek the reported source, but these potential recruits cannot see the dancer in the darkness of the hive, and they are deaf. They run along with their antennae either being struck by her vibrating abdomen or held against her thorax. Although Alexander (1967) has criticized that the classes of stimuli they receive have not actually been determined, the vibrational sensitivity of their antennae corresponds to the frequency produced by the vibrating wings, and the orientation of the dancer can be assessed tactually by followers touching her thorax. A recipient bee that has followed one or more dances (the number followed varies in part with the recipient's experience as a forager) performs yet another formalized tactile display to obtain a regurgitated sample of nectar: she taps the dancer with her antennae in a particular pattern.

Tactile communication is also important in the communication of nonsocial insects. For instance, tactile stimuli have been implicated at several stages in the precopulatory patterns of cockroaches. In the mating behavior of the Cuban cockroach *Byrsotria fumigata,* although a female sex pheromone (see the next section), initially leads a male to begin precopulatory display, some later stages of the sequence require tactile stimulation (Barth 1964). For example, a male who has been attempting to determine whether a female is fully receptive will perform his full wing-raising display only after the female has briefly engaged him in antennal fencing or at least has lurched forward and gently bumped him. Other necessary tactile components of the performance may not be as formalized.

Although tactile displays are apparently abundant, they can be difficult to assess. First, it is not clear that an observer can see all tactile displays. Further, some of the touching that can be seen during communicative behavior may not be formalized. For example, some touching is done in attempts to monitor or detect scents and less volatile chemical products; this is to be expected when one individual presses its nose against another's axillary, anal, or genital regions in which glands producing these chemicals are often located. The trunk of the Asiatic elephant, for instance, is used in a great deal of social touching, but Eisenberg, McKay, and Jainudeen (1971) caution that "it is extremely difficult to separate potential communication signals with tactile input from those involving olfactory signaling."

Also, some formalized behavior may be tactile without appearing to be. E. Gould (1969) describes several postures of shrews, but notes

that they do not appear to be used as visible sources of information. Instead, they may be perceived only when one animal "actually comes in close tactile contact with another and touches the opponent's vibrissae or gland-secretion on the flanks or neck." A final and less bothersome problem is that some production of substrate-borne vibrations may affect not auditory but tactile receptor organs (see the previous section, Audible Displays), and be wrongly categorized if we fail to understand the sensory limitations of a species.

Displays Releasing Chemicals

The chemical products of animals can be informative. If a chemical released by an animal elicits responses, either from other individuals or even from itself as it later encounters the taste or scent, the chemical is known as a pheromone.

Some pheromones are specially manufactured for their informative functions. Others are instead produced by symbiotic bacteria. The Indian mongoose *Herpestes auropunctatus,* for instance, has two anal scent pockets that contain epidermal sloughings mixed with lipids and proteins excreted by sebaceous and apocrine sweat glands; bacteria produce volatile carboxylic acids from this pasty mass that are highly odorous and are deposited by mongooses in scent marking (Gorman, Nedwell, and Smith 1974); the bacterial populations may differ among individual mongooses, because each animal's combination of concentrations of the six odorous acids can be recognized by others (Gorman 1976). Yet a third class of pheromones may be just metabolites that recipient individuals find informative; for instance, a breakdown product of a hormone such as estrogen or androgen.

Whatever their chemical nature, however, pheromones cannot be said to be involved in display behavior unless the acts of exuding, excreting, depositing, or ejecting them are somehow specialized. Pheromonal display behavior is thus the specialized releasing of chemicals, perhaps only at specific places or only in specific circumstances.

Pheromonal communication, often incorporating display behavior, is characteristic of many kinds of invertebrate animals. That female silk moths release a chemical sex attractant has been known since the late 1800s, and today the study of insect pheromones is one of the most rapidly developing areas of investigation in ethology.

Bombykol, the sex attractant of a silk moth, is like that of many other insects in functioning as an airborne trail of scent to which adults of the opposite sex (males, in this species) are extremely sensitive. A single molecule of bombykol will trigger a nerve inpulse in a receptor cell of a male. (This will not lead him to respond behaviorally. Nerves fire

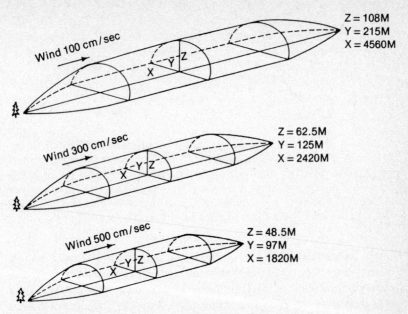

Figure 2.7. Plumes of a sex attractant. The distribution downwind of de-tectable concentrations of bombykol released by a female silk moth is shown for three different wind velocities. The higher, more turbulent winds dis-tribute the pheromone more uniformly; hence the volume of air within which threshold concentrations occur is smaller. The calculated dimensions assume that the ratio of the emission rate of the pheromone to the response threshold of a recipient male (called the Q/K ratio) is 10^{11} molecules per cubic centi-meter per second. (From Bossert and Wilson 1963.)

spontaneously from time to time, so that a field of receptor cells always provides the brain with some "noise." It takes about 200 receptor firings within 1 second to exceed the noise level with which a male silk moth must contend, and to elicit a behavioral response; see Schneider 1974. Still, this represents only an incredibly minute portion of the molecules released by a single female.) Because receptor systems have evolved such sensitivity to the appropriate sex attractants, the pheromones can be effective over enormous distances. Released into a wind of the appropriate strength for optimal distribution, an attractant forms a long plume extending away from the communicator (figure 2.7). The direc-tion of the wind provides some of the information needed by a recipient, who responds by flying upwind as long as it is able to detect the odor, and by quartering back and forth across the wind on losing it (Kennedy and Marsh 1974).

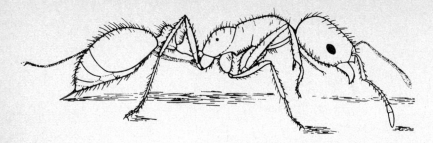

Figure 2.8. A fire ant worker laying an odor trail. The ant moves slowly from left to right and crouches, bending its abdomen downward and extruding its sting so that the tip touches the ground. The pheromone proceeds from the Dufour's gland through the sting. It is applied to the ground like ink from a pen. The sting is periodically lifted clear for short distances, breaking the pheromonal trail into a series of streaks. (From Wilson 1962c.)

Other insects lay odor trails on the substrate. Such trail-laying display behavior is analogous to the dance of honeybees because it provides directional information. In fact, some bee species with less elaborate dances than those of honeybees mark their routes as they return from a food source to the hive by depositing pheromones at numerous points along the way (Lindauer 1961, 1967). On the basis of careful experimental work with fire ants, Wilson (1971a:249) describes their odor trails as the "most elaborate of all known forms of chemical communication." Ants of this and many other species lay trails of a pheromonal product from the abdominal Dufour's gland by dragging the tip of the sting along the ground as they release the scent (figure 2.8). A fire ant does this as it returns to the nest from a site at which it has found food that can be harvested, occasionally doubling back on its own trail (Wilson 1962c).

Most invertebrates have a variety of pheromones available to release in different circumstances (see, for example, the extensive review of social insect behavior by Wilson 1971a). These pheromones differ in chemical structure, in part to be distinguishable, and in part as adaptations to requirements for differential rates of diffusion. For example, the alarm pheromone of the ant *Pogonomyrmex badius* diffuses very rapidly and for about 35 seconds establishes a small active space (maximum radius about 6 cm) of odor emanating from the excited communicator. Rapid spreading is not as useful in the case of odor trail pheromones of ants; these should spread less far than do alarm pheromones, and to be efficient should persist longer than 35 seconds. In fact, ant odor trail pheromones are less volatile than are alarm pheromones and usually last a matter of minutes, still fading quickly enough so that they do not

continue to guide foraging ants to depleted food sources. Still more extreme are the pheromonal trails that individual snails, limpets, and other gastropods each lay in the mucous path it excretes to lubricate its "foot" (Wells and Buckley 1972). The trails of some of these species persist for days, despite the flux and reflux of tides, and guide an individual back to its home "scar" on the rock where it rests when not foraging (Cook 1969; Cook, Bamford, Freeman, and Teidman 1969).

Pheromones are commonly released by vertebrates as well as by invertebrates—snakes, for instance, lay pheromonal trails (Gehlbach, Watkins, and Kroll 1971). In fact, birds may be among the only vertebrates in which pheromones are unlikely to be important sources of information, and even their lack of ability to sense odors is still being questioned; a few species of birds have quite a good sense of smell. Many mammals are known to scent-mark sites, depositing pheromones on environmental objects (discussed in chapter 9). Black-tailed deer, for instance, rub branches around their sleeping sites with exudate from a forehead gland, and in a dispute between adults each may rub its hocks together while urinating, spreading a urine-carried pheromone on its fur (see Müller-Schwarze 1971). This species of deer has at least four other kinds of pheromones that do not appear to be involved in display behavior, including a metatarsal scent released when they are alarmed (figure 2.9). Alarm-correlated pheromones have also been convincingly demonstrated in the case of laboratory mice who, like deer, do not release them through a formalized act (Rottman and Snowdon 1972), and in woodchucks, in which release is done with special marking behavior (G. Haslett, research in progress).

Among the exciting questions now before researchers in this area is the extent to which humans produce and respond to pheromones. Because they are important to other species of primates (see chapter 9), because people react to the scents of other mammals and they to ours, and because we have a variety of appropriate glands (such as the apocrine for producing scents and hair tufts with which to diffuse them, there is considerable reason to test for their occurrence (Comfort 1971). Human pheromones might be involved in the regulation of endocrine cycles (chapter 9), the timing of puberty, sexual attraction, and individual recognition. If so, they are important sources of information. They may not, however, be much involved with display behavior.

In the study of pheromones most attention is focused primarily on implicating chemicals produced by one individual in the elicitation of responses by other individuals. Relatively little attention has been accorded to whether the release or deposit of the chemicals involves formalized behavior, but much apparently does not. Thus most information that is provided chemically is quite separate from the performance

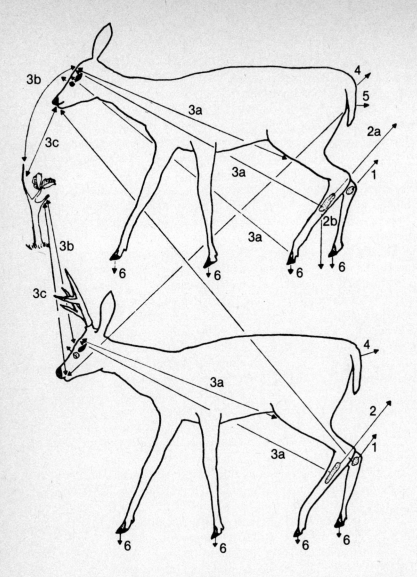

Figure 2.9. Pheromones of the black-tailed deer. Pheromones originate in the tarsal organ (1), metatarsal gland (2), forehead gland (3), anal region (4), urine (5), and interdigital glands (6). All the pheromones spread readily by rapid diffusion into the air except the forehead and interdigital scents.

Deer anoint themselves with pheromone by rubbing their hindleg over their forehead (3a). Twigs are marked by rubbing with the forehead gland (3b), and sniffed and licked (3c) by recipient deer. Scent marking of the substrate is done as the interdigital glands leave pheromone on the ground (6), and as the metatarsal glands touch the ground (2b) while the deer reclines. (From Müller-Schwarze 1971.)

of displays, even though it is often of considerable contextual importance (see chapter 9).

Electrical Displays

Some aquatic animals have evolved special sensitivity to electric fields. American eels, for instance, may be able to detect the electric fields generated by ocean currents moving through the vertical component of the geomagnetic field (Rommel and McCleave 1972), which could aid their fabled long-distance homing. Yet other animals have evolved the capability of producing and amplifying their own electric fields, and use the actively emitted energy at least to detect objects in their environments. The use of ultrasonic echo-location by bats and dolphins is comparable (Griffin 1958; Sales and Pye 1974), although with electric fields reception depends not on reflections but on local alterations of current strength (Lissmann 1958). The emission of electric energy provides opportunities for communication, and although the technological problems involved in studying it are considerable, some recent research has begun to exploit this opportunity.

Some species of freshwater fishes in muddy waters of both Africa and South America establish current fields resembling an electric dipole around themselves. Each species maintains a particular discharge rate from an electric organ of modified muscle tissue near its tail, and apparently uses this both to explore and to monitor its surroundings. The shape and duration of pulse emission vary and function in social signaling. Through numerous, careful experiments with the South American fish *Gymnotus carapo,* Black-Cleworth (1970) has shown that at least three categories of electric discharge behavior provide different information from one another. Its "unmodified" discharge that is kept up largely to monitor the environment does not appear to be specialized to provide information about behavior, although its species-specific pulse characteristics and baseline frequencies may be in part formalizations permitting identification. Cessation of this usually continuous discharge does appear to be formalized: discharge pauses fall into two classes on the basis of duration, each correlated with different agonistic behavior patterns. A rapid increase and then slower decrease in discharge frequency (figure 2.10), called the sharp-increase-decrease (SID) by Black-Cleworth, also appears to be formalized and correlated with biting and mock-biting. Present evidence suggests that other species of electric fish have comparable repertoires of electric signals (see Hopkins 1974). For example, the gymnotid *Eigenmannia virescens* also employs at least the same three classes as does *G. carapo:* unmodified

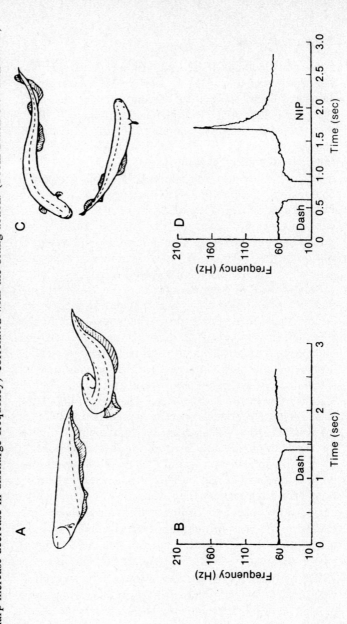

Figure 2.10. Discharge rates of a *Gymnotus carapo* as it dominates another individual. *A*. The dominating fish (on right) dashes toward the other and veers off just short of it. *B*. Electric discharge rates of the dominating individual during this sequence. Note the brief discharge break accompanying the dash.

C. In a different sequence a dominating fish (above) dashes toward another and nips at its tail. *D*. Electric discharge rates of the dominating individual. Note again the brief discharge break accompanying the dash, and also the large SID (sharp-increase-decrease in discharge frequency) correlated with the biting action. (From Black-Cleworth 1970.)

discharges, interruptions, and rises; rises include the SID form, which Hopkins indicates may be universal among electric fish.

From even this rapid survey it is evident that animals have developed displays for reception by each sensory modality that detects external events (with the possible exception of thermoreception); these displays are very diverse. Some of the evolutionary forces that select for this diversity are examined in chapter 12. For the present, the most relevant issue is that no complex animal need lack a range of tools with which to signal; it has opportunities to share information by behaving in formalized ways. The kinds of information made available by this behavior are the subject of the next chapters.

3

Analysis of a Display Repertoire

In studying the behavior of communicating, an ethologist often finds it necessary to observe performance of all parts of the display repertoires of selected species in as much detail as possible. Preferably, this observation is done in the field in natural conditions. This chapter is a summary of the initial results of such a project, "initial" because they suggest much more that could be sought in further, more quantified observations.

The observations permit displays to be described and illustrate correlations between the displays and other behavior of their performers. These correlations permit the analyses that provide the main focus of this chapter, analyses of the information about behavior that the displays make available. A limited number of similar observations of other species is introduced to show the value of interspecies comparisons. The behavior of individuals recipient to displays was observed less thoroughly in this study than the behavior of communicators, and their responses are not treated in detail. The functions of communicating depend on responses and are considered only insofar as they are suggested by the circumstances in which the displays occurred. Displays, messages, responses, and functions are all accessible for study, granted that the right techniques are employed. However, precise determination of responses and functions usually requires experimental intervention, or observations that are extended specifically to seek narrowly defined and recurrent "natural experiments"; the difficult problems that then arise are not discussed until chapter 10.

Naturalistic observations provide the data for this study because, as explained in chapter 1, ethologists try to understand how a species behaves in the kinds of situations that gave direction to its evolution. Every species has become adapted to the ecological conditions of its

habitat with specializations that are in part unique, molded by the action of natural selection. Its adaptations are responses to the kinds and distributions of resources, competitors, and predators typical of the world within which it must contend. These responses have been evolved within the genetic opportunities and limitations peculiar to its ancestral lineage. They characterize every aspect of an animal: its structure, its physiology, and the kinds of behavior it manifests. All aspects of its behavior, including its displays, interrelate to form a coherent, adaptive life-style. Displays should be analyzed within at least a general understanding of the specific natural history that sets this style. Viewed in isolation from it, the adaptive features of each display may not be apparent, and may be misunderstood.

The subject of this chapter is a small passerine bird, a North American tyrannid flycatcher called the eastern phoebe *Sayornis phoebe*. Inconspicuously gray-brown and whitish, it is nonetheless well known to farmers as a bird of barnyards and bridges; in more natural habitats it frequents forest edges, primarily along streams and by ponds. Wherever a phoebe lives it forages primarily by chasing passing insects and catching them in the air, although under some circumstances a phoebe will fly down to the ground to take prey, and berries and seeds are sometimes eaten. Their primarily aerial feeding habits lead phoebes to spend a great deal of time on exposed perches from which they can most easily see and begin to pursue their prey. This exposure may be less risky than it might seem for small birds, because their usually solitary habits, alertness, dull coloration, and partly wooded habitats reduce the ease with which they can be surprised by bird-hunting hawks.

Eastern phoebes do not spend the winter in most of the parts of North America in which they breed, because there are then no flying insects on which to feed. Very rarely some hardy individual makes it through a New England winter subsisting on berries, but most phoebes migrate to the southeastern United States or to Mexico. They spend the winter months solitarily, each apparently on its own territory.

In the very early spring, long before any other tyrannid flycatchers have begun to return north, male phoebes begin to arrive and settle on large territories. Their food supply is then dispersed and only tenuously available. Foraging requires most of their time and attention; a late snowstorm can be fatal. Nonetheless, each male finds some time to proclaim and patrol territorial borders, and to drive out males that intrude on his territory.

After some days females begin to return, and each forms a pair bond with a territorial male. The procedure of pairing, as it is in the case of many species of small birds, is difficult for an observer to follow. It does not appear to involve elaborate ceremonies or special displays, although

females look like males and are initially attacked as they intrude on the territories of eligible bachelors. An unmated female does not respond to these attacks as an intruding male would. Instead of fleeing or fighting back, she elects to remain unaggressively in a male's territory and to follow him and remain in his presence, a social activity called "associating." Within a matter of hours or at the most days after her arrival, the male has begun to associate with her unaggressively.

As the season becomes warmer food becomes more readily available. Each mated female selects a nesting site, sometimes one to which her male has already shown some attachment. Thereafter, the pair spends most of its time in the general vicinity of this site. The female builds a nest, accompanied and watched by her mate. If the weather remains favorable she lays a clutch of four or five eggs. During the ensuing days in which she incubates these eggs, her mate often perches for long periods within a hundred meters or less of the nest. When she comes off the nest to forage he may accompany her, or he may remain in the nest area. At other times, however, he forages alone elsewhere in the territory, with the result that the mates associate less consistently than before incubation began.

After the eggs hatch both parents bring food to the nestlings, although only the female sits on the nest to brood them. Both parents feed the fledglings in the days and early weeks after they leave the nest. The family group then gradually breaks up, the parents more and more often avoiding or repulsing attempts by their young to beg food as the young become more competent and consistent about foraging for themselves. Phoebes sometimes raise a second brood, but with its dissolution no closely knit social organization prevails, and the male may occasionally behave territorially even toward his offspring. By early autumn phoebes begin to leave the north, and the solitary phase of their annual cycle has begun again.

The social life of eastern phoebes is fairly uncomplicated, never involving bonded relationships among more individuals than an immediate family and a neighbor or two. Nonetheless, the species has approximately the same number of displays in its repertoire as do birds with much more complex and enduring social relationships, and it uses these displays in ways that are generally quite comparable to those of other species.

The full detail in which the displays of the phoebe are known is not presented in this chapter (more detail is available: see W. John Smith 1969c, 1970a, 1970b). Instead, the complex array of behavior with which one display correlates is examined closely and the performance of others only in sufficient detail to permit a general understanding of the repertoire as a whole. Some displays are compared with displays of a

flycatcher from another genus, or of the two other species of phoebes, to illustrate the comparative method commonly used by ethologists; no attempt is made to describe the entire repertoires of those species.

The Twh-t Vocalization

One of the most commonly uttered vocal displays of the eastern phoebe can be roughly characterized as "twh-t." Physical analysis of the distribution of this sound's frequency in time reveal a pattern that is easily recognized (figure 3.1*A*). Fortunately for an observer, the twh-t rarely intergrades with other vocalizations and is distinctive to human ears. (Less fortunately, it is impossible for a human to pronounce. A good name for the vocalization is hard to find, however, and the awkward description of its sound does as well as any.)

The twh-t call is uttered in a great variety of circumstances. If we are to understand what is consistent about its performance, we must observe and compare the different activities of the bires who utter it. The first circumstance encountered in the annual behavioral cycle is typical only of unmated males. These can be found early each spring, recently arrived and moving about on their territories as they forage and patrol. Although they are initially solitary, each male must both establish relations with any neighboring males and obtain a mate during this period.

Shortly after arriving from migration each male spends much time in the more wooded portions of his territory, perching low and seeking insects on or near the ground. Because food is very scarce, he has little time to do other than forage; in fact, in inclement weather he forages without ceasing throughout the day. His foraging movements take him to peripheral parts of his territory, and on warm days he periodically stops foraging there and flies to a treetop perch to "sing" loudly for a few minutes. Because his brown back renders him hard to see when he is close to the dead leaves of the forest floor, these elevated singing performances greatly enhance the ease with which he can be found.

He must be found if he is to obtain a mate. Yet he must also forage, just to keep alive. Both tasks are necessary, and his behavior often gives evidence of the difficulty he has in choosing between them. For instance, he may cease foraging but fly only part way to a treetop perch, flying again to approach it in successively higher steps. Sometimes he quits such attempts and returns to foraging; at other times he finally does go all the way, and then sings. These periods of indecision lack *both* foraging behavior and singing: they are times he can be seen to hesitate in choosing between the two activities, temporarily failing to select either.

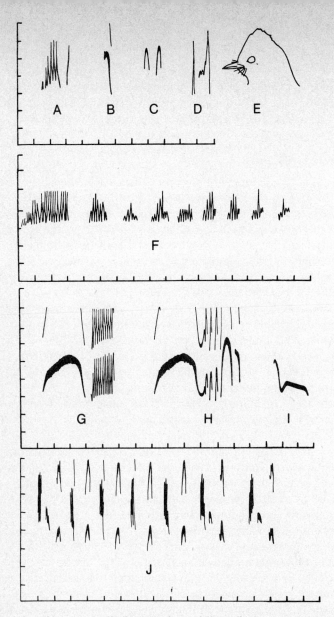

Figure 3.1. Nine vocal displays and a visible display. Examples of the eastern phoebe's twh-t (*A*); tp (*B*); double vocalization (*C*); bi-peaked vocalization, t-keet (*D*); crown ruffled (*E*); chatter vocalization (*F*); fee-bee (*G*); fee-b-be-bee (*H*); and initially peaked vocalization (*I*); and of the eastern kingbird's kit-ter (*J*). The sonagrams will not enable readers to imagine the sound of the vocalizations in much detail (see Chapter 2). Instead, they should be used primarily to visualize differences among them and secondarily to note some similarities (such as the "fee" portion of the two song units, the harsh "bee" of the fee-bee, and the prolonged initial unit of the chatter vocalization). (From W. John Smith 1969c, 1966.)

During these periods of indecisive behavior the twh-t call is uttered. It can be the only call we hear, if he does not go on to sing. Thus a bout of twh-t calling may replace a bout of singing, at least with respect to the apparent function of advertising a male's presence on his territory. Even if singing does occur, the amount of time in which a male is vocal and therefore relatively easy to locate is extended by the twh-t calls that precede and sometimes follow the song bout.

One other feature of the occurrence of the twh-t vocalization in the patrolling behavior of unmated males is very striking to an observer, but makes no obvious sense until further uses of the display have been studied. The twh-t is uttered only when alighting from a flight. Not from all flights—only from flights during the periods of apparent indecision. As it happens, this observed correlation is crucial in enabling us to determine the kinds of information made available by the call.

The utterance of twh-t in patrolling is not confined to unmated males in early spring, although patrolling behavior is uncommon after pair formation unless a male should lose his mate. Patrolling bouts can also be elicited at any time during the breeding cycle by the appearance of an intruding phoebe, but intrusions are common primarily in the early phases. A territorial male usually routs an intruder quite readily, and as he follows him toward the perimeter may begin to utter twh-t on alightings. He does not chase beyond his borders, but may briefly patrol them or stop and sing from a station there before returning to his mate.

Although mated males patrol little, they give display during other activities. For instance, males sing nearly every morning in the predawn twilight, particularly during the early part of the breeding season. Toward the end of such a crepuscular bout of song a female, during the days when she is laying eggs, is likely to fly toward her mate and alight some meters away. Some females utter a twh-t as they alight, others a few song components. The male immediately ceases singing and flies toward his mate. If the mates are evenly matched and do not fight much, copulation may follow quickly and without other obvious preliminaries other than the female's adoption of a posture that facilitates mounting.

In many pairs, however, the female is slightly or markedly dominant over the male, and in most circumstances ready to rebuff him aggressively should he approach closely. In these pairs males are always cautious about approaching their mates, and after a predawn song bout rarely go directly to them in one flight. Usually a cautious male stops short and utters the twh-t on alighting. He may make successive abortive approaches, uttering twh-t with each alighting. Some subordinate males then make tangential flights around their mates, apparently too aroused not to fly, but reluctant to approach closer than a few meters. They may

alternately fly and alight in very quick succession for several minutes, uttering twh-t on each alighting. In addition, they may do something that is otherwise unusual: they utter twh-t in flight and on taking flight, and even when perched (not that they ever perch more than momentarily). When twh-t is uttered on taking flight or in flight the male usually flies more toward the female than tangentially, except when many flights are being made in very rapid succession. When indecisive behavior reaches such extremes a male may fail to get all the way to his mate, and both birds gradually begin to forage. With his first foraging flights, a male once again calls twh-t only in correlation with alightings; soon twh-t occurs only with some alightings, and then the calling ceases.

A phoebe forages by watching for passing insects from a perch, then chasing them in a rapid flight and returning to a perch for more watching. Occasionally after a particularly erratic, twisting chase, an individual alights unusually close to its mate. This appears to be accidental. The bird may abruptly shy away from its mate as it touches down, utter twh-t, and perhaps immediately fly somewhat farther away. If the dominant individual alights, the other is more likely to utter twh-t, making a flight intention movement or taking flight to a safer perch. Very rarely, both individuals call twh-t as if greeting each other. As the season progresses mates gradually appear more accustomed to each other's presence, and this use of twh-t decreases. It reappears when one is excited by a predator, or when a male persists in trying to approach and mount his unreceptive mate.

Yet another circumstance involves predators. When a phoebe discovers a predator or an animal that might be dangerous near the nest it usually does not attack and strike it, but makes abortive approaches and withdrawals or watches it closely while making tangential flights. Its behavior is "agonistic" or "hostile," terms ethologists use to imply the possibility of either attack or escape or both. These are indecisive flights, and twh-t may be uttered on alighting from them.

Twh-t is uttered in further circumstances, but most do not occur often. For example, I once watched a male associate with his foraging mate as she moved farther and farther from the center of their territory. Eventually he turned and flew back about 12 m into the territory, uttering a twh-t as he alit. He stayed there until she returned to him, and then began to work farther back into the territory, often uttering twh-t on alighting from a flight that took him away from his mate.

In all circumstances in which a phoebe utters this display, certain aspects of its behavior remain comparable. For instance, the vocalization comes primarily on alighting from a flight or, much less commonly, it precedes flights or even occurs during or between them if they are short and frequently repeated. These flights come in periods of indeci-

sive behavior. The vocalization is *not* given when there is little or no evidence of indecision, even if flying is frequent: for example, when insects are pursued and either caught or lost, after which further flight is usually unnecessary and alighting is silent. Other aspects of the communicator's behavior are less consistently observable than flights, alightings, and evidences of indecisive behavior, but when not observable can be inferred. In particular, the indecisive behavior always involves interacting or seeking to interact as at least one choice. This is observable when the bird is near its mate or a predator, but not when the bird is patrolling alone. In patrolling it is necessary to make assumptions about why the bird takes high perches (they maximize its conspicuousness and hence facilitate advertising) and to consider what happens if it does encounter another phoebe (it interacts) to infer that it is seeking to interact.

On the basis of these observations we can describe the information that is made available by the display. That is, we can determine what a phoebe could learn by hearing a twh-t vocalization just from the vocalization itself, neglecting everything else such a recipient might know or be in a position to apprehend.

First, this vocalization is only uttered by an eastern phoebe. No other species within this phoebe's range has a closely similar call (it is occasionally mimicked by, for example, some individual mockingbirds, but their mimicry is not difficult to recognize). In fact, the physical form of the call not only identifies the caller, it enables a binaural recipient to determine the location from which the call is coming (see chapter 6). Second, the vocalizing phoebe is indecisively facing behavioral choices in which flight is involved. It is probably terminating a flight as it vocalizes, but the flight is not likely to settle the issue over which it is indecisive. A decisive flight is at that instant less probable than some unspecified alternative that prolongs the period of indecision. One of the caller's choices involves an unspecified form of interacting: it may associate, copulate, or attack, or may be seeking an opportunity to interact in any way. Again, the twh-t by itself does not enable a recipient to know which possibility.

To summarize, a twh-t display makes available information that identifies the caller and indicates its location, as well as information about the probabilities with which it will elect to fly, interact or seek to interact, perform unspecified alternatives to interacting, and be indecisive.

This information is obviously limited. Nonetheless, it must often be relevant and potentially useful. Consider again the main kinds of circumstances in which the display is employed. First, in early spring patrolling, the pattern in which a male repeats the display is largely

peculiar to unmated males. It increases their conspicuousness and provides information needed by a migrating female as she seeks a territorial
male with whom to pair. The same information is needed by a migrating
male as he avoids clashing with settled individuals. Later, when an
individual uses twh-t in the presence of its mate it tends to withdraw, or
to cease approaching, but will associate. Such a communicator is indecisive and, whatever else, poses no threat to its mate. If the pair bond is to
persist and if insemination is to occur, withdrawal should not be required too frequently; the display provides information that must be
useful to a recipient in controlling its tendency to rebuff or flee from
approaches by its mate. Third, individuals who utter this call around the
nest in the absence of their mates are always responding to some
apparent trouble such as a predator. Although phoebes rarely attack
large predators directly they harry them and sometimes attack smaller
ones. By frequently flying about and calling they provide distraction that
may draw a predator away from their nests, or draw in their mates or
birds of other species to "mob" and further harass a predator until it
leaves the area. In short, in the circumstances in which a phoebe
communicates about its hesitancy to fly in a decisive act and its readiness to interact, this information can be relevant to a recipient that deals
with it along with information available from other sources.

For Comparison: The Kit-ter Display of the Eastern Kingbird

The eastern kingbird *Tyrannus tyrannus* is a slightly larger flycatcher
than the eastern phoebe. Although the two species are in the same
phylogenetic lineage (the taxonomic family Tyrannidae) they are not
closely related. Like eastern phoebes, eastern kingbirds lack complex
social organization when not breeding, although they may form fairly
large, loose, and apparently anonymous foraging flocks. Males migrate
north and establish breeding territories before the females return, and
the most complex social organization that develops is the family. This
breaks up when the fledglings are competent to forage for themselves
(for details about the life history and displays see W. John Smith
1966).

Ecological differences between the two species have important consequences. First, eastern kingbirds forage more in the open than do
eastern phoebes, perching in high, conspicuous places from which they
can see in many directions, and flying out to take their prey at considerable distances. They are adapted to such exposure by being vigilant and
by their behavior of attacking and driving away potential predators.
Hawks intruding over a kingbird's territory are harried; if they are
relatively clumsy, slow-flying species a kingbird may strike them re-

peatedly and even ride their backs, pulling out feathers. This certainly deters hawks from remaining and possibly trains them to stay away.

Second, eastern kingbirds build relatively exposed nests. A phoebe's nest is often placed so that terrestrial predators cannot climb to it, and it is hidden from aerial predators by overhead obstructions. A kingbird's nest is in the crotch of a tree, usually commanding a good view and readily seen. Phoebes are cautious in defense of their nests, but kingbirds of both sexes attack readily—some even strike humans. A kingbird's aggression carries over into its social life, and mates are prone to squabble. In particular, and unlike phoebes, a male frequently attacks his mate.

The eastern kingbird has a vocal display that is unlike the phoebe's twh-t in form (it sounds like a series of kit-ter and kit-ter-ter components (see figure 3.1*J*), but very like it in many of the ways it is employed. For instance, patrolling males utter kit-ter vocalizations as they fly from perch to perch during advertising behavior. The prolonged call is usually begun in flight and finished on or shortly after alighting, or finished as the bird at the last minute veers away and flies on. Mates give the vocalization when approaching each other, especially during the early phases of pair-bonding and at any time thereafter when one is excited and likely to attack the other. Unlike phoebes, both individuals are likely to utter kit-ter series simultaneously, accompanying them with a wing flutter display in what amounts to a ceremonial interaction of greeting.

Kingbirds utter kit-ter when responding to predators or potential predators, sometimes on making flight intention movements toward them from a perch, or in flight as one begins to circle back toward a predator that it is repeatedly attacking. I have also seen a male kingbird chase a hawk beyond its territory, then turn back, uttering a kit-ter series as it slowed and turned.

The aggressiveness of kingbirds begins to disrupt their family interactions as the fledglings become largely independent. Fledglings and their parents may fight as the young persistently pester the adults for food; at this stage of the family the young not uncommonly utter kit-ter series as they follow their parents in flight. Thus at a time in the breeding cycle when the twh-t is scarcely given by phoebes, the kit-ter becomes abundantly uttered by kingbirds.

In sum, the kit-ter vocalization of the eastern kingbird and the twh-t of the eastern phoebe correlate with similar sorts of behavior, although the kit-ter is uttered more often and in a somewhat larger variety of circumstances. The differences correlate with the kingbird's much more aggressive approach to pair, familial, and predator situations. Yet the information made available by the two vocalizations is closely compa-

rable. First, their forms both identify the communicators (at least to the extent of indicating their species) and facilitate locating a communicator. Second, behavioral correlations enable us to say that both vocalizations provide information about the likelihood of flight, of interacting or seeking to interact and of unspecified alternatives, and of being indecisive. There is this difference: observations reveal that, on the average, flight is more likely when an eastern kingbird utters kit-ter series than when an eastern phoebe utters a twh-t. This amounts to a difference in the relative probability of flight. Thus detailed observations of the uttering of vocal displays in different species permit comparisons of the information that the displays carry. The information may be of the same kinds for different displays despite some differences in the patterns of their performances and major differences in their physical forms.

Singing

To return to the eastern phoebe, additional vocal displays make up the singing behavior referred to earlier. That birds "sing" is a notion applied popularly to vocal performances that people find aesthetically pleasing, but singing lacks a fully accepted and rigorous descriptive meaning in ethology. It is most often used as a label for continuous and regular bouts of calling. Viewed in that sense, an eastern phoebe's singing has two component displays. One sounds like "fee-bee" and gives the bird its common name. The other is similar, and sounds like "fee-b-be-bee" (figure 3.1*G* and *H*). Some employment of these song vocalizations is not considered singing behavior, because it lacks regular continuity. For instance, at times when singing bouts are uncommon a lone individual may occasionally utter some fee-bee units, a few of which come at about the rate of slow singing, but with pauses of unpredictable length and with very few or no fee-b-be-bee units. In other cases, a female approaching her mate in the predawn twilight at the season when she is relatively receptive to copulation may rapidly utter a few fee-b-be-bee units, but with no regular rate.

Singing regularly correlates with being at important sites in a territory such as at the boundaries or in the nest region. It must provide some information about the singer's reluctance to leave these sites.

A singer's behavior is not complex: the bird is almost always alone, lacking a suitable individual with whom to interact. During territorial patrolling there may be no other phoebes in the vicinity. During the predawn bouts uttered by most paired males, darkness and lack of knowledge of the mate's position may preclude interaction. Opponents who are not within a male's territory are not fully suitable. Thus, when

males countersing—that is, have singing duels from opposite sides of a territorial boundary—neither will cross to engage the other; a male who sings while driving out an intruder will not follow it beyond his territory.

A singing phoebe can almost always be inferred to be seeking opportunities for interaction. Its loud song vocalizations and high perches make it conspicuous. It may range widely along its territorial boundaries while performing. And if an opportunity to interact should arise, should another phoebe appear within the singer's territory, the communicator *ceases* its singing and immediately attempts to initiate interaction: association or copulation with a mate, or attack on an intruder of either sex.

The two song units are generally given in similar events, but there are important differences in some details. When singing is made up primarily of fee-bee units the singer often pauses briefly for a foraging flight, or may preen, shake his feathers, bill-wipe, or perform other "comfort" acts. Foraging flights occur only in the intervals before fee-bee units (154 flights all occurred before fee-bee units in a sample of singing with 6,685 fee-bee and 3,219 fee-b-be-bee). Comfort acts occur primarily in this period, although exceptionally brief ones occasionally precede a fee-b-be-bee (38 before the former and 3 before the latter in a sample including 1,816 and 1,408 of the respective song units; W. John Smith 1970a). When fee-bee units are less predominant, both flights and comfort movements are exceedingly rare during singing.

The fee-b-be-bee unit rarely predominates in the singing of most males, although I have one observation of a male singing several in rapid succession immediately after attacking and chasing away an intruder from near his nest, and another of two birds uttering it in strings while countersinging in an early territorial encounter. Females may utter this unit in irregular strings (not within the definition of singing, although this is not crucial to an interpretation of the two vocalizations) when approaching their mates; they are apparently seeking to copulate, but are also aggressive.

The differences between the activities with which the two song units correlate thus seem to indicate that the fee-b-be-bee is uttered by a bird that is very ready to interact in some way, including attack or copulation, while the fee-bee is uttered by a bird at least somewhat preoccupied with individual maintenance behavior and perhaps less ready to engage in intense interaction, although it will interact at least to associate.

Changes in the proportions of the two units during a singing bout appear to reflect shifts in readiness to interact "intensely," that is, differences in the likelihood that the singer will attack or attempt other contact behavior instead of simply associating. They also reflect shifts in

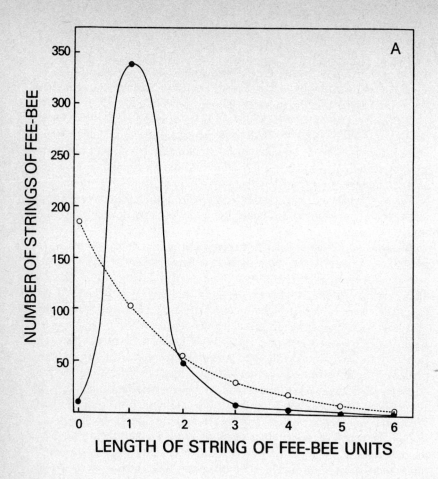

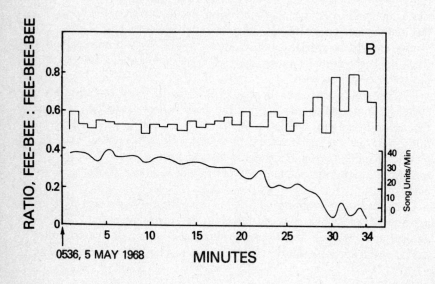

Figure 3.2. Organization of a bout of singing by an eastern phoebe. This is an analysis of a predawn bout of singing recorded on 5 May 1968 from its start at 0536. It was sung by a male whose mate was incubating. The upper curve shows that the modal length of the 414 strings of the fee-bee song was one (that is, the singer usually alternated fee-bee with fee-b-be-bee). He uttered 338 strings of length one, compared with only forty-nine repetitions of two fee-bee songs, fifteen longer strings, and twelve repetitions of the fee-b-be-bee song (the last are scored here as twelve strings of zero fee-bee songs). The dotted line shows the distribution of string lengths to be expected if the male uttered his two kinds of song in random order.

The histogram in the lower part of the figure shows the relative proportions of fee-bee and fee-b-be-bee in each minute: the former tends to increase at the expense of the latter toward the end of the bout. The curve below the histogram (its axis labeled on the right) indicates that the rate of singing dropped markedly as the proportion of fee-bee songs increased. Though no part of the figure provides information about the moment-by-moment sequential patterning of the singing, the male began with five fee-bee songs in a string then largely alternated the two kinds of song for about twenty-five minutes, gradually introducing longer strings as his rate of singing slowed and became irregular.

readiness to engage in such activities as individual maintenance that are incompatible with contact forms of interacting, although not with associating. The prevailing ratio of fee-bee to fee-b-be-bee is thus informative. It is also readily assessed, because sequences of songs within singing bouts are patterned.

In the case of most males, who sing more fee-bee than fee-b-be-bee units, only the former are usually repeated in runs or strings, and each run is terminated by a single fee-b-be-bee unit. When the numbers of strings of each length are tallied (such as strings of 0, 1, 2, . . . n units), strings of 1 fee-bee are shown to be the most common, on the average (a string of 1 amounts to alternation of the two song units, as the string is terminated by a single fee-b-be-bee). Next most common are strings of 2 fee-bee units. Longer strings of up to 15 or more fee-bee units occur, but no string lengths other than 1 and 2 units appear to be much favored; that is, longer string lengths are approximately randomly distributed with respect to the average ratio of fee-bee to fee-b-be-bee units in a bird's singing (see figure 3.2A). Strings of fee-b-be-bee are uncommon, although an occasional very rapid singing bout (well in excess of 30 units per minute) begins with several fee-b-be-bee units in succession. In general, most bouts of singing tend to follow one of three main patterns: they may begin primarily with strings of 1 fee-bee and progress to strings of 2 for awhile, then cease, or they may continue to longer strings (as in figure 3.2B), or they may comprise primarily

longer strings; rare utterance of primarily fee-b-be-bee is followed by longer strings.

The inferences about the information made available by the two song units can be tested in part by the behavior of the minority of male phoebes that regularly sing more fee-b-be-bee than fee-bee units. They too most often alternate the two units, but when they repeat one in strings it is almost always fee-b-be-bee. Yet they do not reverse the correlation of fee-bee with foraging flights and comfort acts. For instance, of 2,826 song units recorded from such males only one quarter were fee-bee units and flights preceded 127 of them (about 1 flight for every 5.5 units). Only 15 flights preceded the much more common fee-b-be-bee units, about 1 for every 142. From a slightly smaller sample of singing with comfort movements the comparable ratios were 1:9 and 1:53. Thus each of the two units probably encodes the same information when it is sung by any phoebe, but individual birds can make either unit the one which more often succeeds itself in strings in their singing. Perhaps the ratio of fee-bee to fee-b-be-bee in the singing of different males falls along a continuum in which a slight excess of fee-bee units is most common, and singing preferences may reflect individual differences in temperament. Exploration of this possibility would require detailed study of the singing of many more individual phoebes than has yet been done.

The Tp Vocalization

A brief "tp" is the most abundantly uttered call of eastern phoebes (figure 3.1B). Tp is perhaps the only vocal display given during much of the asocial behavior of winter, and may be the only call of the earliest spring migrants. Yet it is not common even in their patrolling behavior and does not become common until birds have paired and chosen nest sites.

The tp is less loud and much less complex in form than are the song units, but is used with them to a limited extent. It comes before or after some singing, and in pauses, which are rare, during predawn song bouts. If male eastern phoebes become vocal in the evening twilight when they rarely sing, say within 15 or 20 minutes before going to roost, they usually employ a few bouts of tp calling. It is as if tp replaces song in the crepuscular period in which light is failing and activity must be reduced or curtailed. No kind of interaction is very likely to occur then, and copulation is particularly unlikely.

Within the pair, one individual may utter tp calls when its mate appears to be about to approach and to try to interact. For instance, a female may utter tp when being followed by her mate as she collects nest

materials early in the season. Only when an individual appears apprehensive in the presence of its mate is it likely to give the tp commonly away from the nest site; for example, a male who is often attacked by his mate will forage quietly if alone, but with tp calls if in association with his mate. In a pair that I watched closely, the female often called tp when about to attack her mate or after attacking him. Until the time when their eggs hatched he did not often reply. Thereafter, he uttered tp whenever he was in her presence, whether she was attacking or not, and also whenever he was near the nest site even if she were away.

Either parent calls tp in response to almost any potential disruption of nesting activities, for example while watching or mobbing a predator, or while watching or listening to other birds that are mobbing. In these circumstances a phoebe that simply watches, seldom flying, utters only tp; one that makes many indecisive flights utters both tp and the twh-t display. One absent when its mate begins tp calling may return to the nest area, but whether in response to the tp calls or other clues is not usually clear.

The call is uttered frequently and in many kinds of situations. Some individual phoebes who utter it relatively infrequently usually restrict it to obviously agonistic situations in which they attack or escape, and to behavior like patrolling that can readily lead to agonistic interactions. Whether individuals that utter tp very commonly during the nesting season are simply responding to more events as agonistic, however, is difficult to test. Only some of their employment can be traced directly to concern over possible threats to the nest such as the presence of a farm cat. At other times there is little to suggest the cause of their concern, and the call does not seem to increase the ease with which their behavior can be predicted: they call while perched during foraging, preening, or even apparent inactivity. Perhaps the presence of even an unobtrusive observer sometimes makes them more ready to behave agonistically, but they may not even show consistent orientations toward or away from him, let alone more decisive behavior. For the present, the most promising interpretation is that the tp makes available the information that either attack or escape behavior has become anywhere from just slightly more likely to highly probable, while the bird is likely to be indecisive about either selection.

Three Visible Displays Related to Flight

Three visible displays, tail wag, wing shuffle, and wing flutter, make information available about flying and the probability of behaving indecisively. They differ in the probability of flying associated with each, and only the wing fluttering display makes information available about the

probability of interacting. Frequency of use also varies: tail wagging is common, wing shuffling less so, and wing fluttering uncommon.

Tail wagging, in which the tail moves conspicuously through a large, often eccentric arc, may sometimes help a phoebe to balance—particularly when a tail wag happens to coincide with alighting, as one often does. However, there are usually several additional wags, and a bout of wagging is occasionally performed during prolonged perching. In other words, this movement does not function just to assist balance. The form of the movement has been little altered in the course of its specialization as a display, except that the basic unit has become repeated to an extent not required by its initial (balancing) function. It is formalized in this seemingly excessive use: the excess is informative.

The communicator usually remains perched immediately after a tail wagging display, but the likelihood of flight is greater than when a phoebe is quiescent, as is especially apparent when bouts of the display are repeated. Repetition is common when a phoebe is indecisively vacillating between two choices involving flight such as returning to a nest to incubate versus continuing to forage, or approaching versus withdrawing from a predator. As tail wagging is done in diverse situations, however, the display provides no information about the kinds of alternative acts that may be selected. In short, tail wagging shows that the communicator is behaving indecisively about unspecified alternative acts that may involve flying, although flight is unlikely immediately after a tail wagging display.

In the wing shuffle display the folded wings briefly make small shuffling movements, as if the bird is either settling to a comfortable position or beginning to lift its wings to fly. This shuffle often coincides with the initial tail wag as a bird alights, but tail wagging continues, and it often occurs without the shuffle. Wing shuffling occurs in various kinds of events in which flight is momentarily possible, but in which the probability of flight immediately subsequent to the display is generally lower than in events in which twh-t vocalization is uttered; both displays are sometimes used in the same event.

The wing flutter, in which the wings are held high over the back and fluttered rapidly through a small arc, differs from the other displays in being employed primarily *in* flight. That is, the *immediate* probability of flying is high with this display, although it also correlates with a high likelihood that flying will soon cease or change its course. The display usually comes shortly before flight terminates, and is sometimes coincident with alighting. Although in some cases wing fluttering coincides with taking flight, these flights are rarely sustained. Wing fluttering is done in interactions, or when the communicator appears to be seeking opportunity to interact as in advertising behavior. The display is per-

formed during attempts to copulate, in most close agonistic encounters between mates that do not develop into vigorous fights, and with the nest-site-showing display.

Nest-Site-Showing Display and Chatter Vocalization

The nest-site-showing performance is visible: a male goes to a ledge suitable for nesting (sometimes to the nest itself, before there are eggs) and hovers in front of it or even gingerly alights, with a prolonged wing flutter display. Because a male neither selects the nest site his mate uses nor helps to build a nest, his visits to a potential nest site are not directly functional; rather, they are behavior that is specialized to be informative, and is made the more conspicuous by being combined with the wing fluttering. Further, throughout his approach and while fluttering at the site he utters the chatter vocalization display, a rapid string of harsh calls (figure 3.1F), each resembling the main portion of a twh-t. The compounded display performance (visible plus vocal) appears to make information about his strong attachment to the site conspicuously available. This site attachment is directly functional later, when there is a nest, where males spend a great deal of time throughout the nesting cycle.

Males display this way when alone, both before pairing and after the first nesting of a season when they will leave their families to return to the nest site. A male also nest-site-shows in the presence of his mate, flying from her to the nest or a similar site near it, returning part way to her, then flying back to perform again at the site. The display appears to be done only when a male has to choose between remaining at the site and doing something else: such as patrolling, foraging, or associating with his mate or family as they move about. It thus makes available information that the communicator is behaving indecisively; he may remain at a site or do some incompatible alternative. The wing fluttering component of the performance attests to a temporarily delicate balance between staying and flying away; an individual that either settles on the site or turns from it usually ceases wing fluttering almost immediately.

Females have not been seen to nest-site-show, but have occasionally been heard to utter the chatter vocalization. One chattered while coming to feed nestlings when a stuffed owl was 3.5 meters away, apparently severely taxing her attachment to the nest and young. On a very few occasions brief chatters were recorded from mates as "greetings" in the general nest area; in form these calls were particularly like repeated components of the twh-t. Greetings are not much developed as ceremonies in this species; that is, no single patterned exchange of signals recurs frequently, and most meetings are initiated without displays.

Flight Displays

The simultaneous performance of visible and vocal displays in nest-site-showing appears to represent a fixed sort of compounding that anchors one end of a highly variable continuum of flight performances. The remaining flight displays seem more frantic than most nest-site-showing, and at their extreme involve rapid, erratic, ascending flight, sometimes circling over the point of origin. Rapid and erratic flights that are intermediate between the ascending kind and nest-site-showing are done by a male in the presence of his mate during unusually vehement interactions near the nest.

Only nest-site-showing is closely tied to the nest site itself, but all flight displays are performed within the territory. If one is begun near the nest site it usually passes over it. These displays thus make available the information that the bird will probably remain in the region where it is displaying. They also indicate that it will behave indecisively, although further indications of the alternatives it may select do not appear to be made available by visible aspects of the performances. There are usually vocalizations, however: the ascending flights can in fact be thought of in part as providing a conspicuous platform from which to utter chatter vocalizaitions or bipeaked vocalizations (see next section). Chatter vocalizations correlate with a range of possible actions that can interfere with remaining near a nest site, whereas bipeaked vocalizations appear to correlate with agonistic behavior.

Eastern phoebes give the dramatic ascending performances extremely infrequently. A male that has been uttering tp for a long period and has begun to repeat it very rapidly, or that has been singing a great deal, may launch abruptly into an ascending flight display. On alighting he immediately sings very rapidly, initially with an excess of the fee-b-be-bee song form. Once, in a flight display with bipeaked vocalizations that followed a territorial border dispute, I saw a bird begin the fee-b-be-bee singing while still in flight as it straightened out and made for a perch.

The very few observations and the variable forms of flight displays make them especially hard to interpret. Yet comparison with other species strengthens the case for regarding them as belonging to a graded continuum: the black and Say's phoebes, the other two species in the genus *Sayornis,* have highly variable performances that appear to span similar ranges of form to the eastern phoebe's, and to be performed in similar circumstances. The eastern kingbird and its congeners also have very variable flight performances incorporating diverse vocalizations. At their extreme these include mock aerial fights: the displaying kingbird acts as if he were grappling and tumbling down with an imaginary opponent. Such mock fights may be a further reflection of the more aggressive nature of kingbirds.

The Bipeaked Vocalization and Other Agonistic Displays

The bipeaked vocalization is a loud "T-keet" (figure 3.1*D*), known only from the rare ascending flight displays and from intraspecific fights between an eastern phoebe and either its mate or an intruder it is driving from the territory. Because such encounters can be fast and vicious, it is not usually possible to determine which individual utters each call. The calls are heard while both are fighting, and not when one breaks off to flee. Because most attacks of one bird on another are silent, the bipeaked vocalization is not given by birds that are simply attacking. Its utterers are probably fighting defensively. The same call in flight displays does not correlate with defensiveness in fights, but with being unable or unwilling to interact. What the bird would do if immediately given an opportunity has not yet been observed. It is likely, though, that this vocalization provides information about the probability of attack behavior in circumstances in which attack is not the only possibility.

At least four displays other than tp and the bipeaked vocalization are involved with information about agonistic behavior. The crown ruffled display (figure 3.1*E*), raising the feathers atop the head, commonly occur with tp calls and appears to correlate with indecision between attacking and either withdrawing or remaining at a distance. It is variable in form, and when both mates approach a predator near the nest the one that approaches closer or even strikes (if the predator is small) has the more conspicuously raised crest.

Between approaches to a predator, birds may adopt what appears to be a specialized posture, a crouch display. Though this posture is not very striking in the eastern phoebe's repertoire, eastern kingbirds have a similar but very obviously specialized crouch display that they use in agonistic circumstances.

The third agonistic display is an abrupt whirring sound made by the wings of a phoebe chasing hard after another individual; because similarly hard flights can be made without this whirring, it appears to be the result of a specialized manner of flying. Another audible but nonvocal display is produced by snapping the bill shut loudly in an act that is like biting but done with nothing in the bill; these snaps are made as an approaching phoebe passes close by a target without actually striking.

Vanishingly Rare Displays and Inconspicuous Displays

Two vocal displays of the eastern phoebe are uttered very rarely, and yet both are prominent in the display behavior of the two western species of *Sayornis* to which it is closely related.

The two western species give an initially peaked vocalization as a component of their singing performances; the black phoebe even has

special variants of this unusually plastic display that are uttered primarily in correlation with the act of alighting when it is behaving indecisively. The variable initially peaked vocalization of the eastern phoebe (figure 3.1*I*) is extremely similar to the forms of those of the other species, especially the black. It is uttered when an indecisive communicator is actively making short flights in circumstances when the fee-bee song form and the twh-t vocalization might be expected. However, it is given so rarely that the events seem to be chosen almost capriciously.

The doubled vocalization of the eastern phoebe is uttered even more rarely and capriciously, having been recorded from only two males of the large number of pairs of this species that were studied. Its form (figure 3.1*C*) is extremely similar to that of the chatter vocalizations of the other two species and, like those displays, it appears to be uttered primarily in the nest vicinity by a bird that is doing much indecisive flying.

These two little-known displays of the eastern phoebe appear to be no more than vestiges within its repertoire, redundant with more commonly performed displays. Perhaps, as the sole member of a large and essentially Andean lineage of tyrannid flycatchers to invade eastern North America, the species has undergone fairly recent and important evolutionary changes. Whatever the case, the finding is not uncommon: displays that are apparently vestigial are known from the repertoires of intensively studied species of birds in several evolutionary lineages. To Moynihan (1970b) this comparative evidence jibes with other characteristics of display repertoires and suggests that some evolutionary processes replace displays within species' repertoires; the theories he proposed to account for this continual turnover are discussed in chapter 11.

The eastern phoebe may have other displays that are more commonly performed than present evidence indicates. Some inconspicuous fluffings and sleekings of its plumage may serve as visible displays, yet are sufficiently subtle that a human observer detects them only with difficulty when working at a much greater distance from the birds than they may be from one another. None appears to be very prominent in this flycatcher's behavior, and because it is a somewhat asocial and cryptic bird it may never have evolved much elaboration of plumage signaling.

Information and Displays

Most display acts found in the behavior of adult eastern phoebes are instantly recognizable, and their forms differ from those of displays of other species that live with it. They thus provide information about the communicator's identity. The forms of the vocal displays also provide

Table 3.1. Summary of information made available about behavior by displays of the eastern phoebe.

Display	Behavior								
	Interaction			Seeking[a]	Escape	Flying	Indecisive behavior	Unspecified alternatives to other acts	Remaining in a neighborhood or at a site
	Not further specified	Attack	Copulation						
Twh-t	x	–	–	–	–	x	x	x	–
Fee-bee and fee-b-be-bee	x	–	–	x	–	–	–	x	x
Tp	–	x	–	–	x	–	x	–	–
Tail wag	–	–	–	–	–	x	x	x	–
Wing shuffle	–	–	–	–	–	x	x	x	–
Wing flutter	x	–	–	–	–	x	x	x	–
Nest-site-showing	–	–	–	–	–	–	x	x	x
Chatter vocalization	–	–	–	–	–	–	x	x	x
Ascending flight displays	x	–	–	x	–	x	x	x	x
Bipeaked vocalization	–	x	–	–	x	–	–	–	–
Crown ruffled	–	x	–	–	x	–	x	–	–
Crouch	–	x	–	–	x	–	x	–	–
Wing whirr	–	x	–	–	–	–	x	x	–
Bill snap	–	x	–	–	x	–	x	–	–

Differences among the ways some displays are performed are obscured because the relative probabilities of different acts are not indicated.
[a] The criteria for concluding that a display provides information about the behavior of seeking to act are developed in chapter 5.

information about a communicator's location. And each display makes available information about the kinds of behavior in which the communicator will probably engage (table 3.1). This was learned by finding that the performance of each display consistently correlates with a limited range of other communicator activities, however diverse the circumstances in which the display occurs. By applying similar observational methods to the performance of displays by other organisms, we can make a comparative survey of the kinds of information that are generally made available by this special class of behavior patterns, and gain some appreciation of what displays can contribute to the processes of interacting (see chapters 5 and 6).

Several of the phoebe's displays carry information about the likelihood of flying. Flights differ and serve quite different ends at different times, but the kinds of flights are *not* specified by these displays. Birds are very active creatures and can move rapidly; information about the probability that one will move is, in many circumstances, very useful for others to have, even if other sources of relevant information must be sought before responses are made. The information that a communicator will restrict its activities to certain neighborhoods is also useful and is made available by several displays.

Some displays correlate with readiness to interact, and some more specifically with the likelihood of attacking or fleeing, or of being indecisively balanced between these or other activities. However, although many displays deal with information about agonistic or indecisive behavior or both, none is known that deals with the likelihood of attempting copulatory behavior, even though initiating copulation interaction is often difficult for eastern phoebes. For some pairs in which the male is more timid than his mate, copulation is especially difficult. Nonetheless, it need not occur frequently for reproduction to be assured.

The phoebe's displays are remarkably spare in the detail they provide about a communicator's behavior. Perhaps only some acts are sufficiently crucial—or crucial sufficiently often—for evolutionary forces to have tied one or more displays to the task of making available information about them. The implications of this finding are important in developing the general semantic theory offered in chapter 7.

4

Messages: The Kinds of Information Made Available by Displays

Displays are acts specialized to make information available. Seeking correlates for the performances of displays of the eastern phoebe (see chapter 3) revealed that information of several different kinds, both behavioral and nonbehavioral, can be made available by each display. Other research indicates that the information provided by the phoebe's displays is surprisingly representative of the kinds dealt with by the displays of many species of birds and mammals and perhaps of most vertebrate animals. That is, it appears that the display behavior of most species traffics in the same few kinds of information. The task of this and the next two chapters is to clarify the concepts and procedures involved in studying this information, to give an account of the descriptive categories that best represent our current knowledge, and to provide some indication of how widely it is used by different species. We can then ask why these kinds of information should be so fundamentally important to animals.

Information is an abstract property of things and events. It can be dissociated from them and disseminated without loss to them. The displays that act as vehicles carrying information are not abstract. They have physical reality—for instance, as patterns of waves in the media through which they travel. Their "referents," the things and events about which they provide information, are the communicators themselves and such characteristics of the communicators as their behavior or probable behavior. (In principle, referents could also be external to communicators, just as are the referents of many words. However, most putative cases of displays with such referents are doubtful, and it is not fully clear how to describe the information made available in the remaining cases.) Each kind of information that a display makes available about a referent is designated as a "message" of that display, following W. John

Smith (1965, 1968, 1969a, 1969b; the term "message" was taken from Cherry 1957; see also the second edition of Cherry 1966).

The study of messages is concerned only with the kinds of information that displays enable their users to share. The issue is to learn how the universe of possible referents is divided by displays. Of all the many topics about which animals might adaptively communicate, with which does display behavior deal? What kinds of relationships exist among the kinds of information that are made available? What is the potential and what are the limitations of sharing information by displays?

At this level of analysis the subject of information should be treated only descriptively. Message analyses cannot yield a categorization of the functions of communicating, and whether any individuals make any use of the information that is provided by displays is not even at issue at this stage. Ethologists too often let interpretations of functions intrude, at least when they are dealing with the kinds of behavior referred to by displays. This is not appropriate, and often involves a simplistic linking of two quite different functions: the function of the behavior that is the referent of a display, and the function of the act of communicating about that behavior. The range of activities to which each behavioral message refers can be defined by their function: for instance, the message of escape behavior covers acts as disparate as fleeing and freezing but both share an important function for the communicator. The acts covered by the message of locomoting (such as flying or walking; see chapter 5) are functionally similar in moving the communicator through space, although beyond that their functions diverge widely. Recognizing the functions that unite the kinds of behavior done by a communicator in correlation with a display, however, usually sheds limited light on the functions of *communicating* about that behavior. An analysis of the functions of communicating asks why it is adaptive for an individual to make information about itself available and for another to respond. Procedures for such analyses depend in part on studying the responses made to displays; they are discussed in chapter 10. The study and categorization of messages asks questions only about the kinds of information being made available, and ideally depends on analyses of the behavior of displaying individuals, not of recipients.

General Classes of Messages

The kinds of information made available by displays fall into general classes similar to the journalist's familiar questions: what happens, how does it happen, who makes it happen, and, usually, where does it happen. A journalist would also report on when, but display behavior seems to be able to deal only with the present and its possible implica-

tions for the future. A journalist may also go beyond descripton to determine why by interpreting the causes of an event. This usually requires more information than is available in the event itself. Causes are antecedents—some immediate, some of very long term—and they must be sought with special kinds of analyses. The causes of display behavior can also be analyzed, and at various levels, but displays themselves provide only some of the necessary information. However relevant they may be to a full understanding of display behavior, topics of causation are peripheral to a study of messages. They are considered, in one way or another, in chapters 8 through 12.

A journalist can offer many different perspectives, but the information provided by most displays, in fact I suspect by all nonhuman displays, reflects an entirely egocentric perspective. As Lyons (1972:71–72) remarks, theoretical linguists would describe it as "indexical": it is information about the sender. True, ethologists have not always phrased the information in indexical terms, particularly when discussing honey-bee dances and what are often called "hawk alarm calls" (discussed under separate messages in chapters 5 and 6), but the available evidence indicates that it would be more appropriate if they did. For displays, the "what" and "how" of the journalist are classes of messages that indicate what a communicator is likely to do and how it will do it; the "who" and "where" describe the communicator itself.

Behavioral Messages

Both behavioral (what and how) and nonbehavioral (who and where) classes of information are made available by displays. The what and how subdivision of behavioral messages is described more precisely by viewing the what class as behavioral selections and the how class as behavioral supplemental messages.

First, behavioral *selection* messages provide information about what a communicator is likely to do: of all the acts it could perform, the selections it makes or might make. In practice, each behavioral selection message is a category of acts that, although they may be performed very differently by different species, have comparable basic tasks: for example, they move the communicator through space (locomotion), remove it from danger (escape), or keep it near another individual (association). Second, the *supplemental* behavioral messages provide information about how a selection will be performed in two senses: (1) how likely the selection is to be performed, and (2) in what manner it will likely be performed for instance—partially or fully, weakly or vigorously, and in what direction. Such effectively adverbial information always accompanies behavioral selection information and is essential;

the term "supplemental" is not meant to imply secondary importance.

The term "selection" has been used deliberately to describe animals as choosing among such various acts in their behavioral repertoires as attack or escape, locomotion, or doing something else that cannot be done while locomoting. To say that animals "choose" or "select" among the possible responses open to them in any situation is to imply only that they use one of several alternative "programs" (in the sense of Mayr 1974a, 1974b; see the discussion of teleological terminology in chapter 1). These programs dictate courses of action that have been tested, in many cases by natural selection during evolution, and found to produce adaptive results in definable circumstances. For example, if an animal hears a vocalization given by a fleeing associate, the type of vocalization often called an "alarm call," it may either flee or freeze (become stiffly immobile). Whether it freezes or flees depends on the kinds and quality of cover available, how near it is to this cover, whether it is already high (vulnerable to aerial predators) or low (vulnerable to terrestrial predators) in the habitat, and whether or not it was moving when it heard the call. It may vacillate, unable to settle on the appropriate tactic, but vacillation may be fatal in dealing with a predator. Most individuals will respond immediately to this circumstance by either fleeing or freezing, here described as "selecting" or "choosing" between these two responses (vacillating, when it occurs, is also described as a selection). As a logical extension, it is assumed that animals choose continuously: at every instant they somehow assess the information available to them and, as a result, either continue their current behavior or switch to one of several options (see first section, chapter 8). No particular model of internal causation is implied, and specifically not a model of "intentional" choice. Sometimes choices can be seen to be intentional (the problems in studying this, with respect to signaling choices, are discussed in chapter 10), but this is not necessary to the notion that selections are made—programming can make selecting an automatic process.

Nonbehavioral Messages

Nonbehavioral messages are also divisible into two classes. First, the "who" class *identifies* the communicator, providing information about the kinds of groups of which it is a member: its species, local population, and perhaps its sex. Of course, any information that identifies a communicator inevitably permits some prediction of its behavior: the communicator will remain within the physical limits of what is possible for it, will probably choose its behavior from the normal repertoire of its group, and will direct its attentions toward other individuals of certain

classes. That is, identifying information defines a framework for behavioral predictions. Second, the "where" class of messages provides information about the communicator's *location* through special physical characteristics of some displays.

Procedures for Analyzing Messages

Analyzing messages provides an accurate description of the kinds of information made available by each display. The physical form of behavior that appears to be a display must be described first (in principle a syntactic task, although some problems that can arise in defining which acts are display units cannot be solved by syntactic analyses alone; see chapter 13). For at least audible displays, the physical form determines the extent to which information is made available about the communicator's location. Once an observer can recognize a display, identifying messages can be determined by learning who performs it. Each kind of display is performed only by communicators of certain classes (certain species, perhaps just one sex or age group, and so forth), and those classes are thus among its referents.

Behavioral messages are found by examining the behavior with which performance of the display correlates. No display comes at random moments; it tends to occur before, during, or after some activities and not others, and so provides the information that the former are more likely to be performed than are the latter. The activities with which a display most commonly correlates at the time when observations are made (such as the time of day or the phase of an annual cycle) tend to be described first, and provisional message hypotheses can be formulated to describe them. No assumption should be made that messages known from the displays of other species will necessarily fit; the provisional hypotheses must describe what is observed, and must be modified or replaced to remain descriptive as further behavioral correlations are observed. (When it becomes desirable to compare the messages of displays from different species, procedures for categorization must be developed; see the next section of this chapter.) As a pattern becomes clear, an observer can seek two kinds of tests of the adequacy of the provisional message hypotheses. First, the display is sought in circumstances in which it has not been encountered, but which are either expected in terms of the hypothesis or contrary to them. Second, additional examples of its occurrence in complex circumstances are sought to clarify details that may enable the hypotheses to be made more precise.

If a display were performed only in correlation with environmental events that bore no consistent relationship to the behavior of the com-

municator, then those events (perhaps defined by the presence of things in the communicator's environment) would be its referents. Display communication would then be less egocentric than is claimed earlier in this chapter. The claim that most displays do not have referents external to the communicator is empirical, because all kinds of correlations are sought during research. It is also a claim that some ethologists dispute— for example, in discussing alarm calls and the dancing of honeybees (see chapters 6 and 7)—and it should be recognized as provisional. If detailed study does reveal cases of displays with referents external to the communicator, the finding will be exciting but will pose no serious problems for the theoretical approach suggested in these chapters. (Perhaps the best cases now known are displays performed only when communicators are remaining steadfastly at or within some particular site such as a nest or a territory. These displays can be said to have sites as external referents, but the issue is equivocal. There is a difference between a display performed only in correlation with behavior at a specific site and an alarm call that might refer to a hawk as yet unseen by the call's recipients. The recipients of a nest-site-showing display can see both the display and the nest or potential nest site; any information that is made available by the display about the presence of the site is entirely redundant. And displays performed only in territories in reality refer to behavior, because territoriality is not property of space, but of the way an animal behaves with respect to space.)

In practice, analyzing the patterns of communicator behavior that correlate with performance of a display to determine its behavioral messages demands a great many observations and extensive sampling of the range of a communicator's activities. For example, suppose we find a pair of eastern phoebes in which the female has completed building her nest. We should recognize the twh-t vocal display discussed in chapter 3, and find that the male utters it when approaching his mate in the predawn twilight. It would be characteristic of incomplete approaches and tangential digressions, but after flights directly to her he would attempt to copulate without display. We might conclude that twh-t is uttered when a male, ready to approach for copulation, is likely to divert momentarily into an alternative like associating. We could then make the provisional hypothesis that twh-t encodes messages about the relative probabilities of copulation, association, and indecisive vacillating behavior. However, continued observations reveal twh-t in other kinds of events and render this hypothesis untenable. During foraging the male, and occasionally the female, utters twh-t on suddenly alighting very near its mate or when the mate alights nearby in an apparently hasty and unsuitable perch selection. That these calls come with an intention movement or a short flight away, but not with fleeing farther,

suggests that the communicator is prepared to escape, but not far, and will associate. In other observations, twh-t is uttered while mobbing predators near the nest and on alighting from approach and escape flights. None of these observations substantiates the postulated message of copulation behavior: no attempts to copulate are seen in most foraging, and attack is seen to be the alternative to escape from the predator in those cases in which the communicator approaches all the way. (By inference, all flights in which the communicator comes close to a predator are flights beginning as attack, even though they are usually aborted.)

Taken all together, the observations suggest that the communicator will behave indecisively, at least briefly. Further, it may copulate, associate, attack, or flee—although in any one event we cannot know which without further information. We can say that some form of interactional behavior is always one possibility. Thus the display provides information that the communicator may interact in an unspecified way, exercise an incompatible alternative, or behave indecisively. Note that we cannot say that escape is encoded because the alternative to interacting may not be escape but some other form of interaction. The communicator is *not* always ready to escape when it utters twh-t.

A further correlation is especially consistent throughout these observations: twh-t is uttered as a bird alights. The very few exceptions correlate with unusually indecisive behavior involving flight, as shown in chapter 3. Whatever other messages twh-t carries, information about flight is certainly encoded: flight is momentarily unlikely, but is involved in the indecisive actions.

Combining these observations enables a reasonable interpretation of the messages of twh-t, although observations in yet other circumstances can still help. For instance, if a female phoebe falls prey to a hawk, or if we begin to observe in early spring, we would find that males without mates utter this display abundantly when alone. They do not call twh-t when just foraging, nor when making maximal efforts to seek interactions, which they do by singing from high perches. This vocalization is uttered when they vacillate indecisively between foraging and going to a singing perch. Because it continues to correlate with the act of alighting, the messages postulated about the immediate likelihood of flight behavior are adequate. However, the information about interactional behavior is more fully revealed. Not only is the kind of interaction not specified, the information does not even imply that interaction is possible: a bird uttering twh-t may just be seeking an opportunity to interact (inferred from the correlation with advertising behavior), and behaving indecisively with respect to an alternative that involves individual maintenance behavior.

Reworking this example from chapter 3 indicates that the messages of a display are not understood until observations are sufficiently complete to reveal the whole range of behavior patterns with which the display correlates. Anything less may yield too narrow an interpretation.

In addition, the example shows how analyses actually proceed during the study of a species' displays. As the range of behavior correlating with a display is gradually discovered, postulates about the display's behavioral selection messages are made, tested, modified, or replaced. These postulates are not determined by the dozen message categories described in chapter 5 (although in the example each does fit one of those categories), but by whatever descriptions most economically encompass the known range of behavioral correlations. Followed carefully, this procedure will reveal both previously undiscovered messages and further examples of already known categories.

The procedures involve recognizing as messages only the referents that consistently correlate with the performance of a display (for example, in the case of twh-t, flight but not escape). This is because any specialized signal must have constant referents that are known to all relevant individuals; without them it cannot facilitate the ability of participants in an encounter to predict, and hence to accommodate to, each other's behavior. Certainly Humpty Dumpty's whimsical use of denotation did much to contribute to Alice's impression that he was a very unsatisfactory person. The egg might be able to translate the poem *Jabberwocky,* but he could scarcely talk without having to explain himself after every utterance. His words did not, as he claimed, each do a lot of work; instead, a great deal had to be done to stipulate his referents and permit his words to function at all.

Humpty Dumpty's words represented a dysfunctional extreme of a phenomenon that is not limited to Wonderland, but must be universal among species in which aspects of the use of signals are left open to be learned. Each individual in such a species constructs what Huxley (1966:259) described as "his private ritualizations," becoming idiosyncratically attached to particular connotations and using his signals at least slightly differently from most other individuals. His constant companions may come to know the patterns governing his use of frequently occurring signals, and thus to this extent share his understanding of their referents even if they do not perform the same signals in exactly the same ways. Among strangers, he will be less well understood. Even within his usual social group, however, an individual cannot afford the extremes deliberately chosen by Humpty Dumpty. For the most part, we understand each other not to the extent that our words or other signals mean just what each individual chooses them to mean, but rather to the extent that we all employ them in the same ways, making the same

information available (this is often described by saying that we must share a common code). Because an individual's signals are functional to the extent that each represents a "societally standardized concept" (Carroll 1964), he must either accept socially defined referents for the signals and match them approximately to his individually formed concepts or modify his concepts to correspond to the ways in which members of his society use signals. An individual's private conceptions of referents are important primarily for pragmatic research because they can hinder communication, making the process less effective. The public or shared aspects of referents are the main subjects to be described by semantic analyses.

However, one communicative phenomenon partially escapes the need for a set of referents known to all the users of a signal, but must take root in this need. The phenomenon is described in human communication as the use of metaphor. When no signal in a repertoire fits a situation exactly, one that fits approximately may be useful if it is not taken literally. The attributes that its referents have in common with the situation at hand will, taken in context, convey a relationship that is significant to the communicator. Thus, although the word "box" may have as its referent a rectangular receptacle with four upright sides, we extend the term to describe three-sided cañons that end steeply and even metaphysical spaces that enclose (one can become "boxed-in" by one's premises). Metaphors aid communication to the extent that they do convey information about referents (particularly if other existing signals are less adequate), and to the extent that recipients grasp the implied relationships—something a recipient cannot do without knowing the usually accepted referents of the terms. The metaphorical use of terms thus trades on resemblances between characteristics of the established and novel referents.

Metaphor is a creative device for communicating about novel referents, including novel aspects of established topics. It is also involved in the evolution of language by either a process of changing a term's referents, often as the term also changes (the English *moon* derives from the Latin *mensis,* which referred to a month, and the French *lune* from the Latin *lux,* which referred to light; see R. Brown 1958) or by expanding the use of a term so that its referents, strictly speaking, become those attributes shared by diverse topics. In the latter case the term becomes broadly used, and ambiguous in the absence of information from contextual sources.

Important as the use of metaphor is to humans, however, we have as yet no idea of the extent to which it is involved in the communicating of nonhuman animals. To speak of displays as having "fixed" relations to their referents should thus not be taken to imply that no creative use can

be made of them in situations in which a communicator has no signal that fits closely. And, even if species other than humans do use metaphor, fixed referents necessarily underlie the phenomenon.

Also, it must be emphasized that to say that a signal has fixed relationships to its referents—that is, it carries the same messages whoever performs it—is very different from saying that the functions of making these messages available are also fixed. A message can influence recipient individuals very differently in different situations: responses to messages are labile. But this is because the circumstances of displaying, not the messages made available by a display, are variable. The twh-t vocalization is uttered by lone birds as they patrol, by birds who are making sexual advances or just associating with their mates, and by birds flitting about in the presence of predators. A recipient can attend both the display and many other sources of information in any event, and the influence of those other sources can greatly alter the way it responds.

One practical consequence of this, already discussed in chapter 1, is that we can much more readily study messages by seeking correlations between a communicator's behavior and its performance of a display than between the occurrence of a display and the behavior of recipients. Recipients may not respond at all or may respond largely on the basis of information from sources other than the display. However, when a display does elicit important features of their behavior, these responses do provide clues to the display's messages.

Even though the communicator's behavior is the most direct and almost always the best place to seek clues about the behavioral messages of a display, its study can present problems. Some arise because the circumstances in which a display is given may change before the communicator follows through with an appropriate action. The communicator then does something else, leaving the observer with a spurious correlation. Such apparent anomalies account for a portion of the observations of any display. Many adhere to no pattern or are marked by communicator inactivity when activity would have been expected. Sometimes the behavior of individuals who were recipient to the display provides the most useful clues. For instance, if one kind of display is almost always immediately followed by an opponent fleeing it may be reasonable to infer a message of attack behavior. An observer will not often see attack, but probably because the recipient's usual response obviates it. A provisional rule might be offered: an observer may have to take recipient behavior into account in a message analysis if recipients of a display respond quickly with acts that appear to forestall, cope with, or cooperate with the communicator's behavior, because the recipients' acts may immediately affect the communicator's selection of

behavior. Of course, such a rule leaves far too much to the discretion of individual observers, and it will need to be limited once its pitfalls become evident.

Even many displays that do not lead immediately to obviating responses can present an observer with problems. These displays are performed by communicators who are behaving indecisively and probably will continue to do so, but who may become decisive if circumstances permit. Difficult though it may be, observing the behavior of the communicators themselves is usually the best way to seek clues to the messages that refer to the decisive acts. These problems are considered more fully in the discussions of the "indecisive" and "probability" messages in chapters 5 and 6, respectively.

Comparisons of Messages among Species

Comparing the kinds of information made available by displays of different species is useful. It permits the study of each species to be made more complete, because features discovered in species where they are obvious can then be sought in other species where they may be quite subtle. As more and more species are compared, patterns emerge in the kinds of information in which the displays of different species traffic. It can then be asked why features should be obvious in one set of species and yet subtle or absent in another. This opens inquiry into the principles governing the distribution of messages among animals, which is done by attempting to correlate patterns of message distribution with ecological, social, or phylogenetic differences among species. The comparisons thus lead to pragmatic research into the ways in which the messages of display behavior can be adaptive. In addition, broad comparisons of messages reveal the range of informative contributions that display behavior can make to the task of interacting, providing an idea of the scope within which displays can operate as sources of information, as well as of the general bounds to the opportunities they offer.

Classifying Messages

Comparing messages immediately presents serious problems of categorization. Even in dealing with studies of single species it is necessary to compare one message with another, and categories are helpful for descriptive sorting. Behavioral and nonbehavioral categories have already been mentioned in the analyses of phoebe displays, and behavioral information has been further categorized as referring to interacting, attacking, or flying. This categorization has required decisions about

how to classify observations. For example, attack behavior has been viewed as a subclass of interacting, and the message referring to attack includes both some approach behavior (aborted attacks, discussed above) and actual striking behavior (a phoebe attacks a snake with its bill, and attacks another phoebe with its bill, carpals, and feet); the message referring to flight behavior has been interpreted as including both acutal flying and what appear to be preparatory movements of taking flight.

In analyzing the messages of phoebe displays, decisions about the behavior to be lumped within each category were guided by the kinds of behavior observed to correlate with each display. In the displays of other tyrannid flycatchers a similar procedure has produced similar results, as well as a broadening of the behavioral categories. For instance, flying behavior is not an adequate category in a species that employs its vocal analogue of a phoebe's twh-t not only in relationship to flight, but also in the same relationship to hopping and walking. Because flycatchers make no distinction between modes of locomotion in using their displays, the behavioral category was broadened from flying to locomotory behavior. Further, when displays of Adelie penguins were studied by David Ainley (see chapter 5), patterns were found that are closely comparable to those of tyrannid flycatchers even though penguins are flightless and flycatchers are not aquatic. Two penguin displays correlate with walking, diving, and swimming. The category of locomotory behavior is useful for describing their messages.

Ainley was unaware of the flycatcher studies while he was in Antarctica, and his observations were unbiased by any intent to seek tyrannid behavioral categories in penguins. The categories were found to fit on the basis of a posteriori analyses of his data, which is important. In principle, fitting messages into categories must not be done on the basis of a priori postulates or intuition because these may guide (and limit) observation. The messages of a display must always be defined a posteriori on the basis of careful observations that seek anything that might correlate with the performance of the display. In practice, some postulates always guide observing and categorizing—an observer is not a *tabula rasa*. For instance, I have come to expect displays to provide information about a communicator's behavior, and doubt that they provide information about external referents (for example, I doubt that a display of a prey species provides information about a class of predators); these expectations bias me to seek evidence for *both* kinds of information. More limiting are biases based on ignorance. It is always easier to test for the presence or absence of a message that has been previously recognized than one that has not: at least some criteria for the former are known, but the existence of the latter message may be

suspected primarily because some of the display's correlations fit no existing expectations. Describing the correlations may be difficult until some feature is seen that ties them together. Indeed, observation probably favors recognition of that which is readily classified, a bias to note the old more readily than the new (unless the new is blatant).

It is necessary, then, to seek unexpected correlations for displays and to be able to categorize these as appropriately as possible. Among the serious problems is that there is no simple way of deciding how much detail must be described before behavior can be assigned to a category. Appropriate categories, however, must describe what has been observed. When observations do not fit existing categories it is necessary either to redefine the inclusiveness of the categories or, if that renders the categories uselessly broad, to recognize new categories. Categories are only the tools of analysis, not its end.

Categories must not only describe, they must also facilitate comparison. The comparison of diverse animals requires inclusive categories, so that a bird's pecking and hitting with its carpals (wrists) might have to be lumped with the clawing of a cat and the butting of a goat. Certainly, such procedures involve some arbitrary decisions and the errors these portend. It would not do to forget this risk, even though the effects of small errors at such gross behavioral levels may not usually be significant.

Despite the inevitable risks, display behavior cannot be studied unless the raw data of observations are classified. Nothing can be studied without sorting initial descriptions into categories; classification is necessary to organize data in the successive steps of any scientific endeavor. But classifying need not be a sinister act if it is not an end in itself, and if it is kept in mind that any classification is limited by the kinds of similarities and differences that are used in applying it.

The use of categories leads to two kinds of danger for the development of theories: first, that the processes of classifying will come to guide the nature of the comparisons that are attempted, freezing theory into too narrow a mold; and second, that the desired comparisons will too freely guide classification, producing highly arbitrary categories. The risks can be minimized by keeping the criteria for categorical inclusion and exclusion as explicit and objective as possible. The assumptions underlying the criteria should be reexamined repeatedly and changed as new data warrant. Classifications should be revised or replaced when they are not suited to the task at hand; no classificatory scheme should be expected to handle all data or serve all kinds of comparisons equally well. The classification of messages offered in this book should not be taken so seriously that it is not changed as further discoveries warrant, or it will hinder rather than facilitate comparisons.

Studies Contributing to Message Comparisons

To analyze just the messages of the displays of any one species is time-consuming. Detailed interspecific comparisons require message analyses of whole display repertoires, and not many devoted specifically to message interpretations have been completed as yet.

The procedures for determining messages were first devised during research on birds: kingbirds, phoebes, and other species in the tyrannid flycatcher family. Detailed accounts of display repertoires have been published for species of both kingbirds and phoebes, which represent two tyrannid genera that are not closely related within the taxonomic family (W. John Smith 1966, 1969c, 1970a, 1970b). Less detailed accounts are available for several other genera (W. John Smith 1967, 1970c, 1971a, 1971b), in which species range from foliage-gleaners in tropical forests to lark- and shrike like birds living in treeless regions of the high Andes. Field work on other key genera from all sections of this large and diversified family has been completed, and some analyses are far advanced. These studies of flycatchers have been used as the basis for a broadly comparative project, designed specifically to investigate the kinds of messages made available by the displays of species with similar and divergent social habits and ecological characteristics, close and distant evolutionary (phylogenetic) relationships, and different degrees of complexity in their patterns of displaying.

The messages of the relatively large display repertoire of the Carolina chickadee *Parus carolinensis* were studied by S. T. Smith (1972). Chickadees represent a phylogenetic lineage quite different from that of tyrannid flycatchers, and the annual cycle of social behavior in this species is more complex than that of any tyrannid yet studied in detail: territorial pairs nest and raise families, and then join organized over-wintering flocks of about five to ten individuals. With Susan T. Smith and other collaborators I am studying members of other taxonomic families of birds, including the Vireonidae in which males give long continued singing performances that present special opportunities for encoding many details of their behavior (W. John Smith, Pawlukiewicz, and Smith, in press). We are also studying mammals, and have reported on a long-term study of communication in a captive colony of complexly social ground squirrels known as black-tailed prairie dogs *Cynomys ludovicianus* (W. John Smith et al. 1973; and W. John Smith et al., in press 1 and 2). Other studies of both wild and captive prairie dogs are being completed by Penny Bernstein and Sharon L. Smith. One display that is in the repertoires of both humans and gorillas (the latter studied in captivity) has been examined: tongue showing (W. John Smith, Chase, and Lieblich 1974). This study provides a start for

message analyses of primate displays and of the facial expressions in which contributing displays involving the mouth, nose, eyes, eyebrows, or forehead can be variously combined and recombined. Other work on the facial displaying of humans and gorillas is in progress.

Messages of many other species' displays can be interpreted from studies published by other workers, although with varying degrees of success. Some studies require no fundamental reinterpretation, because the authors provide message analyses based on procedures essentially comparable with those outlined earlier in this chapter. For instance, Ainley's detailed analysis (1974a) of the Adelie penguin's display repertoire, the work of Howell, Araya, and Millie (1974) on the gray gull, of Mock (1976) on the great blue heron *Ardea herodias,* and of Schreiber (in press) on the brown pelican *Pelecanus occidentalis,* and Rand's studies (ms.) of the tropical frog *Engystomops pustulosus* make possible detailed comparisons between the animals listed above and representatives of four unusually interesting families of birds and a nocturnal amphibian. On the other hand, most published studies require considerable reinterpretation, and some are not well suited for it. The aim of such research has been to learn not how interactions are managed, but how the behavior of an individual is controlled. That is, most studies have been designed to permit interpretation of the motivational states of displaying and responding individuals, which requires using operational procedures and assumptions that are not fully pertinent to the study of interaction and its components. Another kind of problem arises in some research on captive animals kept under conditions that greatly limit opportunities for social behavior: message interpretations can be incomplete when many of the kinds of behavior with which displays might correlate cannot be observed. The net result is that the basis from which to compare messages among diverse species is less extensive than might be expected even after decades of research on display behavior. Still, enough is known to suggest some extremely interesting generalizations.

Results of Comparing Messages

Comparisons have revealed that the displays of diverse species carry the same few categories of messages. Each of these widespread messages has been found in the displays of at least several species of birds and some mammals, and in addition often in other vertebrate and invertebrate species. Some are probably all but universal: displays encoding a message of escape behavior, for instance, are found in almost all the vertebrates that have been studied, and in many nonsessile invertebrates. Other message categories are universal: for example, information about

the probability that the communicator will act is inevitably encoded along with each message specifying a behavioral selection (see chapter 6).

The displays of most species of vertebrate animals carry few or no messages not belonging to these few widespread categories. To put it another way, most or all displays of any species may provide only information of kinds also carried by the displays of most other species—despite different life styles and evolutionary histories. Many additional messages are in fact known, each encoded by the displays of a relatively few species (see chapter 6), but such parochial messages are usually employed only by species (at least among vertebrate animals) that have minimally complex forms of social interaction.

That a few message categories should be of widespread distribution among species and even dominate their display communication implies that there are basic constraints on message evolution; these are considered in chapter 7. Limitations as fundamental as these in turn imply that comparative studies can tell us a great deal about what to expect of the communicating behavior of animals as yet unstudied and of animals such as prelinguistic hominids that are no longer available for study.

The next two chapters review messages, concentrating on those that now appear to be of widespread distribution among animals. These widespread messages are of most interest in any general account of the behavior of communicating. They are the most pervasive and, for many species, the most frequently significant in the business of managing interactions. However, firm criteria for deciding that a message is widespread do not yet exist and cannot be established until many more species have been studied by standardized techniques. These criteria must be based on not just the number of species that employ a message, but also on measures of the ecological, social, and phylogenetic diversity of the species.

The messages described in the next chapters are understood to different degrees, and our descriptions of some will be altered as we become more familiar with them. Differences, none unanticipated, already exist between this listing and those first published (W. John Smith 1969a, 1969b), and are discussed in these chapters. Despite further expected changes, however, the considerable similarities among many and diverse species suggest that some messages now recognized must be fundamentally important in animal communication. Many of them are perhaps important even for the displays of our own species.

The dozen behavioral selection messages that are almost certainly widespread in the display repertoires of adult vertebrate animals are presented in chapter 5. Examples of those that may not be or are not widespread are discussed in chapter 6, along with the behavioral selec-

tion messages known from the displays of dependent infants, a very special social class.

In describing any behavioral selection message it is impossible to eliminate mention of the probabilistic nature of animal communication. Thus, although the behavioral supplemental message designated as "probability" is not itself treated until later, it is used in discussing each behavioral selection message. It will serve adequately if interpreted in its simple, common usage; that is, as providing an indication of how the probability that a specific kind of behavior is being or will be performed. Probability and other widespread categories of supplemental information about behavior are considered in chapter 6, which concludes with a description of nonbehavioral message categories.

Chapters 5 and 6 contain a great deal of descriptive material with examples of the range of behavior encompassed by each message category, and the diversity of species that have displays encoding it. The annotated list included here can be used to scan and compare widespread messages. Its highly condensed descriptions are somewhat simplified, however.

Behavioral Messages

Widespread Behavioral Selection Messages

Interactional behavior: attempting to interact or to avoid interaction of any of several kinds that may include attack, copulation, association, and other possibilities; this message does not specify which kind of interaction is relevant in any particular event

Attack behavior: attempting to inflict injury

Escape behavior: attempting to withdraw or avoid; can include "freezing" behavior

Copulation behavior: attempting to inseminate, be inseminated, or to carry out the functionally equivalent act of fertilization characteristic of each species

Associating behavior: remaining in the company of another individual

Indecisive behavior: vacillating or hesitating between other selections

Locomotory behavior: walking, running, flying, swimming, etc., with no specification of particular functional classes such as running to escape or to join

General set of behavior patterns: unspecified kinds of activities that usually or always are alternatives physically incompatible with other behavioral selections encoded by a display

Behavior of remaining with site: restricting movements to a particular neighborhood

Seeking behavior: attempting to gain the opportunity to perform some other selection such as interaction or escape

Receptive behavior: prepared to accept some other selection such as copulation or care-giving (the latter may not itself be a widespread behavioral selection message)

Attentive behavior: paying attention to a stimulus; monitoring

Widespread Behavioral Supplementals

Probability: the likelihood that a given behavioral selection is being or will be made

Intensity: the forcefulness or rapidity with which a given selection will be performed (not a unitary message category; subdivision will be required.)

Relative stability: a measure of the expected persistence of a given selection

Direction: the direction a given behavioral selection will take; for example, toward or away from some other individual

Nonbehavioral Messages

Identifying Messages

Population classes: species, subspecific populations, individual
Physiological classes: maturity, breeding state, sex
Bonding classes: pairs, families, troops

Location Message

Location message: information that enables the source of a display to be pinpointed

5

Widespread Behavioral Selection Messages

Every individual has a repertoire of behavior made up of all the many kinds of acts it can perform. It can be thought of as continuously choosing among these acts, even at times when its behavior is unchanging (among the choices available at any instant is to do whatever was done in the previous instant). Any choice can be called a "behavioral selection."

Each kind of display has a consistent and specifiable relationship to certain choices. It is performed in correlation with some kinds of behavior and not others. Thus, to know that an individual is performing a particular display is to learn something about the behavior it may select—every display can thus be described as encoding messages about behavioral selections.

Performances of almost every display correlate with more than one category of behavior. Each display therefore encodes more than one behavioral selection message; whenever it occurs, it makes available information about more than one possible choice. Many displays, in fact, indicate that a communicator may behave in one of two or more incompatible ways, and while making that choice is behaving indecisively.

Each behavioral selection message described in this chapter is illustrated with examples of displays known to encode it. These displays are by no means exhaustively interpreted. It should often be obvious that a given display correlates with more kinds of behavior than those referred to by the message under discussion. This does not necessarily imply an incorrect analysis. Instead, in most cases the analysis is simply incomplete, carried only as far as need be for the task at hand. To lose sight of this would lead to a grossly oversimplified view of the informative potential of displays.

There are other potential pitfalls in a chapter limited to the abstract topic of behavioral selection messages. Thus it must be reemphasized that analyses of these messages are very different from attempts to understand what functions displays can serve by making such information available. This was explained in the preceding chapter but, because ethologists have for so long been preoccupied with functional interpretations of display behavior, experienced readers may tend to stray toward those familiar kinds of interpretation. Functions of communicating are not analyzed here, although occasionally some are mentioned to provide perspective. Functions are not treated systematically until chapter 10; the analyses to follow are concerned with *kinds* of information.

A dozen behavioral selection messages are described and discussed in this chapter. They are all of those now known to be widespread; that is, they are encoded in the display repertoires of diverse species. The chapter begins with the message of interactional behavior, and proceeds to messages specifying particular classes of interacting: attack, copulation, and association. Escape behavior is not interactional (it is the opposite, in fact), but is considered along with attack behavior because the two are often encoded together as alternatives for a communicator. None of the remaining widespread messages specifies a subset of the interaction message, and the order in which they appear is simply one that is convenient for describing them.

Message of Interactional Behavior

Some displays correlate with attempts by the communicator either to interact or to shun interaction. Because they correlate with various sorts of interacting they do not usually specify any one kind. Instead, they primarily provide information about the communicator's readiness, or lack of readiness, to join in acts that involve other individuals.

Readiness to Interact

The eastern phoebe's twh-t display is uttered by birds that are trying to interact, although they may instead do other things such as flee or forage, and their behavior is indecisive. If interaction follows twh-t, it can vary from associating or copulating with a mate to attacking a territorial intruder or a predator. If the bird is alone and cannot interact (a problem taken up in the next section) it may patrol, making itself conspicuous as it seeks opportunities to interact by means of either territorial defense or mating behavior—but, again, it does not specify which. Many other passerine birds have comparable displays. For instance, a male chaffinch *Fringilla coelebs* may utter his "tchirp" courtship call while in association with his mate and continue associating or

Figure 5.1. The bow display of the Adelie penguin. In their initial meeting, a female (left) is bowing after having been attracted to a male who was calling on his territory. He has adopted a related display, the oblique stare bow, in response to her approach. He is accepting her, but still somewhat ready to attack in defense of the territory and may peck her if she approaches too quickly or with her head insufficiently bowed. A male does not often perform the oblique stare bow toward a female after they have established a pair bond, but bows and is not likely to attack her. However, both mates may perform the oblique stare bow toward each other if they have been intensely disturbed—for example, if they have just had to drive an intruder from their territory. (From Ainley 1974a.)

attempt to copulate; as an alternative he may flee from her but, like a phoebe, is unlikely to attack her (Marler 1956a, 1956b).

An Adelie penguin that may interact in some unspecified way does a bow display: it arches its neck and lowers its head, pointing the bill downward, then very slowly raises its head (Ainley 1974a). During pair formation and precopulatory behavior a female Adelie bows to a male who shows her a nest; mated penguins both bow as they move about, arranging stones or changing places, while in close association on a nest site; a male approaches in a deep bow to copulate with his mate; and in their initial meeting and a variety of other greetinglike situations both mates usually bow, often also performing a mutual head-waving display. Even attack interactions occur with bows, although infrequently unless the bow is shallow and combined with an oblique stare display (figure 5.1).

The variable tidbitting compound of displays done by some galliform (chickenlike) birds also initiates a wide variety of social interactions. In at least the chukar partridge (Stokes 1961, 1963) one form of tidbitting occurs both during aggressive encounters between males and in courtship overtures, when it may be followed by associating or attempted copulation. Adults also tidbit over special foods to which their offspring then come. (Young galliforms do not beg for food, but seek their own; adults help them to encounter and thus learn some of the rarer kinds of

foods.) Adults even tidbit over significant food discoveries at some seasons when the performance attracts other adults (see Williams, Stokes, and Wallen 1968).

Primates have many displays that appear to make available information about readiness to interact. Some encode this along with an alternative of escape behavior. These are often performed by an individual that solicits from a more dominant animal reassurance that the latter will permit the interaction. The extended, upturned palm of a supplicant chimpanzee *Pan troglodytes* (van Lawick–Goodall 1968) and the squeaks of night monkeys *Aotus trivirgatus* are examples considered below under the message of receptive behavior, which they also appear to encode. Other displays done by primates in these circumstances are employed in both soliciting and offering reassurance, such as kissing, certain forms of mounting, and various touching patterns of chimpanzees, or just in offering reassurance, such as embracing or patting by chimpanzees (van Lawick–Goodall 1968). It is unlikely that these displays provide any information about escape behavior. They may, however, indicate that of all possible ways of interacting, attack is improbable.

Absence of Opportunities to Interact

The repertoires of most species contain displays that communicators perform when predisposed to interact but need partners or opponents. The song of phoebes and many other bird species (see review by Andrew 1961b) is customarily uttered when no other individuals are present at the site in which the communicator is remaining, usually its nest or territory. Singing individuals of species as phylogenetically diverse as the chaffinch (Marler 1956a), European blackbird *Turdus merula* (Snow 1958), green-backed sparrow *Arremonops conirostris* (Moynihan 1963a), Carolina chickadee *Parus carolinensis* (S. T. Smith 1972), and eastern phoebe (see chapter 3) behave as if seeking other birds (the message of seeking behavior is treated later in this chapter): they take high, conspicuous stations or sing in special display flights, and may actively patrol. Song ceases immediately if a suitable recipient becomes available, and some form of interaction is usually attempted. This can range from territorial defense against neighbors and intruders to associating, copulating, or otherwise interacting with a mate. (Singing also ceases immediately if a predator appears on the scene, but the singer then attempts to avoid the predator; that is, singing does not appear to be performed in seeking attacks from predators, even though it may attract them. Responses by predators are among the costs, the

inadaptive consequences of singing behavior, and they interrupt the business of seeking opportunities to interact.)

In many species, what is commonly called "singing" actually includes a number of displays with different patterns of occurrence, and not all encode the message of interactional behavior without further specification. For instance, Carolina chickadees use their song display on territory before and while breeding and, less often, within the flock range at other times. Singing individuals are prepared both to defend territories and "to interact suitably with mate and family" (S. T. Smith 1972), that is, to enter a broad range of interactions. However, an individual predisposed to interact primarily with its mate or offspring, or patrolling but unlikely to defend its boundaries, utters instead the very closely related faint song display. Its faint song encodes a weighted selection: the communicator is ready to interact, but unlikely to attack. Many other species have similar loud-faint pairs of song displays (Andrew 1961b), and males give the quiet song forms when with their mates.

Carolina chickadees also have a song variant display that does not correlate with seeking a full range of interactional behavior—it is uttered only when seeking an opportunity to attack (S. T. Smith 1972). Many species of birds have much more complex singing repertoires, but detailed message analyses remain to be done.

Many animals other than birds advertise their readiness to interact by means of loud displays. For instance, the description by Golani and Mendelssohn (1971) of howling by the golden jackal *Canis aureus* shows how comparable this display is to the loud songs of birds: "Usually, when one jackal howls, all the jackals within hearing distance join in successively . . . howling acts as a general releaser of social interaction. The act which accompanies or follows howling depends on the jackals present in the area. In all cases, howling is accompanied by a decrease of the distance between jackals. If two pairs are present not far from each other, a fight ensues. If the pair is followed by additional individuals, these individuals run to one of the pairmates" and make submissive overtures; strange jackals are attacked by pairs; cubs approach their mothers and may play-mount; and if no other jackals are in the region, mates perform elaborate precopulatory sequences. In short, after howling jackals seek and join each other, then engage in whatever sorts of interaction are appropriate to the situation.

Shunning Interaction

Individuals also perform displays when shunning interaction rather than seeking it. Such communicators behave as if hesitant to interact and avoid encounters if easily possible. If it is not easy they may accept

interaction, but sometimes in ways that appear at least slightly inappropriate.

One such display, known to be done by humans and young captive gorillas, is tongue showing (W. John Smith, Chase, and Lieblich 1974). A young gorilla often protrudes its tongue as it withdraws or remains apart from an interaction. The tongue is kept visible and makes no licking or other motions (see figure 11.2). The human form of the display is very similar, and people of various age and ethnic groups all appear to show their tongues in comparable circumstances (the range of circumstances available to humans is, of course, greater than that available to gorillas). In general, a communicator tongue shows when in or faced with interactions that disrupt preferred activities, cause delay, or are inherently aversive: for example, when the communicator is reprimanded, attended by a stranger, loses his opportunity to speak in a conversation, or loses or is misunderstood in an argument.

Other displays correlate with more forceful avoidance of interaction than tongue showing. For instance, individual black-tailed prairie dogs utter the chittering bark as they actively reject attempts by others to interact in any way from play to copulating (W. John Smith et al. in press 2). At least one other rodent provides similar information through an ultrasonic vocalization. Laboratory rats utter prolonged calls at between 22 and 30 kHz, less than half the frequency of their other ultrasonic calls and readily distinguished from those (see review by Sales and Pye 1974). They are given during "submissive" behavior in agonistic situations (Sales 1972; Sewell 1967), and in various circumstances when in "a state of social withdrawal" (Barfield and Geyer 1972) such as males in refractory phases between the repeated mountings characteristic of copulatory behavior in this animal (see chapter 2: tactile displays); females resisting attempts by overly attentive males to mount: and males after defeat by another rat or after electrical shock or rough handling by an experimenter.

Some birds may also have displays that correlate with shunning interaction. A high likelihood both of locomotion and of avoiding encounters appear to be encoded in a vocal display of the New World tropical flycatcher *Myiozetetes similis,* uttered by individuals flying away (although not in precipitous escape) from interactions with nestlings, fledglings, mates, or neighbors, by individuals passing others without joining them, and (rarely) by individuals perched near an interaction and remaining apart from it (W. John Smith ms.). The Carolina chickadee's tseet display may also encode hesitance to interact, although the available data do not certainly rule out an escape message (S. T. Smith 1972:58). Escape, of course, is simply a limited set of procedures for rejecting interaction, albeit one that is distinguishable in a categorization of messages.

Making Interpretations

In some displays the message of interactional behavior does not in any way specify the kinds of interactions the communicator will seek, accept, or reject. As long as the range of kinds of interactional behavior correlated with the use of a display remains broad, the display can be said to carry this general interactional message. If, instead, a display correlates only with some kinds of interactional behavior (such as attack and associating, as alternative possibilities) these are recognized as its messages. Those known to be widespread are considered below. Some displays, however, are performed in correlation with diverse ways of interacting when some kinds of interactions are either especially likely or especially unlikely. For example, attack behavior is especially unlikely to correlate with the reassurance displays of primates and the faint song of chickadees, mentioned above: communicators usually interact in friendly ways following these displays. Each class of interactional behavior that is especially weighted in this fashion must be recognized as a message that is encoded in addition to the general message of interactional behavior. How many such classes will be found to be encoded by the displays of diverse species—that is, to be widespread—remains to be seen.

Messages of Attack and Escape Behavior

Displays are said to encode either or both attack and escape messages when all their occurrence is correlated with a range of attack- or escape-related behavior. Behavioral indices of attack differ among and even within species, but include acts that, if completed, will harm another individual. Escape behavior can be any appropriate form of avoidance, ranging from headlong fleeing to turning aside, and including even freezing and other ways of hiding. (Some species have different displays used in correlation with different escape tactics; they effectively subdivide the message of escape behavior into references to more narrowly specific behavioral selections; see chapter 7 for examples. None of these messages appears to be nearly as widespread among animals as is the undivided escape message, therefore none is treated in this chapter.)

Widespread Messages for Common Problems

The message of attack behavior is very widespread among species, the escape message even more so, and many species encode these messages in more than one display. This should not be surprising, if we anticipate chapter 10's topic of "function" and consider that agonistic situations

usually must be dealt with immediately, with some risk entailed, and they recur frequently in the lives of most animals. The messages are widespread because the need either to behave agonistically (take the offensive, defend, or flee) or to forestall agonistic behavior is widespread.

Most forms of social organization characteristic of vertebrate animals involve their participants in situations that elicit agonistic behavior, whether in support of the social structure or to its detriment. For instance, in pair-forming territorial species of birds, the male has boundaries that require defense against intruders, and the female may aid him. Mates will quarrel with each other when forming the pair bond or subsequently when one or both is aroused by the presence of a conspecific intruder, predator, or other outside disturbance. Agonistic interactions occur within families between parents and offspring, for example, when one approaches and surprises another, or when the offspring persistently beg food from the adults past the time when they are capable of being independent. The pair and the family can be stable only if neighbors learn each other's claims, if their confrontations become limited to infrequent, standoff encounters between equals, and if agonistic events within the group are avoided or settled without violence. Displays with attack and escape messages contribute to both these stabilizing conditions. Within such larger, organized social groups as flocks, packs, or troupes, stability may require that agonistic confrontations be almost continuously forestalled. In fact, only newly met opponents are likely to chase, flee, or grapple, and even in their encounters more time is usually devoted to inconclusive moves and to display behavior than to actual fighting. Outright attack and fleeing are not commonly seen when individuals of the same species contest with one another.

In addition to interacting agonistically with each other, individuals of most species will also attack, flee from, or pester ("mob") potential predators that endanger them, their families, or the larger social groups to which they belong. The groups of some species must be at least partly mutual protection adaptations whose members alert one another to dangers. The displays with which black-tailed prairie dogs (W. John Smith et al. in press 1 and 2) and Carolina chickadees (S. T. Smith 1972) make available the information that escape behavior is likely to spread warnings of approaching predators, and must be crucial to the ability of these species to live exposed in open habitats (see chapter 10).

Displays Encoding Both Attack and Escape

Displays can encode both attack and escape messages, even though the behavioral selections themselves are incompatible. Such a combination is

characteristic of many intensively studied displays of birds, mammals, fish, lizards, and even invertebrates.

For example, Tinbergen (1959b), reviewing the work his group has done with gulls, concludes that at least four postures—the oblique, forward, pecking-into-the-ground, and some versions of the upright—each correlate with both attack and escape. These several gull displays differ; each is correlated with different relative and absolute probabilities that attack or escape behavior will occur. If interacting individuals are prone to wrangle, as gulls are, it appears to be essential to provide information about these two classes of behavioral selections simultaneously.

Shifts in probabilities can also be encoded by variations of a single agonistic display. Baboons, for instance, threaten with a head-bob display that has a down and forward attacklike movement followed by an up-and-back withdrawal. Individuals who are relatively ready to attack will emphasize the former, those more ready to flee the latter (Rowell 1972:94).

Black-Cleworth's experimental studies (1970) of the electric fish *Gymnotus carapo* revealed an unusual display that occurs when an individual may retreat or become inactive, but never when it attacks. For some seconds the fish ceases all electrical discharges, even those normally employed for continuous environmental monitoring. (It is still detectable because its opponent maintains its own monitoring electric field.) Discharge cessation is a formalized signal done in correlation with escape or with immobility; the latter may be freezing behavior, elicited by the lack of opportunity for escape within the confines of an aquarium. That the negative correlation with attack behavior is also informative is discussed in the section on interpretations.

Other Message Combinations

The behavioral selections of attack and escape are not always paired; many displays encode either attack or escape with yet other behavioral selection messages. For instance, Blurton Jones (1968) has demonstrated experimentally that attack is encoded with a large range of alternatives, including escape, that are incompatible with it in certain displays of the European great tit *Parus major*. Among the displays of Canada geese (Raveling 1970) the bent-neck and forward postures appear to encode attack and a broad but otherwise unspecified range of alternatives, while the vigorous rolling display encodes messages of attack and an alternative that appears to be the behavior of associating. The submissive posture display of this goose, on the other hand, provides information about escape and association.

Black-tailed prairie dogs have a spectacular jump-yip display (figure

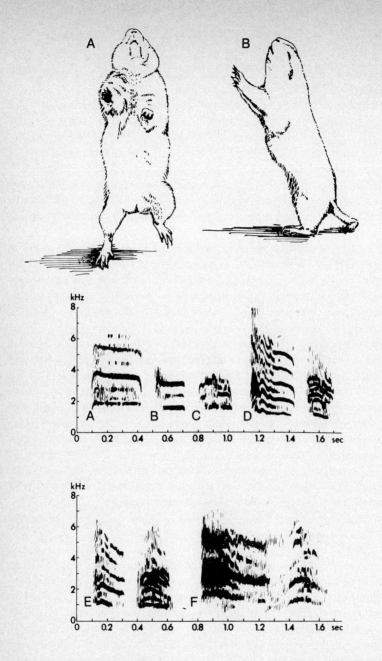

Figure 5.2. Jump-yip display of the black-tailed prairie dog. Front (A) and side (B) views of prairie dogs at the maximum upward stretching of the jump component of the display. Below, sonagrams of common variants of the vocal component "AH-aaah." In D, E, and F there is a break in the aaah part of the call. The harsh-sounding forms (E, and especially F) correlate with a higher probability of attack behavior. (From W. J. Smith et al. in press 1.)

5.2) that is done when escape could prevent or interrupt a large range of incompatible alternatives that it does not specify further. These alternatives include attack, nonaggressive interaction, and such individual maintenance activities as dust-bathing and foraging (W. John Smith et al in press 1). The jump-yip is performed by an individual that has been startled, involved with territorial defense, or is being cautious for other reasons. Whatever the case, it is less likely to flee after jump-yipping than it was before, and usually soon turns to its alternative behavior.

Making Interpretations

On the whole, the attack and escape messages can be readily assigned to displays despite the variability of attack and escape behavior patterns and despite differences among displays both in the relative probabilities of referent acts and in message combinations. Two related message terms that I have employed in past years are less felicitous and contribute nothing to understanding the information encoded by displays. They should be abandoned. The first, "anxiety" (W. John Smith 1966), I subsequently reinterpreted as part of the second, a message of nonagonistic behavior (1969b). The displays that were being interpreted appear to encode messages of escape behavior (perhaps with a rising probability; see S. T. Smith 1972:56), or of seeking an opportunity to escape (see following discussion of seeking behavior), in either case combined with a message referring to a broad class of incompatible behavior such as interaction. Some of the displays in question appear to provide no information about attack behavior, although those termed nonagonistic indicate that attack or attack and escape are unusually unlikely: the acts have neither elevated nor usual but *depressed* probabilities of being selected by a communicator. For example, with the discharge cessation displays of electric fish described previously, the probability of attack is depressed to zero. (Similarly, in the preceding discussion of the message of interactional behavior, displays like tongue showing that are done in shunning interaction can be interpreted as encoding a depressed probability of interacting.) Because description in terms of probability is sufficient, a negatively defined message category is unnecessary.

Message of Copulation Behavior (or Its Equivalent)

Some displays are performed only before or during the social interactions in which eggs are fertilized. These interactions involve either copulation or some behavioral analogue such as the amplexus behavior of frogs. However, although all species depend on fertilization for their

continuance, not all have such displays—possibly because the social cooperation required by the act of fertilizing need not occur frequently nor (in many species) at a moment's notice.

Solitary Animals

Displays that make available information about the behavior involved in fertilizing are known from many species in which individuals are usually solitary and come together primarily or only to accomplish fertilization. They must interact without a basis of interpersonal familiarity, avoiding or controlling each other's agonistic responses and making plain when they are ready to join in a fertilizing act that brings them into unaccustomed proximity.

These problems are faced by many species of invertebrate animals. For instance, fruit flies (species of *Drosophila*) are "not social insects" (R. G. B. Brown 1964): "they meet only when two or more are attracted to the same food source, and there they mate, lay eggs, and soon after die." *D. pseudoobscura* lacks even aggressive behavior, and the several displays with which it courts provide information about copulation, avoidance (effectively escape), and indecisive behavior patterns. One display, vibration, is an "essential prelude" to copulating (Brown 1964) and encodes this message.

Mammals with very limited forms of social interaction also use the copulation message. For example, in South America rodents called cavies (subfamily Caviinae) lack cohesive social groups and have no long-term bonds between individuals (Rood 1970, 1972). Male cavies perform several displays when showing sexual interest in females: the tactile chin-rump follow display, the visible rearing display (in the genus *Galea*) and rumba display (in the genus *Cavia,* see figure 5.3). Vocalizations are uttered with these patterns: "churr" and the "rumble." From Rood's descriptions of the behavior with which these displays correlate, the messages encoded appear to be those of copulation and escape behavior.

More Social Species—Examples from Birds

Many species in which individuals form pair bonds and remain together for at least part of a breeding cycle also use the copulation message.

For example, in precopulatory interactions of the Tasmanian native hen *Tribonyx mortierii* (not a chicken but a large flightless rail) both mates cooperate in performing a sequence of displays (Ridpath 1972). Of 262 observed interactions, 79 percent began when the female trotted near the male with her neck feathers slightly expanded, head slightly

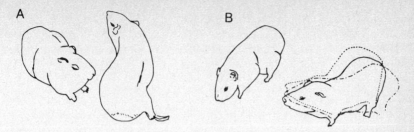

Figure 5.3. The rumba, a visible precopulatory display of *Cavia* males. The male is on the right. As he hesitantly approaches a female (*A*), with his head stretched forward and held parallel to the ground, his body assumes a curve He may circle her. He may stop approaching, but continue to step in place shifting his weight from one hind leg to the other (*B*). This shift causes his hindquarters to sway rhythmically from side to side, and the amplitude of the oscillations may be conspicuously increased, as in this illustration. (From Rood 1972.)

raised, wings expanded slightly above her back, and tail cocked and spread (figure 5.4). She would repeatedly turn her head, watching the male. A similar posture, with somewhat more extensive feather ruffling, is often done while "pacing" by one bird when approached by another in nonsexual interactions, yet males nearly always recognize the trotting, head-turning females as sexually receptive and respond accordingly. A male pursues a trotting female with a similar, high-stepping gait, his neck sleeked and upright, his tail cocked but not spread, and they fall into a rapid prancing run. Finally the female slows and bows her head, then stops and crouches, depressing her tail. The male climbs onto her, lays his head along her back, and pecks her neck. Then he raises his foreparts, makes cloacal contact, and they mate. Of the many displays in this interaction, at least the female's terminal crouching posture and the male's high-stepping prance as he follows her make available information about readiness to copulate; probably the mounted male's touching and pecking of the female's neck do also. Some features of the female's trotting performance also appear to provide such information; if not, the information is inherent in other aspects of the circumstances in which she chooses to trot (that is, in sources of information contextual to the trotting).

Females of many other bird species indicate readiness for copulation by soliciting with a posture that facilitates mounting by a male. For instance, when a female duck is ready to copulate she lies quietly on the surface of the water (see, for example, accounts by Lorenz 1941; Myers 1959; Lind 1959; Dane and van der Kloot 1964; and Johnsgard 1965). She assumes this prone display posture after her mate begins to repeat

Figure 5.4. Precopulatory displays of the Tasmanian native hen. *A.* The female begins the sequence by trotting in front of the male in a specialized posture; he then follows with a similar gait but different posture. *B.* They develop this gait into a prancing run. *C.* Finally, the female slows her pace, and bows. *D.* Then she stops before the male and crouches, depressing her tail. (From Ridpath 1972. Reproduced by permission of the author and CSIRO.)

various displays before her, and maintains it while he continues to display. Its specialization for communicating is evident only because it is adopted before the male tries to mount, or is even in a suitable position from which to try: her display may precede his attempts by as much as 20 minutes (Myers 1959). (In some species, such as the eastern phoebe discussed in chapter 3, females have been observed to adopt the facilitating posture only at the last moment; it may not be formalized as a display in such cases.)

In many other bird species, females solicit mounting with postures that are also used in soliciting other kinds of interaction and that, therefore, do *not* provide a message of copulation behavior. In many songbirds fledglings use these postures, movements, and calls in begging for food; adult females perform them both to beg during "courtship feeding" and, later, to solicit mounting for copulation.

In some species of birds, males have displays that are performed only in correlation with attempts to copulate—the high-stepping prance of a male Tasmanian native hen, for instance. Phillips (1972) found that a male killdeer *Charadrius vociferus,* a North American plover, may make high-stepping movements while standing in place beside his mate before mounting. Once mounted, a male killdeer treads (makes a repetitive tactile display) on his mate's back before copulating; males of various other bird species are also reported to tread. Males of at least some species of gulls have a copulation call (Moynihan 1955a, 1962a; Southern 1974; Beer 1975). Generally, however, male displays with a copulation message have been described in fewer species of birds than have female displays with this message. Among weaverbirds (songbirds of the subfamily Ploceinae), for instance, Crook (1969) found females of at least four species to have vocalizations employed in soliciting copulation, but described for only one species vocalizations used by males during mounting and when trying to entice unreceptive females to permit mounting. In many duck species, the displays performed by a drake before mounting are simply a selection of those he employs much earlier in the process of pair formation when copulation is not attempted, and thus probably do not have a copulation message. There are exceptions: the final three displays of a male goldeneye duck about to mount his mate are entirely restricted to his precopulatory performance, and even compounded into a fixed sequence (Dane and van der Kloot 1964). Similarly, male downy woodpeckers, but not males of other woodpecker species, have a hovering flight display with which they approach a soliciting female (Kilham 1974).

Many pair-forming species, like the phoebe and many other tyrannid flycatchers, seem to lack displays that encode a message of copulation behavior. They may display before copulating (phoebes usually employ

both song and twh-t in precopulatory interactions) or may not. Once mates have become familiar with one another as individuals some pairs can apparently copulate without preliminary displaying, although it remains possible that females have weakly formalized postures that solicit mounting. This is not to say that attempts to copulate usually proceed simply; in fact, in most instances one or both mates may attack the other. Despite the strain this must place on the pair bond, phoebes and other tyrannids are not exceptional: in some gulls (see Moynihan 1962a), shorebirds (for example, some *Charadrius* species, Simmons 1953), and perhaps in many other kinds of birds, mates may display before copulating primarily with the same postures and movements employed in nonsexual agonistic encounters.

Making Interpretations

Although copulation messages are found in the displays of fewer species than are the messages of attack and escape, they are nonetheless widespread among diverse species. Attempting to assess distribution is difficult, however, for several reasons.

First, displays that are reported in the precopulatory behavior of a species often are not adequately sought in other circumstances. If other employment of such a display occurs only briefly in an annual cycle of social behavior, as does the courtship feeding or many bird species, for instance, it may easily be missed. (As mentioned previously, females of many songbird species perform the same displays whether soliciting copulation or soliciting feeding; because their readiness to copulate cannot be known from the displays alone, such displays do not encode a copulation message.)

Second, interpretations of displays have sometimes assigned "sexual" implications if the form of the displays appears to be developed from the act of copulating. For example, in many primates a display soliciting mounting is performed in some greetings, in resolving minor agonistic conflicts, affirming dominance relationships between individuals, initiating grooming, and in other events that may involve or lead to agonistic behavior between two individuals who differ in rank but who do not usually differ in sex (see Hanby 1974). The same display used in precopulatory behavior may also be given by a mother to entice her infant to return to her (Hansen 1966:120). Obviously, the display does not make available the information that the communicator is ready to copulate, despite its form (see the message of receptive behavior).

Third, displays with a message of copulation behavior may be unusually difficult, all but impossible for an observer to recognize or to detect if they are performed only when two animals are very close to each other. For instance, the same species of primates that use a posture

soliciting so many different kinds of mounting appear to have more subtle displays that encode copulation messages. Michael and Zumpe (1970) described three gestures used by captive female rhesus monkeys in soliciting copulation: hand-reach, head-duck, and head-bob. We do not know what these gestures correlate with in free-ranging populations, but the encoding of a copulation message in a tactile display like the hand-reach would not be surprising in a primate in view of the considerable use of tactile simulation in human sexual foreplay described by D. Morris (1971). Inconspicuous tactile, orientational, and postural displays that can function adequately at very close quarters could be much more widespread than we now know.

Recognizing that a copulation message is encoded by display behavior can be difficult when the display forms used in precopulatory behavior are highly variable. Some species appear to encode a copulation message in little more than well-marked variants of displays encoding other messages, presumably risking some recipient confusion (discussed in chapters 12 and 13). A male chukar partridge, for instance, utters a copulation-intention call in running to approach to mount a female (Stokes 1961, 1963). This call is a member of an intergrading group of tidbiting vocalizations, most of which are uttered in a wide range of events—correlating, in fact, with diverse forms of interacting (see previous interactional message section). A male chaffinch risks confusion in much the same way: before he mounts and while mounted and copulating he utters what is called "congested song," a special variant of the species' song display (Marler 1956a).

Message of Associating Behavior

Some displays correlate with the behavior involved in remaining with another individual. When individuals so associate they remain together because one, both, or all will follow, will not leave when the other might not follow, and because each permits the others to be nearby (W. John Smith 1969a, 1969b). Usually they maintain at least some minimum individual distance between each other characteristic of their species, sex, and age. They do not seek contact through fighting or copulating, although members of some species do huddle together while resting, or lie in contact while sunning ("passive contact," Rosenblum and Kaufman 1967:34).

Displays providing information about associating behavior are apparently widespread in persistently social species in which mates or members of larger groups spend much time together even while foraging, resting, or engaged in other essentially individual concerns.

These displays are not commonly performed when animals can maintain their association with ease, but used primarily when other behavior

may disrupt the group. For instance, disruption may result when an individual has just attacked a companion or an outsider, or flees from an approaching predator before the rest of its group reacts, or even when an individual that has been absent approaches to resume peaceful associating with its group. Foraging behavior can also separate members of a group, and displays with a message of associating behavior are given by an individual about to move some distance from its group in seeking another foraging site, or by an animal able to maintain contact with its associates only auditorily.

Agonistic Disruption of Associating

Fighting and fleeing can disrupt associating very abruptly. Many herd-living ungulates display conspicuously when the probability of fleeing from a predator suddenly exceeds that of simply continuing to associate with the herd. Some, like the pronghorn, can erect white rump patches when alarmed, and many deer raise their tails when fleeing, exposing a white undersurface (see chapter 9); many African antelopes employ a very stilted gait when beginning to flee (Guthrie 1971). Such displays may both warn the herd of danger and enable it to remain together while fleeing.

The likelihood that a group will remain together after one or more have fought with each other or with outsiders can also be increased by the performance of displays encoding an association message. For example, in the first phase of the gray-lag goose *Anser anser* triumph ceremony, a gander attacks, orienting directly toward a gander from another group. In the second phase (Fischer 1965), he returns to his family, still in the threatening display posture but carefully orienting obliquely to his mate and offspring. The stereotyped oblique orientation encodes the association alternative to attack, and may be adopted by the whole family. Newborn goslings lack the oblique orientation; when they greet with this ceremony they orient directly at each other with their bills. By the time they are two days old this orientation leads to fighting—and the fighting enables a rank order to develop as weaker individuals begin to orient away. In the Canada goose *Branta canadensis* this procedure is retained as the adult triumph ceremony, but the gray-lag replaces it with the more formal, mutual adoption of oblique orientations during the approach phase (Radesäter 1974, 1975).

Disruption by Foraging or Other Maintenance Behavior

Just before a foraging green-backed sparrow *Arremonops conirostris* flies away from its mate it may utter a plaintive note; when separated, both mates give this vocalization. Following or joining usually reestablishes association immediately (Moynihan 1963a).

Such vocalizations are very widespread. They have often been called "contact calls" by ornithologists on the assumption that they are used in maintaining the coherence of a group during foraging and comparable activities that can largely preoccupy the individual members. The term implies association, not physical contact.

For instance, pairs of zebra finches *Poephila guttata* give brief vocalizations while foraging actively, the mates remaining continuously with each other and often calling in alternation; if separated they switch to louder, more prolonged vocalizations and the male actively searches (Butterfield 1970). Most other species of weaverbirds also remain in pairs or flocks, and utter brief vocalizations at frequent intervals during active foraging (Crook 1969).

Among the vocal displays of the Carolina chickadee (S. T. Smith 1972) the lisping tee is uttered only by birds that are either temporarily severing association, actively maintaining it with some individuals (such as by following) while severing it with others, or reestablishing association with individuals with whom they have bonded relationships. Associating behavior is almost continuous in wintering flocks of this species: individuals remain near one another virtually all day, flying to follow, or occasionally flying, hopping, or sidling slightly farther apart when foraging activities bring them unusually close together. The lisping tee is usually uttered when in flight, typically by the first individual to fly from the flock toward another foraging area and by some of the next to follow; the final members join up silently. In spring and summer, the lisping tee is given by a male foraging with his mate, especially in flights that separate them, and by a male approaching a perch station outside the nesting cavity within which his mate is incubating. Later, both parents utter many lisping tees when accompanied by their dependent offspring while foraging.

The green and solitary sandpipers *Tringa ochrops* and *T. solitaria* provide examples that are phylogenetically distant from the songbirds just mentioned. Each species has "contact calls" that are uttered by both mates when separated by a short distance in an activity such as foraging. They are also uttered by an adult as its eggs hatch and until the young are dry, and continued after the adult flies from the arboreal nest and waits on the ground for the young to jump to join it; parents continue to give these vocalizations as "follow calls" until the young have grown up and fledged (Oring 1968).

Displays of this sort are by no means restricted to birds. For instance, many species of Old World monkeys utter vocal displays when a troop makes a trip to a foraging or sleeping site. In the common langur *Presbytis entellus* of north India males initiate and determine the direction of group movements, gathering the animals together with a "whoop" vocalization (Jay 1965). Male lutongs *Presbytis cristatus* in

Malaysia initiate and control group travel with a "kwah" call repeated at intervals (Bernstein 1968). The much more terrestrial chacma and olive baboons *Papio cyanocephalus* give a doglike bark display (Hall and DeVore 1965) when individuals are about to rejoin after a temporary separation, and a male hamadryas baboon *Papio hamadryas* leads his troop with a "contact grunt" (Kummer 1971). Dense vegetation limits visibility with the large troopes of Japanese macaques *Macaca fuscata,* presenting a problem as they move over wide areas. Itani (1963) described nine vocalizations in three intergrading series: his A-1, A-2, A-3; A-7, A-8, A-9, A-10, A-11; and the discrete but variable A-13 sounds. *All* these appear to encode an association message. Some (A-10, A-11, and A-13) are given by leader individuals, while others are given more widely, like the avian contact calls. Between each difference in form that Itani recognizes there appear to be slight shifts in the messages of escape and seeking behavior that are encoded along with association.

Some New World monkeys also have vocalizations uttered in group travel: for example, Carpenter (1934) heard a "deep, hoarse cluck" only from leading male howler monkeys *Alouatta palliata* at the beginning of and during group progressions. In other species the association message may be encoded in a display given when other animals are being sought. A night monkey *Aotus trivirgatus* utters hoots when alone, ceasing when joined by another individual (Moynihan 1964). A rufous-naped tamarin *Saguinus goeffroyi* utters long whistle displays when separated from its companions or roaming alone, and two bands may face each other across a gap between trees and whistle persistently, then withdraw; long whistles indicate separation from others whom the communicators cannot join either because they do not know where to go, or because the others would not readily permit them (Moynihan 1970a).

Apes, too, have vocalizations encoding association. Carpenter (1940) described a "chatter or series of clucks" uttered by an individual leading a moving gibbon *Hylobates lar* troupe. Two vocalizations are uttered by scattered gorillas (Schaller 1963, 1965, further studied by Fossey 1972). Neither occurs primarily during leading behavior, but one, the hoot bark gets group members to cluster in various circumstances. Silverback males give it along with a special posture, standing motionless and facing in a fixed direction; the rest of the group crowds around, and then all move off (Fossey 1972).

Message of Indecisive Behavior

When a single display encodes information about such actions as attack and escape or perhaps attack, escape, and copulation, none of which has

a preemptive probability of being selected, the communicator most often does not do any one of them immediately. Instead, it may adopt a static posture that could facilitate any of them, may alternate intention movements of approach and withdrawal, may move laterally with respect to the pertinent stimulus (thus locomoting without either approaching or withdrawing; the hesitant approach of a male phoebe to copulate with his mate was described in chapter 3), may redirect a selection such as attack onto an inappropriate stimulus, or may do acts that appear to be irrelevant in the circumstances (such as eating or grooming, which ethologists would then call "displacement" behavior; see chapter 11).

All such heterogeneous behavior can be described as indecisive, and makes up a class of behavioral selections; the communicator at least temporarily carries out none of the incompatible selections that would be decisive and, in fact, often hesitates and vacillates irresolutely until the circumstance has changed and decisive acts are no longer required or even possible. The communicator often continues to display throughout such an incident.

There are at least three ways of knowing that specific behavioral selections are encoded in addition to the readily apparent selection of indecisive behavior. First, if the communicator does commit itself it does so in a fashion that can be predicted for each display, selecting consistently from the same two or three incompatible alternatives. For instance, a rhesus monkey *Macaca mulatta* that threatens with open mouth, thrusting lower jaw, and jerking its head toward its opponent may eventually lunge in attack, or may turn or even move away; it thus reveals attack and escape messages. Second, among the indecisive acts various intention movements reveal something about the incompatible alternatives. A threatening rhesus monkey usually does not attack or flee, but makes intention movements in the appropriate directions; there is a "long period of ambivalence during which the animal jerks head and body . . . without making a decisive move in either direction" (Hinde and Rowell 1962). Third, one of the alternatives may be redirected: for example, in the pair-formation interactions of black-headed gulls *Larus ridibundus* a male unable to select between associating with an intruding female and attacking her may briefly associate, then suddenly fly away from her and from his territory and redirect attack behavior toward a distant gull in a formalized swoop and soar display (Moynihan 1955a).

Some displays encode two messages that do not refer to incompatible alternatives. For example, they may encode a message specifying the behavior of seeking an opportunity of some sort (see below), and another message specifying what behavioral selection will occur if an opportunity is found. However, few or no displays encode only one behavioral selection message. Probably most displays, including most of

those just cited in describing the various messages, encode some probability of indecisive behavior in addition to other possible selections. Because it is so abundantly used, the failure to recognize a message of indecisive behavior was the most serious omission of my initial description of messages (1969b).

Message of Locomotory Behavior

The eastern phoebe's twh-t vocalization is usually uttered on alighting, very occasionally on taking flight or making a flight intention movement, and in one extreme kind of event during brief flights. The eastern kingbird's kit-ter (described, along with the twh-t, in chapter 3) is also uttered on alighting, and somewhat more often during flight in correlation with slowing, turning, or veering. Both displays provide information about a communicator's use of flight behavior, but *not* about functional categories of flight such as approach, withdrawal, attack, or foraging. The displays correlate with all of these acts and more. Other species of flycatchers have vocal displays with comparable behavioral correlates, and some extend the performance of their displays to correlate with hopping or running when they forage on the ground. Thus, the behavioral category is viewed as "locomoting" rather than as "flying," as stated in chapter 4.

Parallels to Flycatchers in Vocalizations of the Adelie Penguin

Ainley (1974a) found that at least two vocalizations of Adelie penguins make information available about locomotory behavior. His observations of the performance of these displays provide striking comparisons for flycatcher displays because penguins are flightless. The comparisons are also valuable because penguins are phylogenetically distant from flycatchers and are persistently social, which phoebes and kingbirds are not.

Ainley named one penguin display the locomotory hesitance vocalization (LHV), adopting the term I used for the kingbird's kit-ter to emphasize the comparable employment of the two species' displays. An Adelie penguin displays with its LHV when moving or showing intention movements of moving, as when approaching its nest territory after a long absence, or approaching another bird: its mate, chick, opponent, or a wandering intruder. It also gives the display when it is being approached (by another penguin, a predator, or a person) and may need to locomote to respond, or when it is being forced away from some site or situation. As in eastern kingbirds, pairs of Adelie penguins greet each other with their LHV, individuals utter it when they hear their neighbors use it, and it is uttered in stalemated fights and when a bird has to

choose between starting a fight or finding some alternative (with Adelies, this can be as birds space themselves on a small ice floe).

When the LHV is uttered the communicator always has both reasons to approach and reasons to take up an alternative: for example, coming to the nest requires giving up feeding behavior, joining a mate risks a quarrel, leaving the nest unattended for any reason (to join a nearby fight or to repulse an intruding creche of chicks) exposes the eggs to predators such as skuas, and fighting involves choices between approach and withdrawal.

One performance seems odd: a parent whose eggs are hatching will direct LHVs toward its nest. In many species the chicks become vocal within the eggs two or three days before hatching, and the incubating parents become very restless. In mallard ducks *Anas platyrhynchos,* for instance, considerable parent-offspring vocal interchange develops in which the incubating female utters various displays (see Hess 1972); new social relationships are being established. This is the only case in which observations do not reveal clear-cut locomotory behavior correlated with the display; extremely detailed analyses of the parental restlessness might, and for the moment this behavior is inferred to include some incipient attempts to leave or to approach the eggs more closely.

The second Adelie penguin display in which Ainley found a locomotory message is in the aark vocalization. It is uttered only when a communicator is locomoting or likely to locomote: while a penguin is walking between its nest and the beach, standing or walking at the water's edge (particularly as individuals begin to dive from a flock into the water), swimming at sea, and when being chased.

Behavioral Correlations

Current evidence indicates that displays encoding a message of locomotory behavior are performed by animals when choosing between incompatible alternatives involving locomotion, or when indecisive behavior affects locomoting.

The social Carolina chickadee has three vocal displays in its repertoire that illustrate this (S. T. Smith 1972). Chickadees utter their chick display when making difficult decisions between fleeing and some incompatible alternative, and are likely to behave indecisively. For example, one will call chick on leaving or making intention movements of leaving a bird feeder when another chickadee comes, and will return quickly if it does leave; chick is repeated rapidly by a chickadee in highly indecisive zigzag or tangential flight while approaching another chickadee or avoiding a larger bird such as a blue jay *Cyanocitta cristata* at a feeder. A less indecisive chickadee, more often flying than not, will

utter a high tee display when either escaping or selecting an alternative while still likely to flee. The high tee, the chick, and another display not providing information about locomotion are often combined in the sequence high tee-chick-dee as a communicator alights, the chick coming either at the instant of touching down or fractionally before, the high tee while still in flight. Finally, the lisping tee is uttered primarily by flying chickadees, and sometimes when flight is very imminent. The communicator is selecting between associating with other chickadees and doing something that would impair association such as flying away from them to a new foraging site, or (less often) flying toward them when they might respond agonistically.

The timing of utterance in correlation with locomotory behavior differs among the chickadee's three vocalizations: lisping tee usually comes in flight, high tee less often in flight but still more often then than when perched, and chick usually on alighting or taking flight, except when repeated rapidly (this can be either in flight or from perch). The timing of the chick reflects that display's strong correlation with indecisive behavior. This correlation makes it difficult to predict the immediate likelihood of flying at the instant when chick is uttered, although it can be said that flying is likely during the incidents in which chick occurs. Comparable problems recur in describing other displays that provide information about both locomotory and indecisive behavior. For instance, the phoebe's twh-t comes on alighting, on ceasing flight. Yet although flight is momentarily improbable, it occurs frequently in incidents when a phoebe's indecisive behavior leads to repetition of twh-t. Short-term reversals are typical of indecisive behavior in which there is flying.

Although displays that provide a message of locomotory behavior differ in how their performance is correlated with locomotory acts, each does have some sort of distinctive timing. The phoebe's twh-t and the kingbird's kit-ter exemplify correlations that are primarily with the termination of locomotion. Examples are presented here of displays that correlate with (1) the period before locomoting and (2) more diverse choice points in locomotory behavior.

DISPLAYS PERFORMED BEFORE LOCOMOTING   Many birds of species that live in flocks have preflight displays indicating that locomotion is becoming probable. These displays characteristically provide other messages about the likelihood of indecisive behavior and of a social alternative to locomoting—often the behavior of associating. Such combinations of messages may enable the displays to facilitate the cohesion of social groups, keeping members informed when other members may be about to leave. For instance, red-legged partridges *Alectoris rufa*

have a flight-intention call which is apparently uttered as individuals decide between flying away from or remaining with other partridges (Goodwin 1953). Gregarious *Chlorospingus* tanagers perform visible wing- and tail-flicking displays as the likelihood increases that the communicator will take flight (Moynihan 1962b). Canada geese live year-round in families that join flocks seasonally; they have calls that are uttered "whenever locomotion (to whatever purpose) is blocked" (Blurton Jones 1960). Two of these vocalizations were described by Raveling (1969), along with a conspicuously visible head-tossing display that must also encode information about locomotion. In a family of geese, members that are showing many indications of being ready to fly head-toss a great deal while waiting for their gander to take the initiative. He is invested with the leadership of the family, and is usually its least indecisive member; when he elects to fly the others follow, and the family does not become dispersed within the larger flock.

Other displays also provide information about the likelihood of forthcoming locomotion but have little or nothing to do with group cohesion. The honeybee's famous dance performance is an example. An individual dances as she prepares to return to a site that is important for foraging or other activities, which helps other bees to find that site. The performance incorporates many formalized components, some making available information about the direction of the dancer's next flight, others about how much energy she will expend in flying (see chapter 6).

DISPLAYS PERFORMED AT DIVERSE TIMES IN RELATION TO LOCOMOTION   Piñon jays *Gymnorhinus cyanocephalus* have a loud kraw vocal display that they utter at times when difficult decisions are being made about flying. For instance, it can be hard for a flock of as many as 250 jays to become coordinated, and the display is much repeated before beginning group flights. The display is also uttered by many jays in flight, if a flock's course takes it out across open ground: the flock then rolls and swirls as the birds in front turn back and are replaced as leaders by birds from the middle, who also turn back (Balda and Bateman 1971).

Indian hill mynahs *Gracula religiosa* have a variable range of um-sounds (Bertram 1970) that may be uttered on taking flight or just before, and in flight when changing direction; the species also has a songlike vocalization that is uttered only in flight, usually shortly after take-off when mates are flying rapidly together, sometimes chasing one another. The "tit flight call" of buntings (songbirds of the subfamily Emberizinae; Andrew 1957) and the "flight call" of the chaffinch (Marler 1956a, 1956b) are both uttered on taking flight and, less commonly, during flight. According to Crook (1969) the "flight calls" of

several species of weaverbirds are uttered "together with preparatory movements for flight prior to take-off, during take-off, and at certain times during flight, for example when a group changes direction, and prior to a group landing."

Red-winged blackbirds *Agelaius phoeniceus* have an unusually rich repertoire of vocal displays that provide information about locomotion. Their "flight" call complex is long and complicated and uttered only when leaving and returning to the nesting territory—and in the latter case is readily followed by singing. Others occur in or just prior to flights at similar times, but also more broadly: for example, during and at the termination of sexual chases, when chasing insects, and when milling around in winter roosting flocks. Still others are uttered only in, or just prior to, the comings and goings of winter flocks (Orians and Christman 1968).

Fully detailed studies of mammalian displays with locomotory messages have not yet been compiled, but the numerous candidates vary from formalized tail movements of the black-tailed prairie dog (P. Bernstein, research in progress), to the chest-beating of gorillas (W. J. Smith and students, research in progress), to the gulps and sneeze-grunts which night monkeys utter throughout periods of high locomotory activity (Moynihan 1964).

Making Interpretations

The function to which communicators apply their locomotory behavior are diverse for each display just described. For instance, the functions of flights correlated with utterance of the phoebe's twh-t include foraging, escape, various classes of interacting (associating, copulating, attack), and taking perches from which to advertise readiness to interact. Sometimes one functional class of locomotion (such as escape, in the high tee of the Carolina chickadee) is specified by a display with a locomotory message, but the alternatives also involve locomotion and are diverse.

It may be that locomotion is necessarily involved in activities represented by other messages encoded by the displays of some species. Thus, a display that encodes information about escape behavior might be performed only when the escape tactics being selected were locomotory and not based on freezing, cowering, or flinching. Such a display should be interpreted as providing information about locomotion, although it would be necessary to make it clear that information about the limited functional class of this locomotory behavior was also provided. Examples of such displays are not known at present, although cases are known in which the escape tactics correlated with the performance of a display are narrowly specified in other ways (see chapter 7).

Message of the General Set of Behavior Patterns

A single display often encodes behavioral selection messages referring to activities that cannot be done simultaneously, such as attack and escape acts in which the motor patterns are incompatible with each other. In a great many cases, possible selections are left very vague and are specified only as incompatible. When the range of possible alternatives appears to include *anything* in a species' behavioral repertoire the message is provisionally, if inelegantly, described as referring to the "general set" of behavior patterns available to a communicator.

It is, of course, difficult to establish whether the available selection is indeed limitless for any display because some activities occur quite rarely. When the range is broad, determining its limits can be very time-consuming, and we have not usually advanced beyond knowing that a given display correlates with a large range of incompatible alternatives.

Displays Encoding Specified and General Set Alternatives

The twh-t vocalization of the eastern phoebe (chapter 3) is uttered when interactional behavior may be selected, but its selection is less likely than some incompatible alternative that can range from solitary foraging to escape, apparently without limit—in fact, both alternatives can be ways of interacting (copulating versus associating). The communicator is usually intent on some single alternative whenever twh-t is uttered, but that alternative can be very different from event to event. Another of this phoebe's displays, the chatter vocalization, encodes a message of the behavior of remaining with site (see discussion of that message, this chapter) along with a message of incompatible alternatives, and it is not clear whether this range of alternatives is as broad as in the case of the twh-t. The chatter vocalization has not been heard during observations of precopulatory behavior, for instance. Nonetheless, it is provisionally assigned a general set message.

Displays of numerous and diverse species are similar to these two phoebe displays in correlating with one specific kind of behavioral selection and a large, incompatible, but otherwise unspecifiable array of alternative selections. Many examples are given throughout this chapter in describing other message categories.

Displays Encoding Only General Set Alternatives

Hailman and Dzelzkalns (1974) have shown that the tail-wagging display of mallards may encode the likelihood of a change between two unspecified alternatives in the duck's behavioral repertoire. Mallards tail-

wag in extremely diverse circumstances, but in 1,085 observed cases always gave the display just before, just after, or both before and after some definable activity such as another display, a maintenance movement, locomotion, or a foraging pattern. That is, tail-wagging correlates with changes in behavior, but gives no information about the activities that are being changed or the alternatives replacing them.

Like tail-wagging, many displays are so abundantly performed as to make precise description of the limits and characteristics of the behavior with which they correlate very difficult. With considerable study, some will likely be found to encode at least one behavioral selection more specific than the general set of behavior—perhaps interactional behavior for many. Those which encode only information about unspecified behavioral selections will probably be found to correlate with a measurable probability of change, indicating, like tail-wagging, that the likely selections are incompatible (change implies incompatibility of alternatives).

Making Interpretations

We may be failing to discover messages less broad than the general set by not exploring the full range of alternatives correlated with the performance of each display to which this message term is assigned. On the other hand, by exploring them we might be tracing out details of little or no significance to the animals under study. What is important to an individual receiving a display may be the information that there are many different alternatives. It may not be important that they include the whole feasible range of alternatives, as long as all are incompatible with the display's messages of locomotion, association, escape, or others, and no one incompatible alternative has a particularly high or low likelihood of being selected.

The range of behavioral selections to which the general set message refers is remarkably broad, but it might be more appropriately renamed the "general incompatible set" of behavioral alternatives. Still ungraceful, the term would reflect our present understanding more accurately. Such a name, however, presupposes that no display will be shown to correlate only with the activity of being intent on some unspecified task, encoding no information that there may be a change to alternative acts. It will not be clear how many message categories deal with what is now called the general set of behavioral selections until more displays that are performed abundantly by communicators have been studied in detail. In the meantime, it is advisable to keep an open mind and a simple terminology.

Messages of the Behavior of Remaining with Site

Displays performed only when a communicator is remaining at a fixed site encode the information that he will remain at a single point, in the vicinity of such a locus, or in an area that allows considerable freedom of movement within fixed boundaries. The behavioral selection referred to is simply "staying put," defined with respect to a site.

Song vocalizations are in many species limited to performance on territories (as mentioned previously). Some species sing while patrolling the boundaries of a large territory that encompasses all of an individual's activities. In species with a very small, nest-centered territory, the owner must leave it to forage and song is uttered only in the defended area.

Birds of species that do not sing may employ comparable displays. The Adelie penguin, for instance, has an ecstatic vocalization that it utters when ready to interact but staying at a site (Ainley 1974a). The site is usually a nesting territory, but occasionally an individual gives the vocalization when stranded alone on an ice floe; it appears to be reluctant to leave such a site, although for reasons that are not clear. The chatter vocalization of the eastern phoebe and the visible nest-site-showing display with which it is used (chapter 3) are other examples of displays performed when showing strong reluctance to leave a nest site. Courting males of such weaverbirds as the African *Ploceus cucullatus* and *P. nigerrimus* (Crook 1963) do a wing beating display as each remains at a little cluster of nests he has constructed, steadfastly refusing to leave even to follow a female who has been attracted to mate; these displays may have evolved from intention movements of flying (see chapter 11) to follow females. The nest-site demonstration display of tricolored blackbirds *Agelaius tricolor* of the western United States is shown in figure 5.5.

A laughing gull *Larus atricilla* that is incubating or brooding performs a display distinguished by posture, movements, and a vocalization when an intruder comes near the nest (Burger and Beer 1975). Facing the intruder, the displaying bird extends its neck upward and forward, tilts its bill slightly up, lowers its tongue bone (bulging out the face below and behind the mouth corners), raises the feathers at the back of its head, and begins repeating a plaintive call. The communicator may attack, but is more likely to remain on its nest. As it becomes more likely to attack its call notes shorten and are repeated at a rate of four to five per second and, with its neck still extended, it begins to raise and lower its head in a forward arc. If it then rises to a semistanding position and begins to extend its wings it is likely to leave its nest and launch an attack. The whole continuously graded performance appears to provide

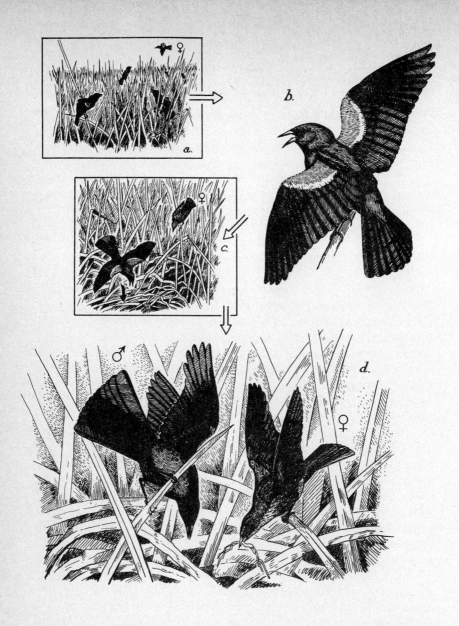

Figure 5.5. Nest-site demonstration display of the tricolored blackbird. *A*. Males on their closely packed territories, each singing and performing the song spread posture as a female flies over. *B*. Having seen the female, a male elevates his wings into a V shape and spreads and depresses his tail. *C*. He then works his way down to a nest site low in the vegetation, still displaying, while the female follows. *D*. At the site he displays and points with his bill. The female also adopts the display and points at the site, and in this case has picked up material for a nest. (From Orians and Christman 1968. Courtesy of the Museum of Vertebrate Zoology, University of California.)

messages about two incompatible kinds of behavior, remaining with site and attacking, and the probability that the gull will do either shifts through the series in the direction of attack.

Temporary Sites

Staying put need not involve as persistent a site attachment as is made to breeding territories; it can have diverse functions. For instance, the "flight seep" call of the European blackbird is often uttered before flights to "any other place which is more than just a few yards away," as well as by territory owners before flying to a roosting place or foraging site away from its territory, and by birds losing territorial encounters (Snow 1958). The communicator chooses between remaining with a site, whether its territory or something much more local and of perhaps ephemeral significance, and departure for any of a variety of reasons. When performing the horizontal display posture great tits would stay put in experiments done by Blurton Jones (1968) remaining on a bird feeder until finished eating and attacking any interfering individuals; the same individuals did not behave territorially over the feeder at other times.

Perhaps more common, however, than just staying put, the behavior includes defense of a site's boundaries: that is, it is territorial behavior. This too can be ephemeral: in early spring in the Andes individual tyrannid flycatchers of the ground-dwelling species *Muscisaxicola albilora* will defend small patches of vegetation for as little as half an hour (W. John Smith 1971a:246). These prebreeding territories are rich foraging habitat, but unsuitable for nesting. Yet even large breeding territories may be briefly held. Tinbergen (1939a), for instance, saw male snow buntings *Plectrophenax nivalis* in Greenland oscillate between holding territories and abandoning them to flock as spring weather shifted, then later stabilizing the same territories.

Displays Combining Two Messages of "Remaining With"

The behavior of "remaining with" can take as its focus either a site or an individual. Two messages can be distinguished, and both may be encoded by the same display. For instance, some encode the messages remain with site and association (that is, remain with a companion) as alternative possibilities. Chickadees, tits, and titmice (family Paridae) are particularly rich in examples. The European blue tit has at least three visible displays that encode remain with site: a hole inspection display, a dance display, and a moth flight display (Stokes 1960). In the first a male, or rarely a female, perches at a nest hole turning its head

back and forth, sometimes poking its head into the hole or entering if the mate is watching. It may fly back and forth between its mate and the hole, but remaining with the site tends to take precedence over associating with the mate. (Similar information is also encoded by the hole inspection displays of related species; see Hinde 1952.) The dance display of the blue tit gives precedence to neither behavioral selection. A male alternates formalized movements toward the nest and toward his mate. The blue tit's moth flight display appears to encode remain with site and interactional behavior (which includes association, but also other such behavior as disputing). In the moth flight a male conspicuously dallies in the air, usually flying toward the nest; about a third of the time he orients from the nest, or lacks relationship to it; that is, the site encoded in the display can also be the territory.

Other displays may encode both remain with site *and* association without the selections being incompatible. As a possible example, in a variant of the tidbitting display, a pair of chukar partridges mutually perform a "nest ceremony" at a nest site (Stokes 1961, 1963). Variants of the choking display performances may also encode both messages: a pair of gulls in their breeding territory align laterally and make scraping movements with their legs and feet as if shaping a nest site in the ground (for black-headed gulls, see Moynihan 1955a; for herring gulls, Tinbergen 1953b). Other choking performances are done in territorial defense with opponents oriented toward each other and not using the scraping movements (unless both members of a pair are defending simultaneously, parallel with each other, and facing their neighbor).

Message of Seeking Behavior

Animals may display while seeking the opportunity to perform some kind of activity during what ethologists call "appetitive" behavior as distinguished from "consummatory" behavior in which the activity is completed. The behavioral selection about which a display provides information if it is done *only* in this way can be termed "seeking." What a communicator is seeking to do is encoded in the same display by a second behavioral selection message. The display is interpreted as providing not just the information that a communicator is ready to do this second selection, but that its behavior includes seeking or preparing to seek an opportunity.

Seeking to Interact

Some displays provide the information that interaction is being sought, without being more specific about what kind of interaction. For ex-

ample, in many species of birds, individuals sing from prominent stations when alone and promptly cease to sing and begin attempts to interact when other individuals become accessible (see previous interactional behavior section). Birds also sing when on opposite sides of a territorial boundary and not, strictly speaking, alone. Yet observations indicate that even when two "countersing" in this way, each communicator lacks access to the other. Neither will cross the boundary to attack; the boundary convention renders them mutually inaccessible, thwarting sufficiently close approach for a more intense interaction than their countersinging. (When boundaries are unstable there is crossing. Songs then cease and fights break out. Countersinging is sustained only when boundaries are mutually respected.) A singing bird does appear to be predisposed to interact and thwarted by either the absence or inaccessibility of another individual.

Singing both carries a message referring to the behavior of seeking an opportunity to interact and is itself as much a part of that seeking behavior as are the acts of patrolling and perching at high vantage points. Singing is a loud, far-reaching advertisement that can call forth recipients to mate or do battle. It thus shares the fundamental characteristic of those linguistic expressions that Austin (1962) has called "performatives." The utterance of a performative is, or is at least a part of, the act to which it refers: in speech, such utterances are "I bet," "I promise," "I thee wed," or "I declare war."

Not all displays correlate with as wide a range of interactional behavior as does song. In other displays the message of seeking behavior is combined with messages that specify particular kinds of interacting such as association, copulation, or attack.

Seeking to Associate

Several species are known to perform displays when seeking to associate. In the songbird genera *Ramphocelus* and *Chlorospingus,* for instance, when an individual becomes separated from its companions both it and one or two members of the group utter plaintive note displays until they rejoin (Moynihan 1962b, 1962d). A similar pattern of performance is characteristic of a vocalization of bobwhite quail (Bartholomew 1967), of a bell-like "hoo" call of the African wild dog *Lycaon pictus* (Estes and Goddard 1967), and of vocalizations of at least four species of New World monkeys: the night monkey and rufous-naped tamarin (see displays described in the association message section), and *Cebus capucinus* and *C. nigrivittatus* (Oppenheimer 1968; Oppenheimer and Oppenheimer 1973). The California bushtit has a complex call that flock members utter as they forage in association, and

another that apparently intergrades with it but is uttered only "by lone individuals suddenly finding themselves separated from one another or from the main flock" (Grinnell 1903). Zebra finches also give two vocalizations in these circumstances, but they do not intergrade (Butterfield 1970).

Only if a display is performed in the absence of an opportunity to associate, correlates with attempts to find another individual, and is not also performed *during* association does it encode the message of seeking behavior in addition to the association message. A display like the plaintive note of the green-backed sparrow (Moynihan 1963a; see also association message section) that is uttered both during and when seeking association provides information only about readiness to associate; a recipient cannot know from the display alone whether or not the communicator has the opportunity. A similar example is the decrescendo call of female mallards and other pond ducks, uttered both when with their mates and when their mates fly away, or by unpaired females when they see other conspecific ducks fly past overhead (Lockner and Phillips 1969). In fact, most displays that encode a message of association behavior may also be performed when seeking association, and cannot be said to provide information about seeking behavior: seeking will occur if necessary, but often is not required.

Seeking to Copulate

If a male lovebird (an *Agapornis* parrot) attempts to mount his mate and she is unreceptive, providing him no opportunity, he may squeak-twitter and rapidly move toward his head the foot that he had lifted in trying to mount. He then uses the foot in a stylized movement like scratching (Dilger 1960). He may even adopt a posture comparable to the pose with which his mate could solicit mounting, a form of "pseudo-female" behavior or role inversion shown by male birds of other species and even by fish (Morris 1955).

Seeking to Attack

The opportunity to attack can be absent if an appropriate opponent cannot be found; not all potential opponents will do. For instance, a territory owner will seek opponents, but only within the limits of the site to which he restricts his activities. Only an intruder is appropriate for attacking, and a neighbor is simply threatened. Displays performed at such times may indicate that an opportunity to attack is being sought, although an observer may have to infer this as opportunities rarely materialize.

For example, at least three species of kingbirds have a spectacular tumble flight display that a male may perform periodically at twilight when it is too dark to have encounters; two neighbors may even go to their mutual border and fly and display in parallel, each on his own side of the border that makes the other inaccessible. Each repeatedly engages in a mock tumbling, grappling, aerial battle alone—shadow-boxing, as it were (W. John Smith 1966). The circumstance is comparable with that in which a male Carolina chickadee utters the song variant display (S. T. Smith 1972): this display is performed when approaching boundary points where he has disputed with neighbors who have not, in this instance, responded to his song display. These chickadees never utter song variants when an opponent is present (hence the interpretation of a message of seeking behavior), or when with their mates and thus prepared to interact in other ways than by fighting (hence the attack message).

Seeking to Escape

The opportunity to avoid aversive events can also be blocked. Many species have displays they perform in correlation with cowering to avoid blows, freezing for various reasons, or struggling to escape when restrained or cornered. Screams are occasionally heard in the field from a prey animal struggling in the grasp of a predator (see, for example, observations reported by Stefanski and Falls 1972), and frequently from birds that are caught in a net or live trap. Infants are especially likely to be unable to escape, as in many species they only gradually attain the motor ability to flee. When able, they scream; nestlings of the European blackbird will scream on being handled once they are about nine days old, and their screams bring their parents to their defense (Snow 1958).

Making Interpretations

Study of the seeking message can present special problems if an observer cannot tell readily when an animal lacks an opportunity. Can lack of opportunity always be discerned from the indecisive behavior of an individual that cannot select among opportunities that are incompatible but available? Perhaps not. And perhaps the message of seeking behavior and the message of indecisive behavior are not always separately encoded by displays. For instance, Blurton Jones (1960) reports that a captive Canada goose that is prevented from attacking by *either* a physical barrier *or* by indecision about fleeing or behaving in any other way incompatible with attacking will perform the bent-neck and forward

displays. The appropriate message interpretation is not apparent; we need to know in detail of more such cases, and preferably in free-ranging populations. Interestingly, however, motivational analyses would view both lack of opportunity and inability to select one course of action from incompatible choices as related; they are aspects of "thwarting" (see chapter 8).

Problems in interpreting this message are in no way diminished by the first term I attached (1969b) to its behavioral selection referent: the behavior of "frustration." That term is more heavily freighted with motivational connotations, more ambiguous, and much less descriptive of behavior than is "seeking."

Message of Receptive Behavior

Some displays indicate the behavioral selections that a communicator will *accept,* not those it is prepared to perform. At least two behavioral selection messages must be provided by such a display, one indicating that the communicator will behave receptively and another indicating the class of acts to which it is receptive. Effectively, the communicator adopts the role of soliciting acts from another individual; it does not offer them.

The contrast between soliciting a class of behavior and offering to perform that class is readily illustrated by the formalized trophallactic (food exchange) behavior of honeybees. To solicit a food exchange a honeybee attempts to thrust its proboscis between the mouthparts of a nestmate; to offer food it moves its folded proboscis forward, often readying a drop of regurgitated liquid at the base (Wilson 1971a:290).

Receptive to Copulation

Soliciting is often obvious in sexual interactions. A female bird or mammal ready to copulate adopts a posture that facilitates mounting by a male; that posture may also be formalized to solicit his mounting.

Although a display of this form encodes information about behaving receptively, the additional behavioral selection that is encoded is not necessarily as narrowly specific as copulation. During the evolution of these displays, in many species the performance of a posture initially specific to soliciting copulation has been generalized to events in which a communicator is receptive to other nonaggressive forms of interacting.

Receptive to Interaction

Various species of monkeys perform a single display posture in soliciting mounting for copulation and in resolving agonistic situations in accor-

dance with status prerogatives (either sex solicits, most often a member of its own sex); at least in rhesus macaques, a mother also adopts this display when summoning an infant to return to her back and be carried (Hansen 1966). The communicator is receptive to being mounted and, more generally, to diverse kinds of interactional behavior to which the mounting is a prelude.

Chimpanzees, like monkeys, solicit reassurance from each other by presenting to be mounted. They also perform displays incorporating bowing, bobbing, crouching, kissing, and hand-offering gestures when soliciting reassurance that interactions typical of their bonded relationships can proceed as usual, with a low probability of attack. (Reassurance, in turn, is offered by mounting, touching, embracing, and kissing displays; see van Lawick–Goodall 1968.)

Even birds indicate receptivity to nonaggressive interaction by soliciting mounting. For example, when a strange individual approaches a pair of the peculiar, storklike African hamerkop *Scopus umbretta* at their nest, the mates alternately solicit and mount each other. Their soliciting employs the same display posture that the female adopts before copulation. Birds also perform other displays in indicating receptivity to interaction in circumstances in which individuals are, or are coming to be, close together. For example, the head toss display of the Chilean gray gull (Moynihan 1962a; Howell, Araya, and Millie 1974) is performed in food begging by young birds and prebreeding females, by both males and females as a precopulatory display (although the female does not feed the male), and occasionally during flight, attack, or escape, or after use of the songlike long call display. Its general function is to solicit reassuring acts, indicative of nonaggressive interaction rather than specifically to beg food or to solicit copulation. (Head tossing is a widespread display among gulls; members of most species perform it or variants of it at least in food begging, pair forming, and precopulatory displays, although adult males tend to use it only before mounting. See Tinbergen 1959b.)

Many species seek evidence of reassurance through displays that solicit allogrooming or allopreening (figure 5.6). In these cooperative activities (see chapter 13), one individual picks over the body surface of another removing detritus and ectoparasites, and continues to pick and search beyond any hygienic functions. Allogrooming is customarily performed when a group is at rest, and correlates with close associating, attempts to copulate, attempts to "borrow" an infant from its mother, and with other essentially friendly acts. That is, it indicates readiness for numerous kinds of interactions and is weighted against the likelihood of attack. Primate species with rigid dominance hierarchies perform a great deal of allogrooming. The comparable allopreening of birds is sometimes done in response to attack behavior (which led Harrison 1965 to

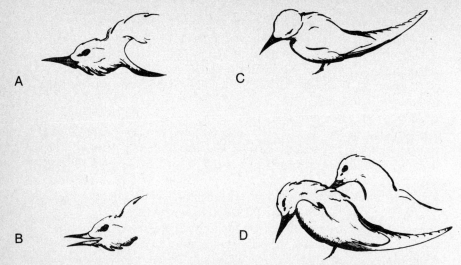

Figure 5.6. Receptivity to allopreening. A white noddy *Anoüs albus* that is receptive to being preened adopts a preening invitation display (*A*), lowering its head and stretching its neck forward, usually while ruffling all the feathers of its head, neck, and upper back. In (*B*) the displaying bird is in addition uttering wheezing notes; it may continue them while being preened. Postures that facilitate the actions of the preener and are commonly adopted during allopreening itself are shown in *C* and *D*. (From Moynihan 1962a.)

interpret it as agonistic), and may also correlate with associating or attempts to copulate (such as in the green heron *Butorides virescens,* Meyerriecks 1960).

The displays that solicit allogrooming present body areas in positions that facilitate grooming. Diverse body areas are presented: adult talapoin monkeys *Miopithecus talapoin,* for instance, have five main soliciting postures (Wolfheim and Rowell 1972). The formalization of these postures is often inconspicuous, limited to taking and maintaining an enabling position before the allogrooming begins (see chapter 11). That is, the formalization is based on performing the position initially to invite grooming, not just to facilitate being groomed.

One of the talapoin positions bows the head toward the petitioned individual. Many species do this, ruffling the fur or feathers of the nape. Species such as the rufous-naped tamarin *Saguinus geoffroyi* even have a special color patch that adds to the conspicuousness of this display (Moynihan 1970a).

Several species of cowbirds (family Icteridae) present the ruffled nape *inter*specifically and elicit preening from members of species with whom they flock or of the host species on whom their brood parasitism

depends (Selander and LaRue 1961, N. G. Smith ms.). In at least the brown-headed cowbird *Molothrus ater,* this head-down display is also performed intraspecifically, sometimes apparently as part of the procedure of dominating another bird (Rothstein 1971 ms.) without attacking it. In intraspecific events the display does not elicit preening, but either an agonistic response or performance of the same display by the recipient; the latter may be followed by the two birds touching and holding their heads together (Rothstein in litt.). In giant cowbirds, some of which also perform their head-bowing display interspecifically, intraspecific performances in agonistic situations do not elicit preening but adoption of the posture by a receptive female will elicit preening from a male before he mounts (N. G. Smith in press). In all cases, the display indicates receptivity to a nonagonistic interactional response from its recipient.

Receptive to Aid or Care

Some species appear to have displays that are performed only in soliciting aid or care in some form. These are often done primarily by dependent infants begging food, eliciting brooding, or shading (see chapter 6). In many species, however, displays performed by begging infants are done by adults when soliciting more than just care. Adult female birds, for instance, may employ the juvenile begging displays of calling, squatting, and wing-fluttering in soliciting aid from their mates—getting their mates to feed them while they need large quantities of food to make eggs—and employ these same displays in soliciting copulation. When it is this broadly given, a display is interpreted as encoding messages of receptive and interactional behavior.

Making Interpretations

To say that the performance of a display which encodes information about the behavior of being receptive amounts to soliciting is a useful description, but the two terms overlap only in part. All attempts to solicit do not involve specification of receptive behavior.

For instance, honeybees are said to solicit cooperative foraging with their dance performances. However, the dancers are not receptive to foraging: they forage but do not receive foraging from their nestmates (whatever that might imply). It might be more appropriate to describe them as soliciting attentive behavior (perhaps true to any animal that displays), an opportunity to provide a nectar or water sample to another bee (they do this for bees who meet them head-to-head after attending their dances), or possibly following behavior—though not they but their

communicated "directions" are followed. In more primitive bees, communicators do solicit following (Lindauer 1967), and members of quite diverse speries of ants that do not use odor trails lead each other with a "tandem calling" display (Möglich, Maschwitz, and Hölldobler 1974). An ant recruiting by tandem calling raises the tip of its abdomen and discharges poison gland secretions from its extruded sting. Nestmates are attracted and, when one makes antennal contact, the recruiter runs back toward the food source. Should the follower lose contact, the leader stops and repeats the scent-based calling display. This behavior does involve receptivity to following; the functionally similar honeybee dancing does not.

Whether the messages of tandem calling be interpreted as "receptive to being followed" or "seeking an opportunity to lead" seems a moot point. Both interpretations are correct. Neither leading nor following can occur without the other, and observation confirms both.

Message of Attentive Behavior

Some displays are performed only when the communicator is attentive to a stimulus to which it is prepared to respond further. For instance, two of the three closely related barking displays of black-tailed prairie-dogs provide information about the communicator's attentiveness (W. John Smith et al. in press 2). A prairie dog will pause either after uttering a bark display or while uttering a continuous barking display and alertly monitor another prairie dog or a potential predator. Along with the monitoring it will avoid interacting with this other animal, and after monitoring eventually turns to some unspecified alternative behavioral selecton. However, when the third display, a chitter bark, is uttered the communicator is sufficiently intent on the alternative to interacting that it is not prepared to stop and be attentive.

Being attentive is a behavioral selection that is more easily discussed along with information about the direction taken by the attending. Information about the directional aspects of a communicator's behavioral selections is considered in chapter 6 among the supplemental messages about behavior, along with various other examples of displays with attentive messages.

6

Further Messages:
Behavioral and Nonbehavioral

In making information available about a communicator's behavior, a display indicates more than the activities that may be selected. It also provides supplemental information about these selections that is, to use an analogy to language, effectively adverbial.

In every display two kinds of information about behavior—selections and supplemental messages—are inevitably yoked. Thus the kinds of information made available by the supplementals must be understood before the potential contributions of behavioral selection messages to interactional events can be evaluated. This chapter therefore begins with a discussion of the known supplemental messages about behavior.

Behavioral selection messages will be considered again after the supplementals are discussed. Behavioral selection messages that do not appear to be widespread among diverse animals are then considered, followed by those known from the displays of dependent infants, a class of individuals with special interactional characteristics.

The final task of this chapter is to consider the kinds of information that displays make available about characteristics of a communicator *other than* its behavior. This information completes the interpretative outline of the kinds of information now known to be made available by displays.

SUPPLEMENTAL BEHAVIORAL MESSAGES

Probability

A behavioral selection with which the performance of a display is correlated could occur every time the display does. However, this correlation is not observed in the case of most displays. Although the behav-

ioral selections interpreted as their messages are much more consistently correlated with the occurrence of the displays than is other behavior, the level of correlated occurrence is far less than 100 percent. Thus a display provides information not about the occurrence of behavioral selections, but about the *probability* of each selection occurring; its messages include both behavioral selections and probabilities.

That the probability of any particular selection coinciding with or immediately following a display must often be less than one is obvious for displays that provide information about two or more incompatible behavioral selections, because only one such selection can be done at a time. For example, either attack or escape behavior may be selected by a rhesus monkey that is jerking its head toward an opponent, but both cannot be done simultaneously; the monkey, in fact, is most likely to behave indecisively (a third incompatible selection; see chapter 5). Even indecisive behavior is not certain, because the attack or escape options are sometimes selected. Thus none of the monkey's options is certain to correlate with any given instance in which the display is performed, although each does so often enough that we can interpret it as a behavioral selection encoded by the display.

ANALYZING PROBABILITY MESSAGES   To be useful to a recipient, probability messages must refer to the average likelihood of a behavioral selection being made as this likelihood exists at the instant of displaying. As a practical matter, however, an investigator can only measure the correlations that are realized in a sample of events (see figure 6.1). This is because the probabilities of acts subsequent to displaying do not remain unchanged once a display has been performed. The communicator's decisions are not frozen from that instant; he continues to respond to circumstances, which often change in ways that cannot be fully anticipated.

The changes can exacerbate, obviate, or otherwise alter the needs for different kinds of action. The display itself leads to some of these changes, which is particularly obvious with threat displays: a gull in a bill-down upright posture or a harshly barking prairie dog may be about to attack, but if their opponents usually withdraw in response to the displays their readiness is hard to measure—they have no further need to attack. Conversely, if their opponents threaten back the communicators may not dare to attack.

Perhaps the probabilities of behavioral selections correlated with the occurrence of displays could be measured only in events in which nothing changes between the performance of a display and the performance of a selection—if such events could be recognized. Even this situation would be suspect, however: if the absence of change is rare

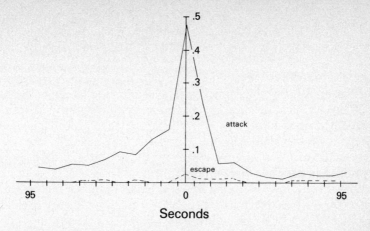

Figure 6.1. Realized correlations of attack and escape behavior with the most aggressive pattern of displaying of the great skua, combining bend, long call, and wing raising. Behavior was scored in 10-second intervals. Attack was seen in nearly half of the intervals in which the displaying was observed (center of the graph, a realized correlation of nearly 0.5), and about one quarter of the observations of the next 10-second interval, then subsequently occurred only about 5 percent or less of the time (realized correlations usually less than 0.05).

A bird that did not attack usually stayed put. Escape, the third most likely behavioral selection to correlate with this displaying, was never seen as often as 5 percent of the time in any interval. Other acts (for example, copulation and nest scraping) did nót occur in the same interval as the displaying and were observed less often than was escape in the other intervals. The information of most use to recipient skuas was presumably that the communicator has a high likelihood of attacking and a low likelihood of escaping shortly after performing these displays. (Because the observations preceding the displaying are also graphed, we can see that attack has been occurring before the displays in many of these incidents. Some of it may have been correlated with immediately previous performances of the same displays, something not revealed by this type of graph.) (From M. Andersson personal communication.)

then it must be unexpected and thus surprising, and the surprise itself is a change.

The result is that it is not possible to measure the instantaneous probabilities of the behavioral selections encoded by displays. The frequency of realized correlations can be measured, but the accuracy with which they reflect instantaneous probabilities decreases as the rapidity of change in circumstances associated with the use of a display increases. Instantaneous probability is thus *indeterminate*. Interacting animals must find it a crucial kind of information, but observers must rely on

frequency measures of realized correlations. These measures are most usefully employed for comparisons and can be stated in relative terms.

RELATIVE PROBABILITIES   Within the repertoire of a single species, more than one display may make available information about the same set of behavioral selections: attack, escape, and indecisive behavior, for instance. Most such displays have different relative probabilities of the selections with which their use correlates. For example, the green-backed sparrow has at least three vocal and three visible displays performed in agonistic circumstances, some when attack is more likely than is escape, others when the converse holds, or attack and escape are about equally probable; for most, indecisive behavior is more likely than is outright attack or escape (Moynihan 1963a).

Lind (1961) compared the occurrence of seven different vocalizations of the black-tailed godwit *Limosa limosa* in a single situation by standing at nests of the species as the parents flew toward him (abortive attack approaches), flew away (escape), or stood at a distance, circled, or flew tangentially (indecisive behavior). He took notes on the numbers of each call uttered singly, but treated brief bouts of calling as single occasions of a call and longer bouts as equivalent to up to four single calls. The effect of introducing this bias cannot be estimated, but assuming that it does not seriously vitiate the results, the calls separate very well in the information that they make available about the probabilities of the three alternatives. The attack-call comes when there is a high probability of flying toward the observer (38 of 45 tallied instances), whereas the staccato-call correlates primarily with indecisive behavior (164 of 246 tallied instances). If a staccato-calling godwit acts decisively, especially if not already close to the observer, approach is more likely than withdrawal (63 and 19 instances). However, as 32 of 37 tossing-flight calls were uttered while flying away, and the remaining 5 while circling at a distance, this vocalization indicates escape as the most probable behavioral selection, with indecisive behavior a poor second. Because these data were obtained from only a single kind of circumstance, all seven vocalizations do not necessarily encode the same three messages of attack, escape, and indecisive behavior. They may have other correlates in other events but their obviously overlapping patterns of behavioral correlations are distinguished by different probabilistic correlations with these three activities.

Probability can also be considered more generally, relative to what would be expected if a given display were not performed. The question becomes, To what extent does the performance of a display correlate with behavior different from that seen in circumstances in which the display is not given? Suppose, for instance, that a subordinant chimpan-

zee approaches a dominant individual with its arm extended, grinning, and making squeaking calls (van Lawick–Goodall 1968). The dominant chimpanzee may embrace it and pat it on the back. It may not be feasible in naturally occurring events to determine that the average probability of subsequent attack by the dominant is, say, 6 percent throughout the range of circumstances in which such back-patting may occur. And it will not usually be worth the enormous effort of making some such statement the objective of a painstaking experimental program. But it can be determined that the back-patting correlates with a probability of subsequent attack by the communicator that is considerably depressed from what it would have been had he not responded with this display.

In this hypothetical example the probability of a behavioral selection (attack) is depressed when the display occurs. In effect, the display provides information about what the communicator will *not* do, a notion comparable to Marler's description (1961a:306) of displays with appeasing functions as conveying "negative" information about attack and escape behavior. The concept of a negative, however, need not be introduced as a further message class in cases like these. It is more simply handled by noting that the displays correlate with probabilities that are low relative to what would be expected were they not performed (discussed further in chapter 5, in considering the attack, escape, and interactional behavior messages).

PERSISTING PROBABILITIES   The performance of some displays is inveterate. Many postures, for instance, are chronic displays. For example, in a troop of rhesus monkeys, individuals differing in dominance status also differ in their customary postures. A sitting rhesus of high rank either relaxes and droops its head or maintains a stiffly upright posture; rhesus of middle rank hunch more; those of low rank slope their spines forward but keep their necks and heads upright (see figure 6.2, which also shows walking postures). Goffman (1963) has described analogous postural differentiation among human individuals differing in rank in hospital staffs; in fact, such postural displays are probably widespread among vertebrate animals.

Postural displays could be said to provide information about status, but they do more because they can be adopted by individuals who have not established their rank relative to one another (status is not a fixed attribute, but is always relative; see Goodenough 1965). These displays are more accurately viewed as consistently predicting certain kinds of interactional behavior. For example, a dominating male rhesus differs from other monkeys because he may suddenly attack without preliminary threat display (other than his chronic postures). He may also

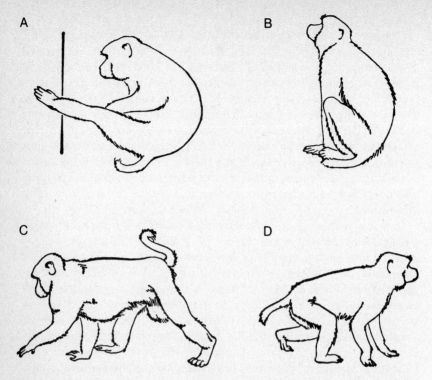

Figure 6.2. Customary postures of rhesus monkeys. The bodily postures assumed by members of a caged group of rhesus macaques differ in accordance with the rank of each monkey within the troupe, particularly the line of the back, the angles of the limbs, and (for walking individuals) the set of the tail. *A*. Relaxed sitting posture, with feet up, of a high-ranking male. *B*. Cautious sitting posture of a low-ranking female. *C*. Confident walk of a high-ranking adult male. *D*. Cautious walk of a low-ranking female. (From Rowell 1972. Reproduced by permission of the author and Penguin Books Ltd.)

terminate attack rushes simply by changing to a stiffly upright posture, whereas a lower-ranking individual that did not carry through an attack would change to threatening displays. Males who dominate are thus relatively instable individuals in at least aggressive actions and more subject to mercurial mood changes than are other members of their troops; they have both a different baseline for attack stimulation and a different ordering of responses once moved to attack (Hinde and Rowell 1962). The postural displays of such males correlate with their behavioral peculiarities, which can be expressed as special weightings of the probability of attack behavior.

However, attack is in most circumstances obviated by the nearly continuous readiness of other individuals to behave deferentially (for a

discussion of the importance of deference in maintaining social order see Rowell 1966, 1972). Patterns of interacting may be so stable within the troop that the significance of chronic postural displays is rarely demonstrated, and yet the displays may imply a very high and persisting probability of certain kinds of behavior being triggered by the removal of the obviating circumstance of deference.

MAKING INTERPRETATIONS   The probability message provides a kind of supplemental information that is inevitably encoded along with each behavioral selection message of a display. We cannot, however, measure information about instantaneous probabilities precisely, and must make do with measures of realized correlations. Also, probability information can be very difficult to interpret for chronic displays, if these displays usually elicit continuous compensating by recipients. Nonetheless, probability information is present in all displays and must be estimated in interpreting them. Such estimates are useful in comparing displays with each other, or in comparing situations in which displays occur with those in which they do not.

Intensity

Many displays provide indication not only of the likelihood that an action will be selected, but also of how intensely it will be performed: how forcefully, with movements of what amplitude, how completely, how rapidly, how long sustained, and other comparable measures.

The appropriate measures of intensity vary with the type of activity. Escape, for instance, can range from a nearly imperceptible movement of withdrawal to a headlong, precipitous dash, whereas the intensity of association behavior may be related more to the average distance apart sought by associating individuals and their failure to grapple with or flee far from one another even if agonistically aroused. Describing these diverse measures as information about intensity emphasizes a fundamental relationship; it is analogous to lumping walking, swimming, flying, and the like as the single behavioral selection of locomotion.

A message category of intensity enables description of kinds of information that are comparable among the displays of a wide range of species. Even though comparable, however, the measures provisionally grouped here as features of intensity do not all vary concordantly; for example, a movement may be performed with great vigor and yet be incomplete. Further study of this message should yield a means of splitting it into more coherent categories than are yet fully evident.

RELATIONSHIP BETWEEN INTENSITY AND PROBABILITY   Few detailed descriptions of the information about intensity provided by displays are

now available. Usually, when differences in intensity and probability vary in similar ways, descriptions tend to emphasize the latter. As the probability of escape, for example, increases, the likelihood of its being more precipitous may also increase. Conversely, an act that has a low probability of being performed may be seen as incomplete "intention" movements of little amplitude or forcefulness. However, nothing is inevitable about parallel variation in probability and intensity, and they can be inversely related. This relation is obvious in very tense interactions such as disputes over a territorial border, in which indecisive behavior is much more probable than is either attacking or fleeing. The highly probable indecisive acts may be abrupt and vigorously performed, but many will be (by definition) incomplete. Complete acts of attack or escape, although improbable, may be quite violent if they do occur.

INTENSITY INFORMATION IN BEE DANCES The studies of dancing honeybees by von Frisch and his students, one of the high points in ethological research on communicating, have revealed an interesting example of the encoding of information about intensity. The information that a dancing honeybee provides her hivemates is about the next flight she is going to make. In most dance performances, the waggle run (see figure 6.4) is the key part: its angle correlates with flight direction (see following discussion on the message of direction), the duration of a tone of about 280 Hz during the run correlates with the approximate distance to be flown, and that tone's pulse rate correlates with the value that can accrue from making the flight (see following consideration of messages that are not widespread). What is usually described as the information made available by the dance about distance can be more accurately interpreted as a measure of the "intensity" of the forthcoming flight, not of map length.

The duration of the 280 Hz tone has been shown to correlate with the amount of energy the dancing bee will expend when she next flies from the hive to the site where she forages (or collects water, or inspects a potential hive location). The correlation is with this measure of the intensity of flight effort rather than with spatial distance, which is apparent from several kinds of evidence. For instance, if the flight from the hive to the food sources must be made upwind or uphill, the measure will indicate a greater expenditure (the tone is more prolonged) than for a comparable source the same map distance downwind or downhill. Bees experimentally forced to walk even a few meters to a site through a passageway with a low, transparent ceiling report with prolonged tones, consistent with the much greater amount of energy they have to expend walking instead of flying.

The energy expenditure that a bee reports is that to go from the hive

to the site, *not* the return trip, even though the last flight made by the dancer was the latter. The dancer's encoding of the intensity of loco-motory effort therefore "points forward in time" (Haldane 1953). Because she may never have made that outward flight, at least in a minimally circuitous "bee line," this form of report is the less simple way of indicating the site's location: she must make some calculations and extrapolations. However, as argued later (under the direction message), she is not reporting about the site itself. Like other display behavior, her dance predicts her own actions—its information is behav-ioral (an argument also made by Haldane and Spurway 1954).

Wilson (1962a, 1971a) has compared the means by which honeybees and fire ants can make available information that enables their nest mates to find food sources. Both species provide directional information through display behavior, a honeybee in her waggle run, and an ant by laying a pheromonal trail, as described in chapter 2. Yet, although the bee's dance performance provides intensity information that correlates with energy expenditure and hence approximately with the distance between a nest and a food source, Wilson's conclusion that an odor trail provides comparable information is inappropriate. Wilson took the number of ants who ceased following a trail at different points as a measure of the amount of information the trail provided about distance. These individuals apparently turn back because of an assessment of the merit of continuing, based on both the concentration of trail pheromone (a measure of other foragers' decisions; see chapter 10) and their distance from the nest. The trail, however, provides only the phero-monal information; distance information probably comes from a fol-lower ant's memory of topographic features and the route it has just followed. The odor trail is just a guide line between two points; it itself contains no distance information that is assessable along the line.

Relative Stability

Information about the moment-to-moment stability of the information that displays make available about the probability and intensity of behavioral selections is provided when a display is repeated. Communi-cators often repeat displays more or less continuously, a phenomenon labeled "tonic" communication by Schleidt (1973), and although much of their repetition follows no patterns, some is formally organized (see chapter 13). The resultant patterns are specialized to encode relative stability messages.

The durations of intervals between vocalizations in the continuous barking of black-tailed prairie dogs are specialized (W. John Smith et al. in press 2). Different relative probabilities of behavioral selections are

indicated when the intervals increase beyond some limit: the probability of engaging in an unspecified alternative to interacting increases. Information about intensity also changes as intervals do: acts are likely to be more abrupt when intervals are brief. When successive intervals remain similar, they inform that probability and intensity are stable. A prairie dog may sustain bouts of continuous barking for tens of minutes, providing much information about stability and its shifts.

Patterned repetition is not the only formalized way of providing information about relative stability. Sustaining the performance of a display act sometimes does the same job and is particularly evident in such formalized aspects of orienting behavior as gaze patterns. (Gaze duration is important in the behavior of the listener during a dyadic human conversation, as discussed next.) Within limits, the more persistent a gaze, the more stable is the attentiveness of the communicator, although a fixed gaze unaccompanied by indications of responding to changes in a recipient's behavior is not necessarily convincing. Sustained attentiveness can be more obvious if prolonged stares are periodically diverted from and then returned to the recipient (Kendon 1967 and personal communication, with reference to unpublished work of Wardwell 1960).

Direction

Among the most important things to know about behavior is the direction it takes: in what direction will an individual attack, from what direction is it fleeing, to which of two or more individuals is it directing attempts to interact. An individual that specifies the direction of its attention and its actions makes available information that effectively indicates the intended recipients of its actions, the whereabouts of a predator, or other things with which it is involved; it does so without requiring messages specific to these external referents.

Information about direction comes from the orientation of an individual's body, limbs, head, or eyes (in gaze patterns). The sources of much of this information are not formalized but rather are simply orientations that enable other activities: orienting toward something facilitates approach, and orienting away facilitates withdrawal; gazing at something is necessary for visual monitoring. Yet the range of orientations that could conveniently facilitate behavior is often much greater than the angles that an individual will take and maintain to some stimulus. Restricted orientations are often the results of formalization that has narrowed the serviceable range to distinctive, obviously informative angles. In some species, formalization of restricted features of oriented behavior has been incorporated into pointing or leading displays.

Unfortunately, it does not follow that formalizations incorporating directional information are always readily studied. First, although some formalized orientations are rigidly held and obvious, others are so variable that they can be revealed only by statistical analyses. Shifting orientations are so often the rule that it is difficult to detect and measure the formalization that may exist, and ethologists have generally tended not to deal with them as display behavior. Second, though formalized patterns of gaze direction and its shifts may be readily apparent to individuals interacting at close quarters, they can be very difficult for a more distant observer with a restricted angle of sight to detect, let alone follow. Nonetheless, useful observations from a very large number of species suggest that formalized orientations are widespread and are important in making available information about direction. The following examples are selected first from a variety of nonhuman vertebrate species, then from human behavior (for which techniques based on frame-by-frame analysis of cine film are revealing great detail), and finally from invertebrate animals.

NONHUMAN VERTEBRATE ANIMALS   Fixed lateral and parallel orientations are obvious examples of formally restricted angles. They have been called "typical compromise" displays by D. Morris (1957), the communicator facing neither toward nor away from a recipient. In many fish, for instance, a threatening individual orients laterally (broadside) to its opponent. In others, such as in *Badis badis* (Barlow 1962), two challengers align head to tail. A similar head-to-tail alignment develops in territorial challenges of the wildebeest *Connochaetes taurinus,* a large African antelope. First one male intrudes on another who responds with a fixed lateral presentation display, then both individuals move alongside one another in a formal reverse parallel position; they may mutually circle while maintaining this orientation (Estes 1969).

Comparable orientations occur in greeting and courting interaction. When greeting each other while "uneasy" (behaving at least somewhat agonistically), paired or pairing eastern kingbirds adopt parallel or reverse parallel orientations from which to perform their wing fluttering and "kit-ter" displays (W. John Smith 1966). A pair-forming male and female gull "often stand or run in parallel with each other" (Tinbergen 1959b); in pairs of the Chilean skua, a close relative of gulls, mates may walk around one another in the upright posture in a standardized "circular parading" (Moynihan 1962a). Both skuas and gulls may suddenly turn their heads away from each other in such events, an orienting movement that has become formalized into a head flagging display in many gull species (Tinbergen and Moynihan 1952; Moynihan 1955a, 1962a).

Both male and female ducks have displays that are performed only

when a communicator can achieve a specific, restricted orientation to a recipient individual. The information these restricted angles provide about the direction of attention is undoubtedly necessary in the milling groups of several males that cluster around and court a female. For instance, a male green-winged teal has seven different displays to perform when he can get close to the female and can orient himself broadside to her, and one display done when swimming directly away from her; two of his other displays are performed with more variable orientation during positional jockeying, one while facing some of the other males who are trying to court the female, and a nod-swimming display done when maneuvering to get parallel to her while trying to avoid the other males (McKinney 1965a). Male mallards will perform their grunt whistle and head-up–tail-up displays to a female only with their bodies oriented at right angles to her. They do considerable maneuvering to achieve this orientation, and the dominant male usually gets the position in front of the female that makes it most likely that he can subsequently attempt to lead her away from his competitors (Weidmann and Darley 1971). McKinney (1975) noted that most of the displays of courting male ducks of this genus (*Anas*) are performed with a "deliberate orientation component"; he proposed that adopting formalized orientations may be the main method by which a male can specify which female interests him. Female ducks also use formalized orientations. A female mallard will display with nod-swimming to a group of courting males, but will perform her inciting display only on taking a fixed bodily orientation with respect to one chosen male (Weidmann and Darley 1971). With both these displays she appears to indicate a readiness to interact, but the inciting conveys information about a greatly narrowed direction of this interacting.

Facing orientations are taken up by communicators of many species in agonistic events in which the communicator may select attack behavior. In the triumph ceremony of the gray-lag goose (see chapter 5, message of associating behavior), for instance, attack is pressed when the postural displays are oriented directly toward another goose, but is stayed when the same displays combine with an oblique orientation (as they usually do among the members of a family). Similarly, a head-forward posture is performed with a direct orientation to an opponent by many species of finches when threatening, and done without directly facing the partner during courtship (Hinde 1956). In these examples, both directly and obliquely facing are formalized orientations that perhaps correlate with the directions to be taken by attack behavior—in the former toward and in the latter deflected from the display's recipient.

In the rhesus macaque, two different threat displays employ nearly opposite bodily orientations with respect to a recipient, and yet both

involve facing and staring, and both may be followed by attacking (Hinde and Rowell 1962). One involves head jerking with orientation toward and gaze fixated directly on the other individual. In the other, often performed during intervals of pacing in encounters between groups, the communicator orients three-quarters away but faces back over his shoulder (a compromise posture). Combined with other displays, staring directly at an opponent can be important in intimidation. For instance, when two tule elk fight one can rear higher, box harder, and still lose the encounter if it is less able to sustain direct staring than is its opponent (McCullough 1969).

Because facing orientations and direct staring are very apt to figure in threatening, members of social groups in most species carefully avoid them at most other times. In flocks of domestic hens, for instance, preliminary measurements by McBride, James, and Shoffner (1963) suggest that directions of facing are not random and that orientations toward the frontal apsect of a neighboring individual are not taken unless that individual comes very close. The resultant oblique orientations are not necessarily formalized; oblique angles may often serve as positions for continuous mutual monitoring, keeping various behavioral options readily available. Oblique angles are formalized in correlation with cautious behavior in various other group-living species. For example, young juvenile gorillas (in observations over several years at the Philadelphia zoo), will carefully hold oblique body and head orientations while their gaze, conspicuous because of the whites of their eyes, is turned to another individual (figure 11.3). Often the communicator then charges (usually still slightly obliquely, figure 11.3B) in an attempt to initiate a play interaction. With slightly different poses, comparable oblique orientations and less rapid approach were often done by these youngsters in coming to a human or to a larger gorilla to cling. Even when several individuals are present, these awkwardly oblique orientations, plus the gaze patterns, can relate quite unambiguously to a single recipient of the communicator's attention.

The fixed orientations adopted with some displays often amount to pointing behavior—a threatening gray-lag goose effectively points at its opponent with its beak, for instance. A parent oystercatcher *Haematopus ostralegus* brings food which it drops on the ground in front of its chicks, and in about 8 percent of the events then stands with its bill held just above the food, pointing to it (Norton-Griffiths 1969). Various quail and related galliform birds point at food with their beaks during some performances of their tidbitting displays (see, for example, Williams, Stokes, and Wallen 1968; Stokes and Williams 1972). The tricolored blackbirds shown in figure 5.5 are pointing with their beaks at the site where the female will build a nest, and birds of many other

species (for gulls and skuas, see Moynihan 1962a:39; for parids, see chapter 5: message of remaining with site behavior) have displays in which they point at prospective nest sites. Some of these displays are elaborate, although many others are little more than sustained orientations.

The orientations of yet other displays can serve to lead a recipient to some special site. For instance, the "ceremonial flight" display of a male black-tailed godwit carries him up and away from his territory, but finishes with a conspicuous swoop back to it, which an unmated female may follow (Lind 1961). Some nonhuman animals are adept at leading with informal behavior: young chimpanzees will beckon companions to follow them with a wave of a hand or the head, or by walking backwards and even trying to drag a companion along—a very direct approach (Menzel 1971). Pet dogs will try comparable procedures on their human masters.

HUMANS   Human behavior provides many examples of formalizations that make available directional information. Pointing with the extended hand and forefinger or whole arm is the most obvious. The head is also often used to point. For instance, the Cuna Indians of northeastern Panama employ what Sherzer (1973) has called a pointed lip gesture: the communicator thrusts his chin in the direction he is gazing, and while raising and lowering his head opens and then closes his mouth, giving the impression of pointing with his lips (figure 6.3). The gesture is performed in diverse circumstances ranging from simply pointing out locations during questions, answers, or statements to making a comment (often mockingly) upon a previous act or interaction, to greeting. The gesture may carry all the information that the communicator makes available about the direction of his actions or attention, or may be combined with verbal statements that provide similar or additional information.

In pointing with the hand or head, the object of attention is itself often being indicated. In human speech, however, referents can be largely within the verbal content; that is, physical objects of attention need not be present during the speech. Gestures that accompany speaking still provide directional information.

The directional movements that correlate with speech have been dealt with by Birdwhistell (1970) as part of body motions or "kinesics," in particular as a special, functionally defined class he calls "kinesic markers." These "markers" do not differ in form from members of his other kinesic classes; in fact, one act may simultaneously or at different times perform both as a marker, and as a part of the developing kinesic structures that parallel the unfolding patterns of speech (see chapter

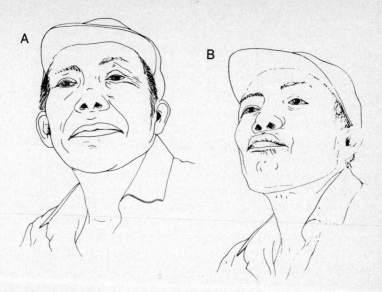

Figure 6.3. The pointed lip gesture. Tracings from photographs made of a Cuna Indian in the San Blas archipelago of Panama when he was asked to indicate, by performing this stereotyped pattern of movements, the direction of something which commanded his attention. *A*. Head position and facial expression before gesturing. *B*. The apex of the pattern: he has turned his head in the indicated direction, and raised his chin; while raising the chin his mouth opened and then closed, and is seen here just after closing as his lips are pursed forward. (From Sherzer 1973; photographs by W. John Smith.)

11). In the roles for which Birdwhistell named them, however, the form of most kinesic markers provides information about the direction of referents in space or time. Take, for instance, "pronominal markers." As a person talks, *distal* movements of the fingers, arms, or head may accompany references to other persons ("they," "their"), other places ("there"), or other things ("those"); proximal movements accompany references to oneself or one's own group ("I," "me," "we," "our"), or to nearby places ("here"), or things ("this," "these"). "Verboid markers" correlate with tense, an indicator of temporal direction; they are sweeping movements that are forward for the future, backward for the past, and inward for the present. What Birdwhistell refers to as "area markers" tend to be loosely patterned but spatially appropriate gestures such as for "on," "under," "behind," and similar terms.

(Two of Birdwhistell's kinesic marker classes do not provide directional information. "Manner markers" belong with what I have called

the "intensity" message, although the diversity of referents available in speech enables it to indicate more than simply the forcefulness of acts. Manner markers are movements that illustrate—to borrow a term from Ekman and Friesen 1969a who call these acts "kinetographic" and "pictographic illustrators"—some of the content of speech: words like "slowly," "swiftly," "carefully," "smoothly." "Pluralization markers" refer to individuals or objects rather than to directions or acts. By making a lateral sweep at the beginning or end of a pronominal marker a speaker indicates a plural state. Interestingly, the phrase "none is" lacks this sweep, whereas "none are" gets it.)

Speech provides the capacity for one of man's most unusual and important interactional behaviors: holding conversations and discussions. During conversations very complex displaying makes information available about both the direction and the relative stability of the attentive behavior of the participants (see, among others, Kendon 1967, 1973; Scheflen 1964; Argyle and Dean 1965; Diebold 1968). For instance, in the dyadic conversations that have been analyzed the listener tends to sustain a direct gaze at the speaker if open to further speech input (Goffman 1963; Kendon 1967), and the speaker intermittently gazes at and disattends the listener (Kendon 1967, 1973). The speaker's shifting gaze pattern indicates attentiveness alternately to the listener's responsiveness and to his own mental processes as he formulates segments of speech. In addition it gathers feedback information about the effectiveness of speech—eyes are unusual signaling organs because they simultaneously both provide and obtain information. A speaker must avoid obtaining visual information while concentrating on internal processing tasks, and commonly averts his eyes as he begins to "take the floor" and speak (Kendon 1967). This is especially obvious when someone pauses to reflect before answering a question: during the pause his gaze shifts laterally from the questioner (Duke 1968).

Analyzing the informative role of gaze during conversations has often been complicated by assumptions about the kinds of behavioral selection information that might be communicated. Some ethologists have described a direct stare as "threatening," for instance, and many students of human behavior have seized on this. But it is a superficial analysis. Many kinds of animals do stare directly while threatening, as mentioned above, but their staring is only one of several simultaneous displays, and it may provide only directional and stability information. This is not to say that direct staring does not greatly enhance the effectiveness of other displays, but that the information it provides by itself is limited and not specific to the act of threatening. Nonetheless, some students of human behavior have assumed that sustained direct staring implies an attempt to dominate.

Experiments to test this have usually been overly simplistic. For example, Strongman and Champness (1968) sought to test dominating, but arbitrarily defined the dominant individual in a dyad as the one who looked away less. More careful experiments seem to imply that sustained staring during conversations correlates with a heightened intensity of interacting, or perhaps with a greater readiness to sustain interaction beyond the conversation itself (see, for example, Argyle and Dean 1965). In fact, Kendon and Cook (1965) showed that subjects who "look in long gazes" are not necessarily feared; they may be liked more than those who use frequent short gazes. Some inferences about interpersonal relationships, although not necessarily dominance, thus do seem to be derived in part from glances of different durations (see also Argyle, Lalljee, and Cook 1968). As a working hypothesis it may be sufficient to assume that the direction of gaze as such provides information only about direction, and that other information comes from other sources such as the duration of staring and the positions and movements of eyebrows, mouth, head, or body.

Gaze patterns are important in conversations, but there are other formalized means of providing information about the direction of attention based on orientations of the head and body, and patterns of movement. For instance, two conversing individuals very commonly orient obliquely to each other (Kendon 1967). Further, by orienting the body differently above and below the waist an individual can provide information about attending in two directions (Scheflen 1964). The division is unequal: the upper torso indicates the direction of immediate attention; the lower designates an individual who is being assured of future or secondary attention, but who has been "put on hold," as it were. When one member of a group is speaking, the listener most ready to interact may employ formalized acts that provide information about both readiness to interact and direction, mirroring the speaker by taking a congruent position and moving in synchrony (Scheflen 1964; Condon and Ogston 1967; Kendon 1970).

INVERTEBRATES   Some of the most interesting displays providing information about direction are used not by humans but by invertebrate animals. Often such displays facilitate return to a site, for example, where food has been discovered; in the case of army ants, they sometimes instead indicate a direction for exploration. The more primitive displays are formalizations of leading behavior. (For example, in some ant species a leader and a follower engage in "tandem running" (Wilson 1959, 1971a; and see the message of receptive behavior, chapter 5): the leader makes short dashes, then awaits tactile stimulation from the follower before continuing. Other species combine leading with the

laying of an odor trail, and in yet others most of the information about direction is contained in an odor trail laid by a forager returning to its nest (Wilson 1971a). These trails provide information about direction along an axis, but are not polarized; recipient ants follow them in the direction away from the nest. Polarized odor trails have been discovered, however, in some gastropods: for example, the freshwater snail *Physa acuta* lays a persistent trail that, on the Perspex used as an experimental substrate, is polarized for about 30 minutes (Wells and Buckley 1972).

The famous dance performance of honeybees is one of the most remarkable and elaborate means of encoding a message of direction that has ever been discovered (figure 6.4). The communicator dances on a vertical comb surface in the darkness of the hive before returning to a site that she has located. Gravity provides a reference, with the upward direction indicating the direction of the sun. For example, should the waggle run portion of her dance (a portion omitted from the round dances done before short flights) be made upward and at an angle of 60° to the left of vertical, her flight from the hive will be oriented 60° to the left of the sun as seen from the hive entrance (see, for example, reviews by von Frisch 1967, 1974; Lindauer 1967; Wilson 1971a; and the examples in figure 6.4).

The common claim that a honeybee dance provides information about the location of a source of food or water or of a potential hive site is a functional interpretation. That is, bees who follow the information they get from attending a dance will arrive in the vicinity of such sites, and getting them there is the function of the dancing formalizations. The information in the dance, however, is about characteristics of the next flight that the dancing communicator will make. This information is clearly evident in the formalized behavior performed by the stingless bees (Meliponini) who do not dance. Their communicators lead recruits back to sites, the leader in some species following an odor trail she laid down on her return to the hive (Lindauer 1967). The signaling that attracts recruits and gives them their initial information about direction is not a waggle run, but an actual flight of the communicator back partway toward her site. This flight is accompanied by a tone and zigzags like a waggle run, and it is repeated. To imply that this behavior, or the honeybee's dancing, encodes information about sites (about entities in the world external to the communicator) complicates its interpretation unnecessarily. Like other displays, bee dances are most simply interpreted as providing information about the behavior and other characteristics of a communicator (see previous discussion under the message of intensity).

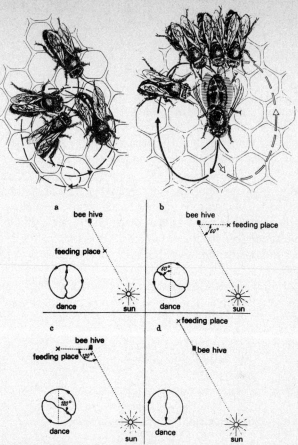

Figure 6.4. Dances of honeybees. Upper left: A returned forager, followed by three hive mates, does a round dance display on the surface of a comb within the hive. The arrows shows that she has made an anticlockwise circle, then turned, and is completing a clockwise circle. She will turn again when she has completed it, and will continue to circle in alternating directions. Upper right: Another bee does a waggle dance display, alternately circling to the right and left after repeating a phase (shown) in which she runs verti- cally downward, waggling her abdomen, while recipients crowd about. This performance is done when feeding (or other important) sites are relatively distant compared to the round dance.

The waggle runs displays are done at different angles with respect to grav- ity. Diagram *A* shows that a directly upward waggle run corresponds to a flight made directly toward the sun from the entrance of the bee hive. Simi- larly, *B* shows that a dance angle of 60° to the left of vertical corresponds to a flight angled 60° to the left of the sun's direction, *C* shows that a dance angle of 120° to the right of vertical corresponds to a flight angled 120° to the right of the sun, and *D* shows that a vertically downward waggle run, as in the upper right illustration, corresponds to a flight directly away from the sun. (From von Frisch 1964.)

LESS WIDESPREAD BEHAVIORAL MESSAGES

Each behavioral selection message discussed in chapter 5 is thought to be encoded by the displays of species of diverse phylogenetic lineages living in a wide range of ecological and social circumstances. Probably other messages with similar distribution have as yet escaped analysis, although it appears as if there may be few of these. On the other hand, certainly many other messages are not widespread in this sense, and some are at least moderately widespread but appear to be limited to special groups of animals. It will be many years before we can categorize with confidence most known messages by their different degrees of distribution among groups of animals, but, as indicated in chapters 4 and 7, it is of considerable theoretical importance that we make such a distinction.

A survey of messages known or thought to be limited in their distribution is not yet warranted. Those well known at present do not appear to involve referents that differ in any fundamental way from the referents of the more widespread messages (for instance, their referents are behavioral, and not semantic categories comparable to such nouns as *food, site,* or *hawk*). They comprise behavioral selections that are relatively narrowly defined.

Messages Specifying Particular Kinds of Interacting

Few messages about particular kinds of interactional behavior are as widespread as is the message of interactional behavior that leaves the kinds of behavior unspecified. Messages of attack and copulating are among these few, as discussed in chatper 5. Postcopulatory displays are performed by crickets, waterfowl, antelope, and many other kinds of animals but usually correlate either with further attempts to copulate or with the behavior of associating; that is, they do not seem to have novel messages. A message of allogrooming or allopreening behavior may be the next best candidate for widespread distribution, yet it can be performed only by species that engage in social grooming interactions, and many species do not.

Social play is another kind of interaction that is often facilitated by the performance of displays, some of which are found only in correlation with play fighting and chasing: the "relaxed open-mouth face" shown by chimpanzees and other primates, for instance (van Hooff 1962, 1967, 1972). Like most displays that appear to provide a play message, its form is a parody of acts used in nonplay fighting. So is a formalized, awkward-looking strike, a sort of "round-house batting" done at the start of play fighting by mammals as diverse as ring-tailed

lemurs, vervet monkeys, and black-tailed prairie dogs (personal observations). Diverse mammals commonly give bounding upward leaps and lolloping running when playing (Bekoff 1974, Symons 1974, and personal observations), behavior typical of a class of formalized acts that Wilson and Kleiman (1974) have called "locomotor-rotational movements": exaggerated head-shaking, body twisting, running, and jumping. When such displays are performed along with other displays that provide information about attack behavior they must help recipients to predict that the attack patterns will be used with restraint. Bekoff (in press) showed this in a quantified study of ninety-two interactions between two captive coyotes: a recipient coyote showed less submissive behavior to threatening displays that were preceded by play signals than to the former alone.

A message of social play behavior may be fairly widespread among species of at least mammals, although the literature is confused on this point. For instance, displays such as laughter are often called play signals, but are performed in correlation with a broader class of behavior than play and thus do not provide information specifically about play. A more serious problem is that play behavior is often highly variable, even innovative, and despite our personal familiarity with it remains hard to define; attempts to define it tend to separate into two approaches, one viewing play as procedures by which animals experimentally test specific kinds of behavior and the other as procedures for generating novel acts and sequences (Fagen 1974). Until consistent definitions are in use, the extent to which this class of behavior is distributed among nonmammalian species will remain unassessable. For the present at least, operational definitions such as that of Bekoff (1972) provide the only practical guide for play behavior research.

Whether social play and allogrooming messages can be considered to be widespread should be left open. However, solicitation of interactions is sometimes very specific, and displays with messages that are limited to a few kinds of animals are easily found. The tactile nudging by which a male gazelle solicits urination by a female is one such display, performed by males who then obtain samples of urine for testing with the *flehmen* response (Estes 1972). The tandem calling displays of ants were described in chapter 5 as providing a message of leading behavior or of being receptive to following behavior. Both interpretations involve a narrowly defined message of interactional behavior that does not appear to be widespread.

In many species, individuals display when soliciting or offering care, particularly in interactions between offspring and their parents (see below). The extent to which messages specific to such behavior are carried from the business of raising young (or being raised) into inter-

actions among adults has not been carefully assessed, however. As indicated when discussing the message of receptive behavior, much of the displaying when adults seek or provide reassurances of mutual support may deal primarily in messages about agonistic behavior and general readiness to interact. It would be surprising if this were the whole story, yet evidence of other messages remains largely inconclusive. For instance, it is often claimed that various bird or mammal species have food-finding calls with which individuals attract each other to share their discoveries of food. Clearly, the calls do attract recipient individuals to food sources in some circumstances (see chapter 10), yet most of these displays are performed in diverse circumstances and on closer examination appear to encode messages about some such behavioral selection as readiness to interact. Still, at least a few species of quail may sufficiently narrow the range of behavior with which they perform one distinctive variant of the basic galliform tidbitting vocalization so that it encodes only messages of eating behavior and incompatible alternatives (Stokes 1967). If so, this variant is basically a food-finding call and encodes an unusual message.

For the most part, the notion that if something such as a call that is uttered on finding food can be demonstrated then the display must make information available about food or foraging behavior reflects a tendency of ethologists to confuse functional analyses with analyses of the information provided by displays. Such conclusions ignore the fact that the functions achieved through the performance of a display are usually very dependent on other sources of information available in the various circumstances in which it is employed. Functions are a poor guide to messages unless the contributions of other sources of information are understood.

Messages Pertaining to the Development of Speaking or Interactional Structures

One class of messages restricted only to humans is found in some "paralinguistic" and "kinesic" displays (see review by Duncan 1969) that are performed along with speech. These displays include variable movements of body parts, modulations of voice and speech pauses, and patterns of gaze and other aspects of orientation that occur in precisely timed correlation with patterns in the flow of speech. Pervasive and crucial, they can function to emphasize, amplify, or even negate the content of speech, often enabling a listener to grasp what has been accidentally garbled or even what has been concealed (Ekman and Friesen 1969b).

The referents of the formalized timing with which many such acts are

performed lie in the active thinking of the communicator: the ways he is relating the ideas being expressed. The results of these relating tasks are usually apparent in the speech flow, and usually the messages of the timed gesturing primarily clarify or elaborate on them.

In the most detailed study to date, examination of a filmed sample of an individual discoursing revealed a five-level "hierarchy in the organization of movement" that matched the hierarchical organization of the speech: each speech unit appeared to have its equivalent, in some sense, in a unit of body motion (Kendon 1972a). Kendon analyzed the speech sample as a "pattern of sound and rhythm," taking as his briefest unit the smallest grouping of syllables over which a complete intonation tune occurred: a "prosodic phrase." These units were joined without pauses into "phrase clusters," and the latter grouped into "locutions." Locutions tended to correspond to complete sentences; they were marked by such formal devices as an increase in loudness in their beginning phrases. Some features of consecutive locutions were shared in "locution groups," one set all having rising intonation, for instance, and the next all falling intonation. These locution groups usually contained what might be called "distinct thoughts." At the final level, "clusters" of locutions were separated from each other by marked pauses, temporizer sounds (such as "um," "uh," "well"), and so on.

Examining body motion, Kendon found a tendency to take up a basic discourse position some time before speaking and to sustain it throughout. Locution clusters tended to have distinctive arm and hand gesticulations and to be separated from each other by shifts in trunk and leg movements. Within the clusters locutions, the principal units, tended to be marked by cyclic head positions, among other things. The form of the cycling movements changed between consecutive locution clusters: the act of using a cycle marked each locution. At the lowest level, each prosodic phrase either had a distinctive movement or a distinctive position that was changed at the end of the phrase. There was much more complexity than this: for instance, very brief locutions such as parenthetical insertions (less integral parts of the whole structure) were not marked in the same ways as longer locutions. Further, with especially important locutions the amount of marking increased; when the point of a whole discourse was reached its locutions tended to be marked by a distinctive hand position such as a boxing-in motion, by the head cycle, and by several vocal devices such as pauses around and stress on the key word.

In some respects the display behavior involved with speech flow is unlike much of that studied in nonhuman species, for example, in being formalized through a precisely timed relationship to other signaling acts of the communicator. Much of the information it provides is perhaps

more appropriately considered within the class of behavioral supplementals than behavioral selections; it might be premature to attempt to present it in message terms. That its timing is relative to speech, a signaling pattern restricted to the human species, strongly suggests that the messages involved are not widespread.

The unfolding of an individual's patterns of speech is not the only performance of formal structure that is marked by display behavior. Many kinds of interactions like conversations, greetings, allogrooming, and social play are the cooperative behavioral products of two or more individuals. The performance of an *interaction* as a structured behavioral unit is a joint responsibility, fundamentally distinct from the performance of individual acts. (The importance of the distinction is not lessened by some acts that, like speaking, are usually found in social interactions, with the result that their patterns often cannot readily be disentangled from those of cooperative interactions such as conversing.)

Formalized interactions are considered in chapter 14. The contributions of individual displays to marking their structural development are of concern here: the indications of pacing, of readiness to remain in or to change roles, of alertness to the deviations of other participants from their roles, and other behavior on which cooperative interacting depends. The messages involved include at least readiness to interact, receptiveness, being attentive, and direction of attending, but other more narrowly defined messages may be involved. How many are widespread among diverse species remains to be learned when we know more about formalized interactions themselves.

Other Messages

Many messages are of adaptive significance for only a few kinds of animals. Their great diversity can be illustrated by a few examples.

In the mating behavior of frogs, males grasp females in an "amplexus" position and fertilize their eggs externally. A male will indiscriminately grasp any other frog, including another male, a female not yet ready to lay, or a female who has laid. Such ill-chosen partners vocalize, and in some species all may utter a single kind of call (A. S. Rand, personal communication). Apparently the only information it carries about behavioral selections is that the communicator will not lay eggs. Such information cannot be pertinent in the mating interactions of many kinds of animals, although there is evidence that in fruit flies, in which matings are also brief, virgins that are too young to breed and older females that have already mated both repulse male advances by producing a loud buzz of about 300 Hz (Ewing and Bennet-Clark 1968). (An alternative interpretation for both frogs and fruit flies is that the communicator is indicating an extremely low probability of

being receptive to copulatory behavior; this interpretation is phrased in terms of widespread messages.)

A message whose referent is the behavior of making a foraging dive may be found in a whistling display of a seabird, the blue-footed booby *Sula nebouxii*. The species forages by flying about 6 to 30 meters above the water, watching for fish. Searching is often done in pairs, sometimes in closely knit groups of up to thirteen members. As one bird begins to dive for a fish it whistles, and the others dive with it, entering the water almost simultaneously and very close together (Parkin, Ewing, and Ford 1970). No other behavior has been observed to correlate with the display. Why simultaneous diving is advantageous is not clear, but it may momentarily confuse groups of fish, making them easier to catch. Further, if the fish scatter erratically soon after, birds not joining in the initial dives could fare poorly. Blue-footed boobies hunting alone do not call, which supports the latter possibility, as does the behavior of related species that hunt in deep water where fish can escape by swimming downward: they neither hunt in tight groups nor mark their dives by calling.

Some species of ants can stridulate, but communicators do so only when seeking to free themselves from physical entrapment, as when picked up by a pair of forceps or, in more natural circumstances, buried by a cave-in of the nest (E. O. Wilson 1971a:243). The message leads to directed digging efforts on the part of other ants. As crocodile eggs hatch, the young produce a vocalization that induces the female parent to return to the nest where she buried them and dig out the young crocodiles (A. S. Rand, personal communication). Conceivably, these displays of ants and crocodiles could encode similar behavioral selection messages, a combination of the widespread seeking message and a narrowly specific version of the escape message (restricted in natural circumstances to correlation with digging behavior), but the latter could be useful only to fossorial animals, and among them only to species in which individuals will aid one another.

Some displays are performed only in correlation with physical exertion, although in diverse sorts of activities. Humans make grunting vocalizations during intense physical efforts. Rhythmic folk songs have evolved culturally for use when groups must cooperate in some task that requires coordination of joint exertion: the "Volga Boatmen" and songs of American railroad crews levering rails into alignment, for instance. Infants of at least some primates and carnivores grunt during exertion (see below). Winter, Ploog, and Latta (1966) reported two sounds of captive squirrel monkeys that they heard only from females straining while giving birth; perhaps a wider range of use occurs in the wild.

As yet, so few such vocalizations with a message of exertion have

been reported even among mammals, which suggests that it is not a widespread message. Indeed, useful responses to such a message may be possible only in species in which cooperation in physically difficult tasks is a well-established part of social behavior. However, at least one other socially cooperative species encodes such a message: the honeybee. As discussed previously (under the behavioral supplemental message of intensity), a dancing bee provides an estimate of the amount of energy she expects to expend on her next flight. The 280 Hz tone she produces during her waggle run encodes a message of exertion. Finding the message in the displays of a few mammals and a social insect does not immediately show it to be "widespread," but it keeps the possibility open and underscores the problem of providing a maximally useful definition of such a concept when we are still working with fragmentary data. The broadest interpretation—that a message has to occur in displays of many animals diverse in phylogenetic histories, ecological adaptations, and social structures to be widespread—will not serve all analytic purposes equally well.

Honeybees encode another message in the tone produced during the waggle run, which appears to have a more restricted distribution among species. The rate at which the amplitude of the tone is pulsed (that is, the loudness, although honeybees probably feel rather than hear it) correlates with the value of making that flight—the benefits that will result from it. Benefits are determined by the richness of the food source or the suitability of the potential hive site she has visited. For sources of equal richness (in experiments, for example, a 1.5 molar sugar solution) at the same distance from the hive the rate of pulsing is the same. If one source is less rich than the other, or farther, upwind, or uphill, it will lead to less rapid pulsing during a longer tone (reviewed by von Frisch 1950; E. O. Wilson 1971a).

Taken together the rate of pulsing (information related to benefits) and the duration (information related to costs) of the tone enable recipients to assess the merit of making any particular flight. Like the cost-related information, what the pulse rate conveys may best be interpreted as a kind of intensity message and its referent some variable characteristics of behavior related to the bee's next flight: for example, how long the flight will be put off while the bee solicits interaction from other bees by dancing, and perhaps the level of resistance that will be shown during the flight to competing attractions. Some of the possible behavioral correlates have yet to be measured, but as it is known that the duration of a bee's whole bout of dancing does correlate directly with the worth of the source, it follows that the pulse rate of the tone provides information about how long the dancer will perform before leaving on her next flight.

Displays with behavioral messages of parochial distribution may be very numerous, yet it appears at present as if most species of at least birds and mammals may have few or none in their repertoires. Because such messages may be very precisely attuned to particular problems, the possibility that they have not evolved abundantly is of considerable theoretical interest, and is treated further in chapter 7.

BEHAVIORAL SELECTION MESSAGES IN INFANCY AND LATER ONTOGENY

Dependent infants of most birds and mammals have interactions with their parents of kinds that do not happen or are uncommon between adults. Infants have only rather uncomplicated behavior patterns to contribute to interactions partly because many are so thoroughly dependent. Typically they also have rather few displays, although these may vary and intergrade in complex ways. From the very limited evidence now available few or no messages seem to be peculiar to infant displays, although some widespread infant messages are probably less widespread in the displays of adults.

Message of Seeking (or Being Receptive to) Care

A comparative study of human, gorilla, domestic cat, lion, and polar bear infants revealed strikingly similar sets of vocalizations that range from faint, low-pitched "uh" sounds to loud, quavering, choking cries of much higher pitch (Ristau 1974). All the vocalizations are uttered when the infants need feeding, warming, or other forms of care, and are attempting to obtain aid or contact, or to continue a care-giving interaction. A small infant can make only feeble searching or clinging attempts, but will usually continue to vocalize if its needs are not met (see also Bell and Ainsworth 1973); that is, the most effective acts with which care is sought are the displays themselves.

Although these vocalizations are very variable, none of their forms appears to correlate with any particular kind of need. That mothers tend to know approximately what is the matter when their infants cry depends largely on the intensity information provided by the vocalizations plus such circumstantial evidence as air temperature or length of time since last feeding, although some other sources of information available are themselves displays. For example, a hungry mammalian infant will often make suckling movements without a nipple in its mouth. Such suckling is not directly functional and is primarily a signal of readiness to ingest.

Messages of Attack, Exertion, Escape, Seeking, and Interaction or Association Behavior

Mammalian infants who encounter insufficient or inappropriate responses may struggle and thrash about in what looks like ineffective attack behavior, and continue to display. Their vocalizations become harsher, apparently supplying an attack message. During unusual exertion they utter grunts, for example, when lifting up their heads, crawling, belching, or defecating. Infants who receive too much stimulation may avert head and eyes and, if they have the motor capability, use special cowering postures that carry escape and seeking messages. If separated from their mothers, infants of at least some primates (such as vervet monkeys, Struhsaker 1967; howler monkeys, Carpenter 1934) and other mammals utter a vocalization that apparently encodes interaction or association and seeking messages.

Birds

Some vocal displays of infant gulls are uttered when a chick is seeking care because it is uncomfortable: it is too hot, too cold, hungry, alone, or being pecked at (Moynihan 1959b; Impekoven 1971). Comparable but less detailed descriptions have been made for some other species of birds, such as the golden plover (Sauer 1962). The open, gaping mouth of a young bird is a display similar to the formalized suckling of infant mammals: both are done when the infants are receptive to being fed. Cowering displays are known from birds as well as mammals in circumstances in which escape is sought.

Infant and Adult Messages

Many behavioral selection messages encoded by these infant displays appear to be the same as those widespread in displays of adults: interaction, association, receptive, seeking, attack, and escape behavior. Some behavioral selection messages that are of widespread importance to infants may be less widely encoded in the repertoire of adults—the messages of caring behavior and exertion and the message specific to ingestive behavior. Close study of infant displays is needed and may well reveal further widespread messages that are more specific to infantile behavior patterns, but the current results suggest considerable accord with the messages of the displays of adults.

Ontogenetic Changes

As ontogeny proceeds, an infant's displays are performed in seeking or maintaining a wider variety of interactions, many of which (in mam-

mals) can at least loosely be called "play." As young animals become less dependent and gradually mature socially, they continue to display (see, for example, Kruijt 1964). The kinds of relationships into which developing animals enter with each other or with adults also change, and different kinds of information become relevant. Their displays may also change, from an infantile to a different adult repertoire, or from infantile forms to adult forms of basically the same displays. In gulls, for instance, infant vocalizations subdivide, diverge, and are elaborated during ontogeny (Moynihan 1959b), and postural and movement displays begin as very variable patterns that are slow to be standardized (Moynihan 1959b; Delius 1973).

Little is known about the message shifts that must occur with some such changes, but some cases have been described. The lost call display of young chicks of the red-legged partridge, for instance, encodes the association and seeking behavioral selection messages; it gradually develops into the adult rally call display (Goodwin 1953), which correlates more broadly with the behavior of associating and loses the seeking message. In some species the calls with which nestling birds beg food from their parents appear to change into the screams adults utter when struggling to escape from a predator. The screams often attract other birds who mob and distract the predator, sometimes enabling escape (for example, see Stefanski and Falls 1972, a paper which includes rare observations of predator events). In other cases, perhaps with little or no change in the form of an infant display, behavioral implications may be broadened. Thus an infant may display when receptive to interaction but be able to engage only in interactions in which it receives care; an adult female may perform the same display when receptive to courtship feeding and copulatory mounting. The change is less in the message than in the behavioral repertoire of the communicator.

Little is known about the behavior correlated with performance of displays by maturing individuals, although it is clear that these are not always identical with adult patterns of performance. Until individuals become both mature and experienced they may display as if practicing, and this displaying is not well correlated with subsequent acts. In the slowly maturing, persistently social Adelie penguin, Ainley (1972, 1974a) found that although young birds display in circumstances in many ways comparable to those of adult use, they have much lower probabilities of following through with or completing the appropriate behavior patterns with which adult performance correlates. When Carolina chickadees are several weeks old they go through a period in which they repeat many of the vocalizations uttered by adults in territorial patrolling and agonistic encounters, but in rapid, jumbled mixtures, often during solitary or semisolitary foraging. The unusual combinations and "amateurish" quality of these vocal displays suggest some form of

practice, and adults have never been seen to respond agonistically to them (S. T. Smith 1972). Practice has usually been invoked as the explanation for the "subsong" of many species of birds, in which immatures (and sometimes prebreeding adults) very quietly utter jumbled, complex sequences of vocal sounds, many of which are not part of their usual repertoire (see, for example, Thorpe 1961:64–70); subsong is not usually done in social circumstances. It brings to mind and may be analogous to the cooing and babbling phases that precede linguistic utterances in human development, and song does tend to develop out of subsong over a period of time.

What are the messages of displays that adolescents perform along with behavior that is characteristic of neither infants nor adults when not engaged in fully matured social activities? Is there a message of practice or play behavior if the form of the display is imperfect or marked in some ways? If the display is in adult form, do contextual information sources enable a recipient to recognize a practice session in which the display's usual predictive relationship to communicator behavior is uncoupled? We cannot yet begin to answer these very significant questions; the ontogeny of communicating behavior is rich with research opportunities.

NONBEHAVIORAL MESSAGES

Every display makes available or is accompanied by some information identifying the communicator who performs it. Without such labeling, recipients would not know to which displays to attend; in any complex habitat they are awash in irrelevant information from the displays of neighboring animals that are usually not of the same species. In many circumstances, a recipient must also know not only *who* the communicator is but *where* it is, and some kinds of displays provide this information.

Identifying Messages

Aspects of the form of displays are often characteristic of particular populations of animals and differ from population to population. Other aspects can be characteristic of particular physiological states or of membership in particular groups of individuals who regularly interact. The form of any single display can provide information identifying a communicator with respect to one or more of these general categories.

Population Classes

Displays commonly provide information about (1) a communicator's species, sometimes also about (2) more local, spatially defined inter-breeding populations, and often they identify even (3) the individual.

1. Species identification The songs of species of birds that breed in the same region are different. These differences are usually particularly striking among species of close phylogenetic relationship who may look alike and in general act more alike than do members of less closely related species (figure 6.5). Throughout a region, each species is clearly and distinctively marked by the form of its song and thus can be identified.

To know with whom to mate and to which kind of animals it is most often worthwhile to pay attention, the birds must be able to distinguish members of their own species. In general, when any animal displays, either the form of the display or a fully available information source coupled to it must identify its species. For instance, in vocal displays, the release of chemicals or electric discharge patterns need to incorporate identifying information because the communicator may not be seen by a recipient. Yet the need for species-specificity is greater in some cases than in others (see chapters 10 and 12): vocalizations uttered at close quarters between mates are often minimally species-specific, whereas most loud vocalizations that are effective over large distances and between individuals who may not be well acquainted are strikingly distinctive. Of course, the communicator can be seen if the display can; visible displays need not be species-specific if other aspects of the communicator's appearance are distinctive (see chapter 9). In fact, many visible displays are very similar among quite diverse species. This similarity is in part to be expected; there are only so many ways in which a bird or any other kind of animal can pose or move (Tinbergen 1959b).

There are exceptions and fringe cases to the rule of specific distinctiveness even among loud vocalizations. Bird faunas around the world, for instance, contain a minority of species that mimic others. The mimicked calls constitute only part of their vocal repertoires, and are usually uttered in distinctive sequences, which makes it possible to identify the communicator. However, a few species that normally forage in mixed species flocks have lost the species-identifying characteristics of one type of call which is uttered when a predator is sighted. These species have converged on a common form of "alarm call" (Marler 1955a); it identifies the communicators not by species, but as members of a limited assemblage of interdependent species.

A comparable convergence, or failure to diverge, characterizes the alarm pheromones of some ants (Blum 1969): at least two species in

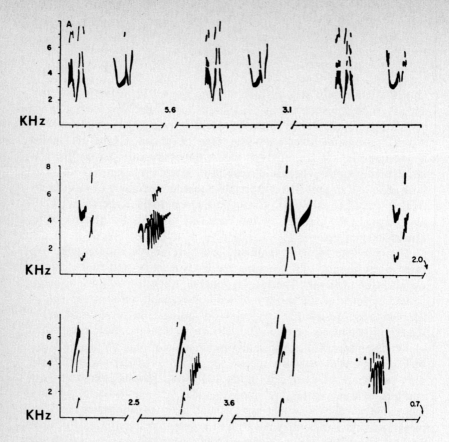

Figure 6.5. Songs of three sibling species of flycatchers. The species are: top line, *Empidonax wrightii*, middle line, *E. oberholseri*, and bottom line, *E. hammondii*. The ordinate shows kHz, the abcissa tenths of seconds; numbers above breaks in the abcissa indicate durations of silent periods that have been removed from the illustrations. The small flycatchers whose songs are shown cannot be distinguished in the field by visible characteristics (hence the term "sibling species"). All have olive-colored upperparts, are whitish below, and have light colored rings around their eyes and two white wing bars. Yet all overlap in the areas in which they breed, and do not form interspecific pairs. Each species uses its readily distinguishable songs in establishing and maintaining pair bonds.

An *E. wrightii* often combines its two song forms in a couplet, as illustrated; it sounds like "sprlat-prit." An *E. oberholseri* may repeat some of its three song forms in a brief sequence before pausing; the sequence illustrated sounds like "prllit-brrltt-peet-prllit." An *E. hammondii* utters its three song forms very emphatically in various combinations; those illustrated sound like "whee-sick," pause, "bzrrp," pause, "whee-sick—bzeep." These onomatopoetic renditions resemble the birds' sounds only very roughly, but help to suggest that what looks different in the illustrations of frequency-time distributions also sounds different. (From N. K. Johnson 1963.)

the genus *Tapinoma* and one in the genus *Liometopum* share exactly the same alarm pheromone, 6-methyl-5-hepten-2-one. But although sharing information about common sources of danger can be adaptive, mating with a member of another species is not, and pheromones involved in mating are more likely to be distinctive. A good illustration is found in two closely related species of gelechiid moths; the female sex pheromones are not only specifically distinctive, but each is also actually inhibitory to males of the other species (Roelofs and Comeau 1969). A male exposed to the wrong one is effectively turned off and cannot respond for a time, even to the sex pheromone of his own species. His sensory receptors are apparently blocked. This is not simply species identification—it is error prevention with a vengeance.

Apart from the effectively cooperative use of predator warning systems or occasional use of social devices for finding food, most species are not usually concerned with identifying animals belonging to classes more inclusive than that of their own species. Identifying phylogenies is not important to them, and they have no reason to share a taxonomist's interest in identifying the species that belong to a category such as their genus. Nonetheless, the species within a genus may share some display features that other species do not have; an ornithologist can often distinguish among the basic song forms of different genera of birds, for example. But the similarities of congeneric songs appear to exist simply because birds have not diverged in all possible ways and thus retain vocal traces of their evolutionary histories. The similarities among songs of species within a genus and the differences among genera (and, for that matter, the differences between bird songs and cicada songs) are not specializations to permit identification. The animals have no need to know. Similarities among the displays of members of a species, aspects of form that enable each species to be differentiated from other species that live with it, identify the most inclusive phylogenetic class that is needed.

2. Identification of local interbreeding populations Within a species population the forms of displays often vary from one region to another. In human language, forms of a single language that are characteristic of different regions have been called "dialects." Dialects are less well studied for the vocalizations of other mammals, and the first case was based on the loud, guttural, vocalizations with which one male elephant seal threatens another (LeBoeuf and Peterson 1969). The elephant seal rookeries that were studied are in a state of flux, however, and the initial dialectal differences are now being swamped (LeBoeuf and Petrinovich 1974). The second case documented for a nonhuman mammal involves short calls of a small lagomorph, the Rocky Mountain pika *Ochotona princeps* (Somers 1973); its dialectal differences may be

more stable. Considerably more documentation of regional differences, however, is available for bird species; and well reviewed by Thielcke (1969) who has done a great deal of work in this area.

Thielcke recognizes different sorts of regional distinctions, from "mosaic" distributions in which the display forms characteristic of an area change suddenly at an abrupt boundary, to "clinal" variation, which is much more gradual geographically. He cautions, however, that samples have too often been not nearly large enough for proper evalua- tion of the true spatial patterns. In describing geographical variation in the songs of the Carolina chickadee, S. T. Smith (1972) noted the same problem, and added that the limits of a dialectal region and the real amount of variation are rarely known because descriptions are com- monly based on disjunct samples from populations that are effectively continuously distributed. She suggested that it is more useful, in the absence of sufficient data, to speak of geographical shifts in display form as being "dialectal" than it is to designate "dialects." Further, Smith pointed out that an unknown but often significant amount of the varia- tion revealed by most sampling procedures may be from nongeographi- cal causes and not recognized because too little attention is paid to the behavior of singing birds. She showed, for instance, that a Carolina chickadee performs each of its four different song displays in correlation with a different set of activities; failure to recognize this confounded Ward's attempt (1966) to analyze geographical variation in this species.

Such dialectal differences among populations suggests a mechanism for assortative mating within a species, a mechanism by which individ- uals with similar genetic adaptations to relatively local conditions can identify each other (see chapter 12). This raises the question of whether or not dialectal differences do in fact correlate with genetic differences among populations (Marler and Tamura 1962). In at least the white- crowned sparrow *Zonotrichia leucophrys,* they do: birds with different regional song forms have been shown by electrophoretic analysis of their tissue proteins to have genetic differences (Baker 1974, 1975).

Dialects of nonhuman species are not fully comparable to language in at least one respect: they are the only known level at which differentia- tion of vocabulary occurs among populations, whereas human popula- tions can be mapped first into groups using different languages, then into dialect groups within language groups. Any geographic subdivision of the form of communicative specialization of a species may promote genetic differentiation, however, and in humans there is some evidence of this effect at least among language groups (Rhoads and Friedlaender 1975).

3. Individual identification The smallest population of all, a single individual, is very often identified by idiosyncratic characteristics

of display form. We can all identify our friends and acquaintances by characteristics of their voices, even though this cannot yet be done unambiguously by current physical methods for analyzing sound (the so-called voiceprint is much less reliable than fingerprints; see Bolt et al. 1969). Other mammals utter much less complexly modulated vocalizations, and sound spectrographs reveal individual characteristics. Using natural and manipulated experiments, Espmark (1971) has shown that measurable vocal differences among reindeer cows and among their calves enable individual recognition. Petrinovich (1974) experimented with played-back recordings to show that elephant seal mothers recognize the calls of their own pups. Many mammal species appear to distinguish among individuals by olfactory clues; for instance, the rabbit (Mykytowycz and Ward 1971; Goodrich and Mykytowycz 1972) and the C57B1 laboratory strain of the house mouse (Bowers and Alexander 1967).

It is again with birds, however, that most ethologists have demonstrated the encoding and use of information identifying individuals. Beer (1970) reviewed both the naturalistic observations and the kinds of experiments that have revealed this message in avian vocal displays. Included is a survey of his own extensive work with adults and chicks of the laughing gull, a species in which the brief notes that begin the long call display differ so much among individuals (figure 6.6) that even human ears can appreciate their ability to act as a "signature" for the vocalization. In some bird species, infants still in the egg learn to identify the voices of their incubating parents (Tschanz 1968; Norton-Griffiths 1969; Beer 1970).

Examples of displays used in parent-offspring interactions, territorial and status interactions, and the interactions of mates have all been shown to permit individual recognition, although at least some displays of vertebrates are exempted. In contrast, no means of personal identification are available in most insect societies, a fundamental difference that distinguishes their behavior from that of most vertebrate societies (E. O. Wilson 1971b).

Physiological Classes

The form of some displays differs among animals of different physiological classes: immature versus adult, in breeding condition versus not, or male versus female.

1. Age Immatures may display differently from adults in many ways; for one thing, frog, mammal, and some bird voices may be higher-pitched because their vocal apparatus is smaller. Animals of many species pass through ontogenetic stages in which they "practice" display-

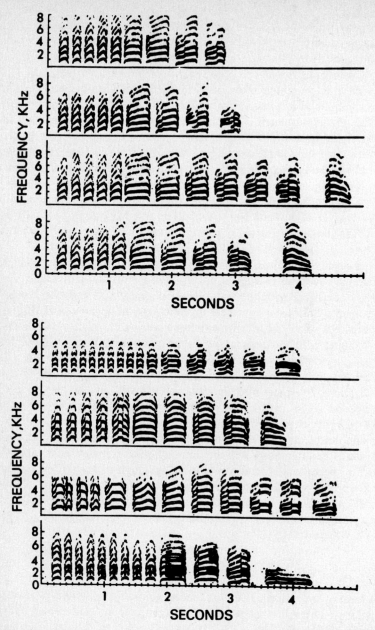

Figure 6.6. Sonagrams of long calls of laughing gulls. The upper four calls are all by the same individual. The first four brief notes are virtually identical in each sonagram, although subsequent notes differ considerably in different events. The lower calls are by four other individuals, and show that the initial brief notes differ in form and number among the gulls. (From Beer 1970.)

ing, as mentioned in discussing the messages of infants' displays. These practice attempts, if peculiar to immatures, can provide information about a communicator's age. For example, most species of ducks have very elaborate courting displays; first-year golden eye drakes perform these with such odd forms that each display is almost unrecognizable (Dane and van der Kloot 1964).

2. *Breeding condition*   The display forms of an individual animal can differ depending on its breeding condition. For instance, a laughing gull's long call display has a hoarse tonal quality outside the breeding season (Beer 1975b) which it lacks during breeding season behavior. However, displays that are performed especially frequently by individuals who are in breeding condition do not necessarily differ in form then. The songs of many species of birds, for instance, although abundantly uttered by breeding birds, are also given in defense of autumn or winter territories, flock ranges, and so forth. Uttering a song provides no information about breeding condition unless it is a form of song restricted to that circumstance.

3. *Sex*   Perhaps the most fundamental distinction between physiological classes is made by displays that are part of the repertoire of only one sex. For example, only females or only males in some invertebrate species release sex attractant pheromones (Jacobson 1965; Ryan 1966) or pheromones that initiate sexual swarming behavior (Brand et al. 1973). In at least one species of electric fish (*Sternopygus macrurus*) the steady state frequency of the monitoring discharges of females is higher than that of males (see Hopkins 1972). In many species of frogs the males gather and sing in "breeding choruses" (reviewed by Blair 1968) where females seek them out for mating; their songs identify the species, sex, and age classes of the communicators.

Bird songs are not always such good indicators of a communicator's sex. There are species of birds in which only males sing, and a few species in which only females are known to sing, for example, in phalaropes (family Phalaropopidae), the females, apart from laying eggs, behave like the males of most other bird species (see, for example, Tinbergen 1935; for information on the hormonal basis of this role reversal see Höhn 1969). In many of the most thoroughly studied bird species, however, females do sing, even if rarely. In fact, in many animals activities customarily performed only by one sex may be shown by the other in special conditions; this has led D. Morris (1952, 1955) to recognize "pseudofemale" and "pseudomale" behavior patterns. When such conditions exist with any degree of regularity, displays cannot be said to identify the sex of a communicator, although the copious performance of some displays will tend to correlate with sex and behavior patterns that are usually sex-specific.

The evolutionary origin of sexual differentiation, the fundamental adaptive advantage of sexual reproduction, lies in its ability to produce offspring with a variety of genotypes by combining genetic material from two different parental individuals (see Mayr 1963:179, 405). In higher organisms this provides the main source of genetic variability needed to deal with unpredictably changing environments. Genetic exchange enables the parent individuals to have more diverse offspring than each could have if it reproduced nonsexually, and thus be more likely to have some offspring that can cope with the unpredictable environments in which they will live (see G. C. Williams 1975). It also endows populations of animals with greater evolutionary plasticity, making them more likely to survive as phylogenetic lineages than are asexual populations (see Maynard Smith 1958).

However the game began, natural selection (which is "opportunistic," Simpson 1949) would have no cause to restrict development of the discovery to functions directly connected with its origins. Sex divides the populations of most species into two equally numerous subpopulations, and natural selection could never pass up such a chance to differentiate, where divisions of labor or the use of physical resources could be exploited. Differentiation has occurred, and abundantly. Some of it has had nothing to do with sex as such, and has no direct or immediate bearing on breeding. For instance, the sexes of a species often differ ecologically, seeking different foods by means of different morphological adaptations (see, for example, Selander 1966, 1972). Other differentiation follows from physiological specializations that are related to breeding. For instance, females in many species have become relatively vulnerable because of their adaptations to produce eggs or fetuses. The evolution of males may then lead to specialization in physical strength or weaponry. Thus, in such open-country terrestrial primates as baboons, females burdened with dependent infants are protected from the attacks of other baboons and to some extent from predators by the actions of fully adult males whose larger size and long canine teeth make them formidable (DeVore and Hall 1965; Rowell 1974). In such cases sexual differentiation affects social behavior in very basic ways with the result that, equality of rights and opportunities notwithstanding, it is often important for a communicator to be able to specify its sex.

The tasks to be accomplished by the behavior of communicating are often divided unevenly between the sexes. Advertising, a task that exposes communicators to predators, is usually handled by just one sex. The risk of advertising is necessary in migratory species, both to enable males to establish and agree on territorial boundaries quickly and to attract females for pair formation. It can also help to ensure the integrity of local breeding populations, each male marking his population of birth

through the dialectal form of his singing (Nottebohm 1969), providing a beacon on which both females and other males can home. Because females in these species have the crucial task of responding to advertising and choosing with whom to pair, the advertisers must provide information about both species and sex. Females can remain inconspicuous throughout pair-formation and subsequently, and thus are more able to assume the tasks of building the nest and incubating. Reversal of roles, with females advertising as in phalaropes, is very uncommon in vertebrates. Probably some factor reduces the physiological demands on such females, who must, after all, still produce eggs. In phalaropes the reversal may be possible because food is unusually plentiful on the tundra or prairies in the spring. A female that can obtain sufficient food to make eggs and has already been able to leave one clutch with a male to incubate should mate again with a second male (Orians 1969), increasing the number of her offspring.

Bonding Classes

Some displays differ within populations so that only individuals who are behaviorally "bonded" to one another share a common form. For instance, the "flight call" of the American goldfinch *Spinus tristis* has special elaborations of form that change and are peculiar to different bonded groups at different times of year (Mundinger 1970). The call's forms always identify bonded goldfinches; the bonds of a breeding pair are enlarged to encompass their family, then later to encompass the members of a small flock. The small flock remains together through the winter, even when within much larger foraging flocks, but eventually disbands into pairs, completing the cycle (see figure 6.7).

Various other kinds of bonds may be marked by the forms of displays. Bonds among males are one example: in Trinidad the males of the little hermit hummingbird *Phaethornis longuemareus* gather together in the forest at singing sites called "leks." Each lek contains subgroups made up of males who sing from stations near each other, sharing song forms that differ from those of other groups within the lek. Within a group, the males appear to have given up peculiarities of their song forms that would identify each individual (Wiley 1971).

Mated individuals of many bird species develop shared vocalizations with which they "duet," a specialization that identifies their membership in a bonded pair (see reviews by Hooker and Hooker 1969; Thorpe 1972; and W. John Smith in press). A duet is a formally patterned mutual performance, and therefore cannot be performed by a single individual; it is an example of a formalized interaction described in chapter 14. Duetting has several functions, not all important to any one

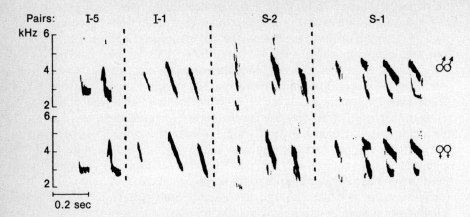

Figure 6.7. The "flight call" of four pairs of American goldfinches. The form of this vocalization is modified on an annual cycle. At different seasons different pairs, families, or small flocks formed of neighboring families each have characteristic forms. Calls of the members of four breeding pairs are illustrated, those of males above their mates. Pairs I-5 and I-1 were neighbors in one field, and S-2 and S-1 were neighbors in a second field. Mates share similar forms, but neighbors do not. (From Mundinger 1970. Copyright 1970 by the American Association for the Advancement of Science.)

species: for instance, it apparently helps mates keep track of each other's locations in vegetation or in a crowded flock; it may enable them to check each other's physiological states (such as coming into breeding condition) and to be ready to breed in synchrony; and it may help them to appease or reassure one another while simultaneously advertising their continued presence on a territory to their neighbors. Each such function depends in part on the information duetting provides about the bond the mates share.

Vocal specializations that identify particular groups are effectively "argots": modifications of some of a species' displays that mark and perhaps aid in the coherence of functionally important subgroups whose membership depends on bonded relations. Membership directly dependent on geographical proximity, on the other hand, is marked by "dialectal" specializations.

I earlier recognized the information about membership in bonded groups as a behavioral selection message, and labeled it "bond-limited" (W. John Smith 1969a, 1969b). This label is not inappropriate, but although individuals who are socially bonded must usually select from the actions that are consistent with the maintenance and functioning of their group, the information is more usefully viewed as "identifying."

Much of what was interpreted in those earlier papers as a bond-limited behavioral subset can be viewed as specification of readiness to interact, encoded in displays that also identify the communicator as being bonded to a certain class of recipients.

The bond that is identified is represented quite literally in a tactile-visible dyadic display of humans, in at least some cultures. Handholding is described by Goffman (1971) as a "tie-sign" indicating a special relationship between two individuals, both to each other (tactually) and to third parties (visibly).

Location Message

The form of some kinds of displays can provide information on "where" (in this sense, the communicator's location) in addition to the "who" of the identifying messages.

Visible displays need no location specializations because if a recipient can see the display it is looking at the communicator (which does not preclude specializations that increase the communicator's conspicuousness). Similarly, the source of a spreading pheromone is found by a recipient that moves up a diffusion gradient, or upwind: the communicator's behavior may be specialized to release the pheromone only under atmospheric conditions that facilitate development of an effective aerial odor trail or only under certain current strengths, if aquatic. Pheromonal marking of sites or of substrate trails with substances that diffuse slowly often provides information about the communicator's future location, because the communicator will likely return to its scent marks. Its reasons for returning can be quite varied: it may be prepared to defend the sites as territorial claims or, in the case of honeybees who mark nearly odorless foraging or water-collecting sites with a Nasanov gland deposit (Free and Williams 1970), to collect at them.

Information about the communicator's location is almost inevitably encoded in vocal displays because their complexity (much of which evolves to carry other kinds of information) makes it possible for recipients to determine the direction of the source by making binaural comparisons of temporal changes in phase and amplitude. Some vocal displays have been specialized to enhance or reduce the amount of information on location that is made available, however. The screams uttered by birds caught by predators are loud, complex, and easily located, whereas the so-called alarm calls that some passerine birds utter when fleeing or likely to flee are much less easily located than are other vocal displays of the same species (Marler 1955a; also discussed in chapter 12 of this book). In a bizarre case, certain bird-eating forest falcons appear to employ species-specific calls that are difficult to locate

as a means of attracting their prey to search for them; the prey apparently responds to a need to know exactly where the nearby hawks are, and the falcons need to see the prey move (N. G. Smith 1969).

A. S. Rand (ms.) has drawn my attention to yet another very complex example. A vocal display of the Panamanian frog *Engystomops pustulosus* is varied in ways that alter its locatability in different social environments. A lone male at a mating pond utters a simple "whine." When it can hear other males nearby it modifies the termination to "whine-chuck" or "whine-chuck-chuck." Despite these descriptions, the modifications are structurally very simple elaborations; they do not alter the duration of the initial call but greatly increase its locatability. Because an individual frog is most easily located when he is calling alone, he utters his least easily located call; presumably it is sufficient to guide patiently searching females but does not overly encourage predators, at least the more active ones. It is less easy to find a particular individual that is calling in a group, although the group itself is easily located. For an individual male to attract a mate he must now be distinguishable, and this is the circumstance in which he gives his more easily located calls.

7

The Origin of
Widespread Messages

That different species of animals have markedly different forms of displays is evident even from the cursory review in chapter 2. However, the kinds of messages their displays make available are very much less numerous and diverse than are the forms of the displays.

Although no exact count of widespread messages is now possible, the evidence suggests that most birds and mammals, many other vertebrates, and some invertebrates encode messages primarily from the widespread referent classes described in chapters 5 and 6, and listed at the end of chapter 4. The list may not be fully complete, and other ethologists might categorize differently in drawing up comparable lists, but the number of referent classes to be considered is small. Messages of more limited distribution are probably numerous on the whole, but most species studied thus far make few or none of them available with the displays at their disposal.

Display behavior thus appears to provide very few kinds of information for use in coping with the great variety of social events that animals must experience. This is all the more remarkable because displays are, for nonhuman animals, the main behavioral specializations for communicating and their messages must be of great importance in interactions.

Not only are few kinds of messages available, but most behavioral messages provide little detail. Of the widespread behavioral selection messages now known, fully half refer only to very broadly inclusive categories of acts: locomotion, the general set of behavioral patterns, indecisive behavior, interactional behavior, seeking, and being receptive. The remainder more narrowly predict a communicator's behavior. For example, even though members of most species can attack in various ways, attack behavior is a much more restricted category than locomotion, which can include movements used in attack as well as in escape, foraging, and many other functions.

The first task of this chapter is to offer an evolutionary hypothesis to account for these message limitations. Relating features of display repertoires (a syntactic topic) to what we know about messages (a semantic topic) is an exercise in pragmatics because it involves considering the functional requirements governing the evolution of display behavior. Another task is to indicate how animals adjust in their evolution to make the most of the limitations on the kinds of information that displays can make available—why some species develop messages that are not of widespread distribution and how others circumvent the limitations, at least in part.

The Limited Number of Displays in Each Species' Repertoire

The syntactic contribution to the puzzzle of the paucity of widespread message classes is that although an enormous diversity of display behavior exists, each species has only a few displays in its own repertoire. Various ethologists have noted both that the total number of displays in the repertoires of most species is remarkably low and that the number is very similar from species to species. Surveys of the numbers of displays of diverse species were prepared separately by Moynihan (1970b) and W. John Smith (1969b), each trying both to compensate for differences in procedures and preferences among numerous authors and to eliminate acts that do not appear from available descriptions to be formalized. We each found an upper limit of about forty displays per species' repertoire, in the traditional sense of "display." Moynihan presented a tabulation of his counts. His hypotheses about the causes of this limit are described in chapter 11, along with his postulate that although new displays are continuously developed in most phylogenetic lineages, they replace obsolescent displays within given populations and do not accumulate.

Such surveys almost certainly do not provide an accurate estimate of the total number of displays available to most species. It has been argued that "biologists have concentrated on highly evolved signals" (Andrew 1969); for the most part, this charge is true. Because they are preoccupied with the evolutionary concept of ritualization and they have seen so many examples of highly distinctive displays, ethologists have tended to name primarily the conspicuously specialized formalizations. We have tended not to count as displays such less conspicuous formalizations as the carpal lifting of gulls that correlates with some probability of attack behavior, even when recognizing that they remain constant and reliable sources of information. (There are exceptions. For example, see the discussion of Stokes's studies of the blue tit, reviewed in chapter 10). Various problems in applying the display concept are evaluated in chapter 13, and it is very likely that because of them we have underesti-

mated the number of displays in each species' repertoire to an undetermined extent.

The counts are not useless because they tally primarily the more conspicuous displays, however. Although underestimates, they are reasonably consistent and they do indicate that the size of a species' repertoire has a real and rather low upper limit. The kinds of acts not included in the tallies tend to be of very much the same sorts for each species: formalized orientations or gaze patterns, formalization of the distances separating interactants, small movements that are performed as recombinable components of conspicuous compounds, and the like.

Students of human communication who have observed such minute formalized actions in great detail uniformly report that they are not numerous. For instance, Kendon (1967) recognized very few categories of formalized patterns of gaze behavior, Scheflen (1964) a limited number of body postures and gestures, and Birdwhistell (1970) a limited repertoire of "kinemes." Kinemes are among the most inclusive behavioral categories of human body motion and positions, and Birdwhistell estimates that there are about fifty to sixty in a given population. Kinemes include many acts that ethologists would readily accept as displays, as well as many acts that may function primarily in relation to speech and thus are peculiarly human, whether displays or not.

Larger estimates have been given in recent checklists of human nonverbal signaling acts in an attempt to make "complete" ad hoc lists available to guide the observations of clinicians (E. C. Grant, 1969, lists 118 "elements," and Brannigan and Humphries, 1972, list 136 "units"). None of these lists should be taken very seriously. When such listings precede detailed behavioral research, considerable oversplitting is inevitable and the criteria for splitting tend to be diverse. The aim of research always influences taxonomic preferences. Birdwhistell, Brannigan, Humphries, and Grant all want discriminating lists, and they emphasize differences—they are taxonomic splitters. Moynihan and I are interested in the evolution of general characteristics and thus concerned with similarities—we are, relatively speaking, "lumpers."

It is not surprising that the lists of Grant or of Brannigan and Humphries should be about double Birdwhistell's estimate, even though they are all concerned with making lists of the same behavior. The current ethological literature indicates that differences in opinions can easily lead one worker to recognize at least twice as many displays as another for the same species of animal; even the definition of display behavior does not provide criteria for recognizing a display unit. Birdwhistell's estimate appears to be about 50 percent higher than Moynihan's or mine, but is, comparatively speaking, even higher because we included vocal displays and he did not.

Granted that our traditional ethological estimates are low and that the human lists are oversplit but incomplete (and based on a species that may have a somewhat larger number of displays, some sets having perhaps evolved along with language), a reasonable guess might be that the upper limit to most species' repertoires is roughly 60 to 80 displays, and probably not more than 150 even for humans. (This total, of course, excludes language itself, here assumed to be an evolutionary elaboration whose complexity exceeds what can be dealt with using the display concept alone, as will be discussed later). The important point is that *all* studies thus far have indicated finite display repertoires well within these projected limits. There is a real limit which is not very high, considering all the kinds of information that might profitably be shared through the use of display behavior.

The Relationship between Numbers of Displays and Kinds and Numbers of Messages

Because each species has only a limited number of displays available to it, and because the set of messages carried by any display is fixed, it is not surprising that each species has available only a limited number of messages. In fact, a species' allotment of displays is even more limiting than is suggested by the gross number of displays in its repertoire because most animals have different sets of displays fit to different sensory modalities, with considerable redundancy among the sets. For example, all the kinds of information that an eastern phoebe encodes in its formalized postures and movements are also encoded in its vocal displays.

That only a minute portion of the great diversity of displays found among animals is available to each species leaves each with a common problem. And it suggests that this scarcity of "message space," of displays to act as vehicles for messages, is in some way responsible for the low limit on the number of widespread messages.

Behavioral Messages

With a limited amount of message space each species has an evolutionary choice between encoding narrowly or broadly predictive behavioral messages. Selection pressures for both must exist. In some circumstances precision and speed are necessary—the only useful response is the right response, immediately. As an example, a bird that sees a predator may flee, uttering a vocalization that carries a message of a high probability of escape behavior, and its flock associates will promptly seek cover or freeze. For most species the number of kinds of such urgent circum-

stances may be small, and most social interaction may take place in an atmosphere that permits both more time to respond and more room for error. Low ambiguity, in short, may often be of secondary importance if the necessary ends are achieved in due time without severe penalties for delay, as in the case of the difficult but ultimately successful attempts of a male phoebe to copulate (see chapter 3). He lacks a message specifying this behavioral selection, but manages nonetheless by advertising his readiness to interact and then (usually) approaching his mate vocalizing about the probability that this locomotion will be terminated as he selects unspecified incompatible acts. Not all his attempts lead to successful copulation, but repeated insemination is not necessary.

The displays performed by the male phoebe in this circumstance are not only sufficient for copulatory interaction but are also useful in other types. This breadth of applicability is the crux of their adaptiveness. Their broadly predictive messages fail to distinguish among various behavioral subsets (for example, the interaction message doesn't distinguish among possibilities as incompatible as attacking, associating, or copulating), but they do permit the displays to be useful over and over again, whenever sources contextual to a display can supply sufficient information to enable a recipient to make these distinctions.

These selective forces, then, lead to divergence in the precision with which messages enable prediction of a communicator's activities. Narrowly predictive messages are relatively precise, but costly: each can be performed in few circumstances, leaving the need for further narrow messages to cover the remaining circumstances a species is likely to encounter. On the other hand, broadly predictive messages are relatively imprecise, but each can be given in many situations and contribute to the selection of much larger numbers of different responses by recipients, with attendant breadth of function. In resolving its evolutionary choice, each species arrives at its own compromise between tying its displays to narrowly or broadly predictive behavioral selection messages.

The cost of "tying up" displays must also lead to the encoding of very similar messages by very diverse species of animals. If selection for message breadth leads to the widest applicability of displays in the greatest number of circumstances, then that we have found only a small number of widespread, broadly predictive behavioral selection messages may be simply because such useful categories of information are so few as to leave little room for more divergent evolution. For example, few message categories are as frequently usable and as able to elicit large numbers of different, adaptive responses in different circumstances as interaction and locomotion messages. Their referents encompass unusually great ranges of behavior. Similarly, the half-dozen narrowly predic-

tive behavioral selection messages that are widespread may be for most species the only ones sufficiently important to be worth their cost in message space (W. John Smith 1969b).

The number of circumstances in which a display can be relevant also appears to affect the encoding of such supplemental behavioral messages as probability and intensity. A species may need to tie down a display to provide the information that there is a high probability of the communicator trying to escape very rapidly, but extremes of probability and intensity are not usually encoded with a message like locomotion. Instead, a display usually indicates a moderate level of probability of such an action relative to that of some alternative: when kit-ter is uttered by an eastern kingbird the communicator may or may not fly or continue to fly in any given case, and on the average is more likely not to. (The probability of flight, although less than 50 percent when the kit-ter is uttered, is still much higher than the probability at an average instant during a kingbird's day—kingbirds do not fly in one-half of all their waking moments. The point is not that the probability of flight is not elevated, but that it is not extreme.) Such a display has a much broader range of employment than a more precise one that could be performed only when locomotion was very probable or very improbable.

Nonbehavioral Messages

The concept of a "space" available for encoding messages has different implications for nonbehavioral and for behavioral messages. Only the number of behavioral messages is limited primarily by the number of displays in a species' repertoire. Nonbehavioral messages depend more on the extent to which the form of each display is complex.

Complexity determines how extensive a hierarchy of classes of identifying information can be represented, because each class is differentiated by a feature of the display's form. For example, a very brief vocalization may have characteristics of form that identify the species and dialectal classes to which a communicator belongs, but lack sufficient additional complexity to encode individual identity or a bonding class (that is, to permit different individuals or groups of bonded individuals to have different forms of the display). More inclusive categories of membership are nearly always encoded, whether or not there is sufficient space for less inclusive categories. Some less inclusive classes of identifying messages may have to compete for space if other selective forces (see chapter 12 and following discussion) restrict the extent to which a display can be elaborated. In some cases it may be feasible to encode information about either individual identity or membership in a small interaction class but not both. In other cases, an individual may belong,

simultaneously or sequentially, to different interaction classes of nearly equal importance but cannot afford the complexity to encode both.

Not all sorts of displays are under pressure to encode the same classes of identifying information, of course. For many visible displays, visible clues available to any recipient that can see the display may preclude the necessity of higher classes and yet favor individually specific idiosyncrasies.

Location is a kind of information that may not usually have to be encoded by visible and tactile displays. For audible displays the complexity of form needed for other ends may often satisfy or exceed the demands of complexity needed to supply information about location. In such cases as the Panamanian frog call discussed in chapter 6, complexity is increased by the communicator in apparent response to a need for increased locatability. The need to conceal information about location may frequently lead to calls of decreased complexity and in the process reduce the amount of identifying information that can be encoded.

Summary

The number of kinds of behavioral information that each species can encode is limited by the small number of displays available to it. Except where responses must be precise and immediate, selection pressures usually tend to operate at the expense of precision, favoring behavioral messages that taken alone do not enable detailed prediction of a communicator's behavioral selections and are not limited by extreme values of such supplemental information as probability and intensity. Each display tends to encode as much nonbehavioral information as is necessary to permit its use in many circumstances, although the extent or precision is limited by the extent of complexity of form available in each display and there are pressures that limit elaboration. Perhaps widely divergent species encode primarily the same sorts of information because relatively few behavioral categories are worth the cost in message space, so that there is little possibility for divergent evolution unless a species has some quite unusual needs.

Kinds versus Quantities of Information

The foregoing proposal relates limitations on the kinds and numbers of kinds of information encoded to limitations on the size of a species' display repertoire. In a somewhat similar proposal Marler (1959:204) suggested that some undefined measure of the *quantity* of information that displays encode is influenced by repertoire size: "given a limited number of signals in the vocabulary, it is most efficient to include the

maximum of information in each one." The concept implied by "maximum of information" was not made explicit but was described as numbers of "items" (different messages) and illustrated by two very different examples. In the first, chaffinch song is said to convey information about species, sex (male, although females also sing), individual identity, and position. No interpretation of behavioral information was made. The second example was the chaffinch's "hawk call," which conveys information about neither species nor location, but was interpreted as telling a recipient "here is something to flee from," a single item of information in a signal of simple form. The interpretation of the latter display is recipient-centered, and very different from message analyses based on the behavior of the communicator; it suggests, however, that such a display may provide information about the probability of the communicator escaping. Whatever behavioral selections are indicated in the "hawk call" display, the precision of information about both location and identification have been adaptively reduced.

Marler's hypothesis is difficult to compare with my proposal, both because we differ in handling what I call "behavioral" messages, and because his hypothesis is phrased exclusively in terms of quantity of information (roughly: number of messages). Basically, I contend that every display must combine kinds of messages that are useful either in a maximum number of circumstances or in meeting an acute and crucial need of the communicators. The consequence of the repertoire limitations on the encoded information are more qualitative than quantitative in tending to favor broadly predictive categories of behavioral selection messages when possible. Nonetheless, quantity is important with respect to the identifying information that Marler described in chaffinch song, because the breadth of applicability of a display is often increased if it encodes many membership classes. An optimal range of applicability is what is important for each display; this range is usually determined more by the kinds of behavioral information encoded than by the density of information of each display. Because Marler's views on this are bound up in a theory of responses they are not considered further until chapter 10.

Message Assortment

Although limited by its small number of displays to providing only a few behavioral selection messages, each species has the evolutionary opportunity to optimize their effectiveness by arranging the particular combinations of messages that each of its displays makes available. Through appropriate assortment of the messages among its displays it appears to achieve at least a limited kind of specialized fit to its communicative needs (W. John Smith 1966, 1969b, 1970c).

The different ways in which the message of escape behavior is made available by a sample of displays from diverse species provide a simple example. The eastern kingbird, a rather aggressive species lacking elaborate social behavior (see chapter 3), combines escape and attack messages in at least one vocalization but has no display that encodes a high probability of escape (W. John Smith 1966). The less aggressive chaffinch, however, has several displays that encode escape messages, including the "tew" of young birds that is "always associated with escape behavior" (and therefore encodes a high probability of escape) and not with attack (Marler 1956a). The Canada goose, individuals of which always live within groups, combines the escape and association messages in what Raveling (1970) calls its "submissive posture." The black-tailed prairie dog, a ground squirrel that is very vulnerable to predators and has a complex social organization, is able to modify its commonly uttered bark displays and barking patterns as the likelihood of escape behavior increases. It makes the frequency of each bark peak more sharply, thereby adding an escape message to the messages already encoded: interactional behavior, the general set of behavior patterns, and attentive behavior (W. John Smith et al. in press 2). The commonly performed jump-yip display of this prairie dog combines only escape and general set messages (W. John Smith et al. in press 1). The kingbird has no such message combination, but many other species of birds do, for example, the Carolina chickadee (S. T. Smith 1972) whose social organization is much more complex than that of kingbirds. This chickadee, in fact, encodes escape in various message assortments in half of its ten vocalizations, reflecting the very frequent need it has to communicate about escape behavior both in controlling intraspecific social encounters and in responding to the predators that endanger its winter flocks as they forage in woodlands largely devoid of leafy cover.

The results of motivational studies of displays of closely related species, although not designed to reveal messages, provide evidence for the existence of widespread differences in what is encoded. They show that displays of very similar form correlate with somewhat different ranges of behavior as they are performed by different species. It is not always clear whether these different correlations imply different supplemental information (for example, about probability and intensity) or different behavioral selection messages, but in many species they probably imply both. Among many examples of differences between species are the frequency with which each species of gull uses such displays as choking or mew calling in similar situations (Moynihan 1962c), and similar frequency differences found by Koref-Santibañez (1972) even among the semi-species of Drosophila paulistorum. (Semi-species are closely related populations that are largely or wholly separated by their geographical distributions; they often represent the earlier stages of

evolutionary differentiation into distinct species and are thus on the borderline between species and subspecies; see Mayr 1963). Bastock's comparative survey of invertebrate and vertebrate animals (1967) indicates that in the "early stages" of evolutionary divergence in courtship behavior between species "large differences in *form* of specific display activities are rare . . . but differences in the frequency of their occurrence . . . are very common indeed."

The homologues of a single form of display sometimes differ among closely related species in the breadth of their messages. For instance, in most New World primates formalized allogrooming occurs in many circumstances as a "general social reaction" (Moynihan 1967:242) and may correlate with the readiness of bonded individuals to interact with each other. However, allogrooming is done less abundantly and only in sexual encounters by two species that are unusual in that they sleep in holes in trees; in these two species it has a copulation message. Moynihan suggests that its restricted performance may be because the physical contacts that are inevitable in the restricted sleeping quarters "provide most of the same social advantages that other species obtain by general social allogrooming."

It would appear that the two hole-sleeping species took an evolutionary opportunity to narrow the use of a display because they had less need of its broader message. Of course, the direction of this evolutionary trend can only be inferred, and the evolutionary broadening of a display's messages is also possible. Indeed, a postulate that Moynihan has made about the evolution of displays within a species' repertoire holds that their employment may often tend to become broader and broader until they are replaced (see chapter 11).

An example that might more likely represent the evolutionary broadening of the message in a series of comparable displays can be taken from what is often called courtship feeding (see, for example, Wickler 1972). In many species of birds the begging vocalizations of hungry young are also uttered by adults. In some species (for example, in the family Paridae: chickadees, titmice, tits) adults may employ the display primarily or solely when they, like the fledglings, are ready to accept food. This use does not alter the behavioral selection messages of the display, although it does broaden the age identification class. Apparently such adults need the food; they are females who either require a vast supplement to their usual intake of food to make eggs or who have too little time to gather food for themselves while they incubate. In other species, paired adults may perform mutual feeding behavior (often with little exchange of food) as a display with a message of interactional behavior, much as allogrooming or allopreening is done by many animals. The begging to be fed has been broadened to indicate readiness for

(or receptivity to) interaction, with little probability of that interaction being attack.

Also, in a comparatively few species, the adult female employs the immature begging performance as a precopulatory display. Here its behavioral selection messages seem to be a disparate mix: receptivity to being fed or readiness to copulate. Such a message combination might evolve in at least two ways. The performance of displays with a message of interactional behavior might become narrowed to correlate with just two kinds of interaction. Or a display that correlated with food begging behavior might become performed by satiated adult females who were no longer receptive to being fed, but who had become receptive to being mounted. In either case, and in all evolutionary narrowing or broadening of a display's messages, there would have to be a stage at which communicators performed the display even though some of its messages did not fit the circumstance. Recipients would then have to be able to recognize this and respond appropriately. That is, communicators and recipients would both have to discard selectively some messages in specific circumstances in which not all messages fit, and both would have to recognize the same implications of the display in these novel circumstances. When such operations were performed, the display would have the properties of a metaphor—it would take on a relation to novel referents. As these referents gradually became associated with regular performance of the display, its messages would be changed. (Of course, if this process can underlie the evolution of shifts in a display's messages, then displays must be able to operate as metaphors without it being overdone. That is, they must occur in circumstances in which they fit imperfectly, and seldom enough that the novelty of their approximate fit is not lost. This procedure would allow some increased use of a display repertoire over that to be expected if displays were performed only when they were fully appropriate, but the extent to which such behavior occurs has not been assessed.)

However the behavioral correlations and hence the behavioral messages of displays are evolved, the display repertoire of each species comes to have a characteristic assortment of messages. Message assortments also come to differ among species and to various extents. To test whether the phenomenon of assortment is adaptive, whether it provides a means of adjusting a limited set of widespread behavioral selection messages to the particular needs of a species, it will be necessary to see if certain patterns in which messages are combined in the displays of a species' repertoire are typical of comparable species. For instance, particular assortment patterns might correlate with such species-specific aspects of social organization as degree of status stratification. Others might correlate with such aspects of habitat as the degree to which

continuous monitoring of other individuals in a group is permitted by the vegetation. If correlations of these sorts cannot be found, or are only weakly evident, the possibility that message assortment results largely from a cyclic evolutionary process of display replacement must be considered. This replacement process is discussed in chapter 11, in reviewing Moynihan's hypotheses for the evolution of displays.

Messages of Limited Distribution

Despite the prevalence of widespread messages, investigations of display repertoires often turn up exceptions—messages that appear to be provided by limited numbers of kinds of animals. A few examples are given in chapter 6 of messages that fit the special needs of only certain animals. Many of these messages are subdivisions of those categories with more widespread distribution: like attack, copulation, or association, allogrooming is one way of interacting. Of these several subdivisions of the information provided by the more inclusive message of interactional behavior, allogrooming is the least widespread; yet other subdivisions of the interaction message are all apparently of limited distribution.

Interaction is a sufficiently important topic for display behavior that it is not surprising to see it subdivided so that some displays provide information about particular ways of interacting. Other message categories are also represented in the displays of some species by more narrowly precise subdivisions. For instance, the Belding's ground squirrel *Spermophilus beldingi* has two vocalizations that are given on sighting a predator, and Turner (1973) demonstrated that these displays correlate not only with different probabilities but also with different tactics of escape. That is, they subdivide the range of behavior implied by the escape message category and refer to more narrowly specific behavior selections. The "churr" is uttered in agonistic situations varying from intraspecific threatening to the sighting of a safely distant predator at ground level. The communicator may stand alertly at a burrow, ready to escape down it, or may flee to a burrow that has two or more entrances, even if it must pass single-entrance burrows. It escape tactic is thus to seek a burrow from which it can get out should an opponent or a predator follow it underground. On the other hand, a squirrel that sees either a hawk or any predator very close by on the ground utters what Turner called its "hawk alarm" vocalization and dashes immediately to the nearest burrow. Seeking a burrow with multiple entrances is not worth the added exposure of more prolonged running in such a circumstance: its escape tactic is to get below ground at once.

Another species, the arctic ground squirrel *Spermophilus undulatus*

also has two vocalizations it utters as it responds to predators. They correlate with differences in monitoring behavior—freezing with head cocked in response to an aerial predator or sitting erect to watch a predator of any sort at ground level (Melchior 1971). In both cases the squirrels are attentive, but the behavioral difference facilitates different scanning procedures and differences in escape-related tactics.

Ground squirrels are not the only kinds of animals that have displays subdividing the escape message. For instance, vervet monkeys *Cercopithecus aethiops* have a "raup" call that correlates with running to or dropping from trees into thickets as a means of escape, and a "chirp" call correlated with remaining in cover or retreating up into the branches of trees (Struhsaker 1967). Yet these narrower messages referring to different escape tactics do not appear to be widespread, even among animals like vervets and ground squirrels that live in exposed habitats and depend on one another for warnings about danger. More commonly the broader escape message is encoded in several displays that make it available with different probability or intensity information and in combination with different other messages.

What ethologists customarily call "hawk alarm" calls, as distinct from calls used in response to terrestrial predators, are rarely if ever so specific that they effectively name a single class of predator. In one classic example, a call of the European blackbird *Turdus merula* has been shown by Messmer and Messmer (1957) to be uttered not only in response to flying predators but also to people near the nest, to mildly frightening prey, and even at times to the mate (see also Curio 1975). Such a display usually encodes escape with a high probability, or escape, attack, and indecisive behavior at high intensities. Unfortunately, most published descriptions of the occurrence of "alarm" displays report little detail or evidence of thorough observation under diverse circumstances. As a result, it is not possible to assess whether performance of some displays correlates only with responding to a particular class of predator. The conservative interpretation at present is that alarm displays have no referents external to the communicators.

Perhaps any of the widespread behavioral selection messages can be subdivided in some way. Heinroth, for instance, gives an example of two vocalizations of the graylag goose that encode different information about locomotion (1911). Its "gangangang" precedes walking or flying in various circumstances in which family association must be maintained, but it "djirb djarb" precedes only vigorous walking behavior. Heinroth found no equivalent to the second in other geese of the genus *Anser,* although their vocal repertoires are otherwise very similar, and he commented on how odd it was that the graylag should have such a specialization.

Even behavioral supplemental messages may be subject to subdivi-

sion. Certainly this might be expected of the intensity message, although we know very little about it as yet. The probability messages of some displays may refer to particular distributions: for example, the probability of a behavioral selection might be remaining steady at the time of displaying, or be increasing, decreasing, peaking, at a momentary low, or be changing in a complex fashion. For instance, it could be momentarily depressed but shortly to be elevated (this could be true of the probability of locomotion when a phoebe utters twh-t, but no measurements have been made yet of the rising phase). When a black-tailed prairie dog jump-yips, the probability of escape usually appears to be decreasing (W. John Smith et al. ms.); the reverse may hold for the loud tee display of the Carolina chickadee (S. T. Smith 1972). However, it cannot usually be economical to devote a display to encoding a shift in the probability of a behavioral selection if this greatly reduces the number of situations in which the display can be used. Further, if the message of escape behavior is important enough to a species to be encoded in two displays at different probabilities, then shifts in either direction can be reported simply by using the displays in sequence.

Most narrow messages that appear to be relatively specific derivations of a broad message category do not seem to be widespread, apart from those already recognized: attack, association, and others (except that all behavioral selection messages are narrowings of the general set message). It would be most useful to understand the nature of the selective forces that drive some species to evolve these specializations or to develop more novel messages. Granted that, first, there is a limited display repertoire available to each species, and, second, that the messages of most displays must function well only when considerd amid many contextual sources of information, it is possible to predict at least three major kinds of circumstances that should foster the origin of messages of restricted distribution.

First, problems of unusual immediacy require appropriate responses quickly and therefore information that can elicit these responses with minimal dependency on contextual sources of information. Messages that must be considered differently in different circumstances are inherently ambiguous, and ambiguity can sometimes be too costly. This is most obvious when a predator endangers a group of animals. Examples of the subdivisions of the escape message with which some species are able to respond precisely to this kind of danger are given above.

Second, species that lack much elaboration of social behavior often have fewer displays than the forty or so that appear to be potentially available. They may not need more, although they could evolve more. If so, they must be under little or no evolutionary pressure to perform most of their displays in the maximum number of circumstances, and

some messages they encode might be expected to refer to very narrowly defined behavioral selections.

Nocturnal nonterritorial frogs, for instance, usually have fewer than ten displays per species (Bogert 1960). At least some of these displays might be expected to make available quite narrow messages. In a very detailed study, A. S. Rand (ms.) has shown that the tropical frog *Engystomops pustulosus* has only seven or eight displays, depending on criteria. Some of these encode representatives of the widespread message set: attack, amplexus (which is behaviorally analogous to copulation), seeking, identification of species and sex, and location. An especially interesting kind of message that may be widespread only among frogs and encoded in their "release call" has been interpreted by Rand as indicating a low to zero probability that the communicator will lay eggs (see chapter 6). Considering the limited amount and kinds of inter-action that most frogs have, this is a very pertinent statement for them to be able to make. Rand has also suggested (personal communication) that a comparative study of messages encoded by various lizard species might be useful in testing the prediction that when a species has evolved some number of displays approaching the maximum available repertoire size it must usually forego most narrowly specialized behavioral selection messages in favor of broadly predictive message categories. The complexity of lizard social behavior is such that they could be expected to approximate this limit.

Lizards do not mark a limit beyond which we should not expect to find simple social patterns. Many phylogenetically "higher" species have quite simple arrangements, perhaps in many cases through evolutionary reduction. For instance, in numerous bird species males organize them-selves at communal sites or leks to which females come for insemina-tion. Social interaction among adults is apparently limited to brief pairings between a male and a female and the continuing territorial or status interactions of the males. The interactions between males and females are so brief that the probable behavior of each bird must be made very clear to the other, as must its species and sex, kinds of identifying information that are obviously very redundantly encoded by such species in behavioral and anatomical specializations. Analyses of the behavioral messages in the complexly patterned displays of lek species have not yet been made, however.

Among mammals, in groups in which heavy mortality is offset by a high birth rate leading to a high rate of population turnover, most species have no cohesive social groups (Rood 1972). Caviinine rodents, for example, engage in few kinds of social interactions. Males of some species appear to encode the narrow copulation message in one or more displays (see chapter 6) but are not yet known to have messages of

more limited distribution. Again, no investigations of these animals have been oriented specifically to the study of messages.

Third, some environments limit the availability of contextual sources of information and must therefore force the evolution of displays that provide more self-sufficient messages. Visual sources of information, for instance, are greatly reduced in deep water and for nocturnal animals. No detail assessments of the messages encoded by animals in such environments have yet been made, and the task will be difficult. However, the effects of the selection pressure to limit ambiguity can be seen in the reduced variability of display form that Moynihan (1964) has detected in the vocal repertoire of the New World night monkey.

The opposite sort of environment may have the opposite effects on the evolution of displays and messages, permitting a further test of the influence of contextual sources of information on the behavior of communicating. That is, where information is continuously and abundantly available from many sources, dependence on contextual sources might be expected to lead to selection for broadly predictive behavioral selection messages. At the extreme, it might even lead to less dependence on displays as information sources, so that the amount of displaying might become reduced.

Circumventing the Limits on Numbers of Displays

Specialized message assortments and the development of narrowly specialized messages are two means by which a species can evolve an optimal repertoire within the limits of its small number of displays. Encoding an increased number of messages, however, ultimately requires finding some means of having more displays. At least two quite different possibilities appear to have been "explored" in the evolution of some species: the use of what amount to clusters of very similar displays, perhaps evolved through the elaboration of minor variants of one initial display; and the development of means of combining displays or formalized components.

Display Variants

Sets or families of closely similar displays are found in some species. An example can be taken from the vocal displays of the black phoebe, a cordilleran relative of the eastern phoebe described in chapter 3. The black phoebe has a variable initially peaked vocalization which it utters much more commonly than the eastern phoebe uses its homologue. The vocalization has three very similar forms (figure 7.1) that intergrade occasionally but are usually recognizably distinct (W. John Smith

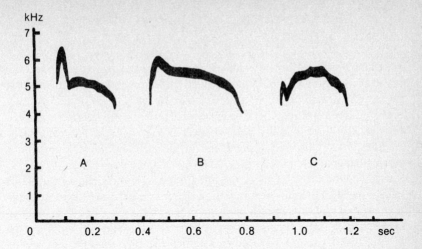

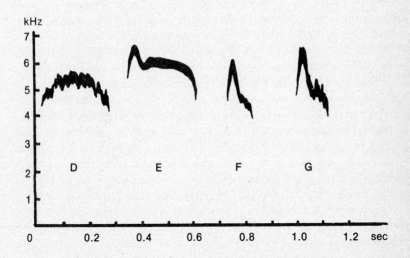

Figure 7.1. Variants of the initially peaked vocalization of the black phoebe. The three principal forms of this display are (A) the chevron-peaked, which is by far the most abundantly uttered, (B) the high-tailed, restricted to within bouts of singing, and (C) the fully humped, uttered in flight and at the termination of flying. Intermediate forms like (E) occur but are uncommon. Other variants exist, such as a quavered form (D) of the fully humped version uttered during flight, and forms (F and G) intermediate between the initially peaked vocalization and a shorter vocal display, the simple vocalization. (From W. John Smith 1970b.)

1970b). Each resembles the other two much more than it resembles any other calls in the species' repertoire, setting the group aside from the other vocal displays. Two are uttered in singing behavior. Of these, one occurs only within bouts of song made up of other vocalizations, and the second, much more abundantly given form, occurs only during the beginning or end of such a song bout, or as a bout itself, entirely replacing the more complex song vocalizations under some conditions. Finally, the third initially peaked form has no special relationship to singing, but occurs during flight and at the termination of flights, both when singing and at certain other distinctive times.

Each of the three forms of the initially peaked vocalization correlates with a partially distinctive range of behavior. Each thus has a message content that overlaps in part with the others, but that also has distinctive features. All three forms appear to encode messages of readiness to interact, and general set. When uttering the first display the bird is usually alone and patrolling or station-calling, or is in the presence of a less aggressive intruding bird but apparently very hesitant to attack (in midsummer such an intruder may be one of the communicator's off-spring from spring nesting). That is, it is seeking to interact or is indecisive about interacting. It shows little tendency to become involved in incompatible activities from the general set, however. When a bird is uttering the second form, indecisive activities may be apparent, but the likelihood of association behavior now surpasses most other kinds of interacting, and the probability of selecting incompatible general set activities is markedly increased. The third display form spans the range in which the first two are uttered, but is in addition highly correlated with flight or its cessation, thus also encoding a message of locomotion. (The behavioral details on which these interpretations are based are given in W. John Smith 1970b, 1970c.)

The three forms of the initially peaked vocalization are distinguishable, but much less easily so than are other vocalizations of the black phoebe. At much distance, or under even slightly noisy conditions, it is relatively quite easy to confuse them or to be unsure of which was uttered. The fundamental advantages of developing such closely related display forms may be that more kinds or assortments of information can be encoded, but with the disadvantage of much less certainty that a recipient will usually be able to distinguish among the signals. Because of the considerable extent of shared information, however, recipient confusion may not be overly costly; it will interfere with precision of response, but should not be likely to lead to totally inappropriate responses. This may be the feature that enables closely related display variants to be evolved in increasing the number of displays in a repertoire.

No messages not already known to be widespread seem to have been developed along with the three display forms just discussed, although much more detailed investigation of the set might reveal correlations with more narrowly delimited behavior than was apparent from the initial study. Presumably, if a species increases the size of its available display repertoire by developing families of closely related displays, then it should be able to afford to tie some of the extra forms to unusually narrow messages and perhaps to evolve message classes that are not widespread among most animals.

This may be illustrated by the tidbitting display patterns found in galliform birds (chickens, quail, partridge, and the like). Tidbitting is a complex performance, comprising a variable, repetitive vocalization, stances, movements, and specialized use of plumage. It varies among species, and in any one species appears to have several patterns that span a range of situations, including care of offspring, courtship, threatening, and flock foraging. From descriptions of its performance by various species (taken primarily from Stokes 1961, 1963, 1967, 1971; Williams, Stokes, and Wallen 1968; Stokes and Williams 1972; and D. Goodwin 1953), the vocal component of tidbitting appears to provide messages of at least interaction and remaining with site and indecisive behavior, and vocal variants may encode escape and copulation. In the bobwhite quail one form may provide the unusual message that the communicator is prepared to ingest food (see Stokes 1967; Stokes and Wallen 1968), combined with directional information and encoded along with some message referring to interactional behavior. Galliforms need to be able to attract their chicks to certain types of food they would not discover alone; although their precocial young forage for themselves they are likely to respond to a wiggling protein-rich worm by fleeing (Hogan 1965, 1966). For chicks to learn not to avoid such tidbits, adults must usually "show" such food to them while using some form of the tidbitting vocalization. Most species do not seem to have a vocal form given only when showing food, but if the bobwhite does the case shows how a special adaptive need can come to tie down one form of a variable family of displays with an unusual message.

The various forms of the tidbitting vocalization in most species intergrade much more freely and frequently than do the forms of a black phoebe's initially peaked vocalization. This must make it all the harder for a recipient in many events to be sure which form was uttered. (It also presents a problem for the ethologist trying to define units of display behavior, and in chapter 13 I suggest that this can be handled by recognizing an additional display whenever any kind of form variation introduces an additional behavioral message.) Some cost, perhaps that arising from recipient errors, keeps most species from employing these

families of related display forms extensively to circumvent the usual limits on the number of displays they can have. Very few species have highly variable display repertoires.

Display Combinations

In a performance like tidbitting several formalized acts are performed in combination, which may reduce the likelihood of a recipient error by providing redundancy, or may add information not otherwise present, or do both. Potentially, however, combinations could generate additional messages not made available by the component displays alone. In human speech, for instance, the feature that Hockett (1960; see also Hockett and Altmann 1968) termed "productivity" or "openness" depends in part on the formalized rules for sequential patterning of the meaningful units that we call words (see discussion by Lyons 1972).

Although combinations of formalized units can potentially circumvent the limitations imposed by the small size of a species' display repertoire, the number of usable concatenations obtained from either simultaneous or sequential recombinations must also be limited. Strictly simultaneous combinations are limited by the amount of sensory information a recipient can take in and discriminate accurately at one time from different sources. Both simultaneous and sequential combinations are limited by restrictions on information processing, such as the capacity of short-term memory, which sufficient organization can overcome.

The most complex example of a signaling repertoire based on sequential combinations is language, although other cases are discussed in chapter 13. The basic, formal components of linguistic utterances are phonemes. These do not carry messages of the same sorts that displays do, but are the minimal sounds capable of changing one meaningful utterance into another. According to Cherry's review (1966) the number of phonemes is small: in English about twenty-eight or, with prosodic features of stress, duration, and pitch included, perhaps forty; only a few dozen are found in any one language. Phonemes, however, combine and recombine grammatically into much larger numbers of stems, affixes, and words, and the working speech vocabulary of most human individuals, although limited, is large by comparison to display repertoires. Finally, however, the fact that words, unlike displays, belong to several functionally different lexical categories that can be recombined and employed in formally structured phrase sequences yields a range of possible statements and messages that is very large indeed—in fact, unknowably large. Thus the organized, sequential recombining of formal elements certainly can influence the limitations set by constraints on basic repertoire size. What is at present problematical is that only one extant species has been able to make extensive use of this potential.

The Pervasive Importance of Context

The displays of most species appear to encode primarily widespread messages: a few broad behavioral selections with approximate measures of supplemental information, plus a few more precise indicators of behavior. These messages alone cannot provide a recipient with sufficient information on which to base its interactional responses. The considerable reliance on information that is only broadly predictive of behavior is adaptive only if the recipient can also make use of sources of information that are contextual to the displays in different circumstances.

Fortunately, abundant evidence indicates that animals do use more than one source of information in selecting a response to almost any kind of situation and that some sources usualy become prepotent, the others supplying varied, informative settings.

Perception itself depends on comparison of information sources, and on treating some as contextual to others. Depth perception, for instance, is based on relative sizes of retinal images of several objects, on the relative effects of haze obscuring different objects, and especially on past experience with objects and with such clues as converging or diverging lines. Even inside 200 meters familiar objects tend to determine the judgment of depth, although in this range stereopsis from retinal disparity is a variable and sometimes strong influence (see, for example, review by Ogle 1962). The perception of a novel shape is simplified in terms of familiarity with similar shapes, and as a shape is discriminated against a background the latter becomes perceived in less detail; that is, it is relegated to the status of context. Perception of one area of color is differently affected by the colors of adjacent areas. Visual, proprioceptive, and labyrinth cues are all compared in determining the perceived vertical direction and interpreted differently when there are cues about motion (see Dichgans et al. 1972). Speed and accuracy of both reading and listening depend on expectations of structure and sequencing of units derived from familiarity with contextual relationships in the language. And in complex scenes such as photographs of the real world, the scanning procedures used in perceptual recognition of objects depend on cues contextual to the objects (Biederman 1972). Even when one stimulus is made especially obvious to the subject in a perception experiment, other sources of information always remain relevant; these cannot be completely eliminated, as many reside in the subject's memory.

Because familiarity (that is, the contextual sources of memory) and other aspects of context are fundamentally important in perception, both primary and contextual sources of information must affect the processes and results of learning. In fact, however, a dispute over the usefulness of

contextual sources of information caused a great controversy in psychology between the noncontinuity theory of discrimination learning and the competing continuity theory. In the extreme noncontinuity view, if an animal learning a discrimination initially showed systematic but wrong responses, it would abandon its wrong hypotheses when it solved the problem and learn *only* the relevant relationships. That is, much of the contextual information with which it had dealt during learning would be discarded. Continuity theory held that contextual information could be learned along with the most salient cues. Tests altering the valence of cues contextual to those chosen by a test animal in its initial responses, before it had developed systematic response tendencies, showed that differential learning occurs even when responses are based on irrelevant "hypotheses" (Spence 1951). Even irrelevant stimuli can be learned, and the learning then applied to a subsequent task in which they are made relevant, as predicted by continuity theory (see Hall 1965). In short, information from contextual sources, whether immediately useful or not, is sufficiently likely to be valuable that it is retained after learning information from primary sources.

MacKintosh (1965) offered a more detailed picture of the role of contextual information sources in learning in his "modified noncontinuity theory": animals learn incidental cues, but they are controlled by an attentional process. MacKintosh claims that continuity theory fails to distinguish between primary and incidental cues, whereas animals separate information sources into most pertinent versus incidental or contextual. They "do not classify their stimulus inputs with equal effectiveness in all possible ways at once." The amount of incidental cue learning can vary as the animal attends to different features.

Even while an animal is attending to a primary cue that it has already learned, it can learn new contextual cues. Church (1957) trained rats to follow leader rats. The leaders employed cues the followers could not learn, because each followed many different leaders that differed in their training. After light cues had been added for 100 post-training trials (that is, at a time when the followers had learned that the important thing to do was simply to follow), the followers used these new cues in 77 percent of the tests then run without leaders. The experiment illustrates "vicarious" learning (see Zajonc 1966:30–33).

All classical conditioning depends on the presence of cues contextual to an "unconditioned stimulus," although to get the animal's attention experimenters usually employ cues that are conspicuous or "salient." The classical conditioning of display responses of the fighting fish *Betta splendens* (Adler and Hogan 1963; Thompson 1966) shows that these fish will pay attention to contextual features during a communication event. That is, when a model of a displaying opponent is paired with a

red stimulus light or a weak electric shock, the latter can come to elicit display behavior in the absence of the unconditioned stimulus of the displaying opponent.

Relatively few experiments on the attentiveness of animals to contextual sources of information have been conducted in nature. Menzel (1969) made a simple demonstration by taking a pink conch shell from the shore where it was ignored by a troup of rhesus monkeys and placing it atop the highest hill on their island, Cayo Santiago. The monkeys treated the shell, which was out of context, very cautiously until it was determined to be safe. His many similar experiments, placing everyday objects out of context, always elicited investigation or caution from the monkeys. In naturalistic experiments with displays, Falls and his students (Falls 1969) have shown that a male white-throated sparrow responds very differently to hearing a tape-recorded song of one of his neighbors if this well-known song is played from the direction in which he customarily hears this neighbor, or from an unexpected direction. The latter elicits a very much more pronounced territorial response, typical of that given not to a known neighbor but to a stranger.

Such experiments indicate how a systematic study of the uses and sources of naturally occurring context could be begun, and it would be useful to have a much more thorough understanding than is now available of just how contextual sources of information operate in social communication. Chapter 9 is a review of ethological literature that bears on sources of information contextual to display behavior.

8

Differing Approaches to the Analysis of Display Behavior

Just because display is a component of interactional behavior does not mean that it can be analyzed only from that perspective, although it does mean that it is not fully understood until its role in interactions is known. Display is only one of many categories of behavior in an animal's repertoire, and, like other kinds of behavior, it requires an individual performer who performs when in the appropriate state. Thus the motivation of displaying must be analyzed.

This chapter describes the motivational analyses that are the central ethological tradition in studying displays, and explores both the differences between and the complementarity of analyses of messages and motives—the aims of the two kinds of analyses are very different and must not be confused. Alternative ethological approaches and their aims are also reviewed, along with a brief account of the traditions and recent approaches of related disciplines.

Traditional Ethological Approaches

Message and Motive

The information-oriented message approach advocated in this book results from an interactional perspective on the study of communicating and is very different from the main procedures that are traditional in ethology for analyzing displays. The tradition focuses on the causations of displays as acts of individuals—that is, the motivational mechanisms underlying display behavior. Yet interactional and motivational approaches are compatible, in fact, complementary: each seeks to explain different aspects of the behavior of communicating.

MESSAGES   As used here, a "message" is a term for a kind of information shared among animals while communicating. To understand the

basic features of a message-oriented analytic approach thus requires both a practical concept of information and an idea of the properties of the information that is usefully shared.

Information is described in chapters 1 and 4 as an abstract property of entities and events that makes their characteristics predictable to individuals with suitable sensory equipment for receiving the information. Information thus enables such individuals to make choices, to select their activities in a given circumstance appropriately for their needs and opportunities. This concept of information as a property that enables choice was formulated by Shannon and Weaver (1949), who made it central to their "mathematical theory of communication" and the basis for measuring not messages but quantities of information. The application of the Shannon-Weaver theory to studies of display behavior is dealt with later in this chapter.

Studying the use of information requires adopting a perspective that has now become almost commonplace. The importance of the concept was argued forcibly by Norbert Wiener (1967:25): "society can only be understood through a study of the messages and the communication facilities which belong to it." He was articulating a powerful point of view that began to develop rapidly in the late 1940's as we became more and more dependent on machinery that facilitates communicating. As Wiener saw, the perspective is applicable to much more than our use of computer and other devices. He called the process of receiving and of using information "the process of our adjusting," and he meant adjusting to all the requirements of our whole environment (1967:27). Organisms, like other open systems that maintain steady states, must continually adjust by choosing responses to their circumstances on the basis of what they know and can perceive. In fact, behavior can be analyzed as if it comprised a sequence of decision points (as mentioned in chapter 4), and different kinds of behavior patterns be shown to have particular "decision structures" (Dawkins and Dawkins 1973).

The information that is most usefully shared among animals by the behavior of communicating might not otherwise be readily available. It is information that a preoccupied, inattentive, or distant recipient might not obtain from other sources. Often it is information that is essentially private to a communicator, inaccessible to other individuals unless made available by special means.

A communicator, like a recipient, is continuously assessing the information available to it and choosing a course of action. Unless an animal signals, however, the ways its assessments bias it toward behavioral selections are not evident until acts are selected and actually performed. When it does formally communicate, its recipients obtain an improved basis for their own processes of selecting, and can adjust in anticipation of the communicator's actions (see, for example, Andrew 1962).

Consider an example. A phoebe may fly and alight, uttering a twh-t, without being seen by a second phoebe. If the latter hears the twh-t it is informed about the flight. Even if it sees the flight the twh-t display may provide it with information it did not have. This information is not just that the probability of the communicator flying is momentarily low (learning that alone is often not particularly helpful for a recipient), but also that the communicator is behaving indecisively and may attempt some form of interactional behavior. When all the messages of its display are taken as a package they are useful: the communicator is being and will probably continue to be indecisive, not immediately making a flight that will lead it directly to or away from a potential encounter, though it may subsequently attempt to interact. Knowing what the communicator probably is and will be doing should enable a recipient to predict more accurately and therefore to cope more effectively with its next acts.

Analyses of messages, the information made available by displaying, are necessary in studying how communication contributes to the orderliness of interactions and the maintenance of social structures.

MOTIVES   The early perspectives of ethology were different—the concept of information was neither explicit nor central. The concepts of motivation and its control by "releasing stimuli" were.

Motivational concepts deal with changes in the internal states that underlie changes in an individual's behavior (Hinde 1970:chapter 8). These internal states are, in the last analysis, the neurophysiological correlates of behavior (Dethier 1964; Hinde 1970—although Hinde argues that when physiological correlates are known, postulates involving motivational or "drive" concepts become irrelevant). The tradition that developed in studying display behavior (described more fully below) was to seek to understand the internal processing that underlies it—the motivation of individuals who perform displays—and the changes in motivation of individuals responding to displays. Those changes were thought to be "released" when displays were perceived, and to lead to rather fixed kinds of response patterns.

The causation of the behavior of individuals is a challenging topic for research. However, it is unfortunate that it became so predominant a focus early in the history of ethology, because reduction of interactional behavior to explanations in terms of units representing individual motivational states is effectively a unidirectional procedure: the properties of interactions cannot readily be derived from the properties of mechanisms internal to each individual participant. This is merely one case of a very general phenomenon in science: it is almost impossible to predict the properties of a given level of organization through synthesis of

components of the next lower level of complexity (van Bergeijk 1967; Wieser 1967; Anderson 1972; see also the discussions of reductionism in Ayala and Dobzhansky 1974).

The procedures of any analysis involve reducing a subject matter into components and studying the mechanisms by which these relate to each other to produce the integration seen in the topic, which is the level immediately superordinate to that of the components. When the subject matter is interactional behavior it should be reduced into the acts of participant individuals, because the course of an interaction depends on what its participants do. The appropriate questions then have to do with what the relevant acts are, what they contribute to the processes of interacting, and how their contributions function to yield orderly (integrated) interactions. Of course, the individuals who participate in interactions are motivated, but to analyze their individual states shifts the focus from the ways by which the flow of an interaction is controlled to the next lower analytic level—the internal mechanisms controlling individual behavior.

COMPLEMENTARITY   The concept of information can be used to describe approaches for studying both motives and messages, even though it was important in generating only the latter. An individual animal can be viewed as having internal states that shift continuously as it processes information from various sources; motivational analyses of communicators are then studies of the information that a communicator is processing when it signals. On the other hand, in studying the contribution of communicating to interacting, we are more interested in the information that a communicator shares with recipient individuals than in the information it is processing. The difference is between the information that a communicator uses in selecting *its* behavior, and the information about that behavior that it makes available to recipients, thereby helping them make *their* choices.

Note that the behavior of an individual communicator is important to both motivational and message analyses. Seen from the perspective used in motivational analyses, individual behavior is the superordinate unit, the level at which integration of motivational mechanisms is expressed. Seen from the perspective used in message analyses, individual behavior is a component of the superordinate level of integration, the interaction: individuals behaving toward and in response to one another make interactions. In human behavior, Goffman (1967) described the difference as between the study of "men and their moments" and "moments and their men"; what an individual does is important from both perspectives, but it is the end point of the first and the starting point of the second.

Thus, although an interactional approach to the study of display behavior diverges from the traditional approach in its aims and many of its procedures, the two kinds of analyses are complementary. Behavioral messages refer to acts performed by individuals, and they can be rephrased in terms of the motivational states that underlie the behavior. When I began to use the term, in fact, I defined a message as being "in some way descriptive of some aspect(s) of the state of the central nervous system" of a communicator (W. J. Smith 1965). Yet, given the strong tradition of motivational analyses, that definition was a tactical error: it invites confusion between the concept of a message as shared information, and concepts dealing with information internal to each individual. Nonetheless, a full understanding of the behavior occurring in interactions depends on analyses of both the mechanisms by which interactions are managed and those by which the behavior of an individual is produced.

Motivational Analyses

ORIGINS AND CHARACTERISTICS OF THE TRADITIONAL APPROACH One of Konrad Lorenz's most significant contributions to the development of ethology was to recognize units of behavior that could be conceptualized and analyzed as properties of a species ("the discovery which I personally consider to be my own most important contribution to science," Lorenz 1974). These "fixed action patterns" comprise relatively stereotyped behavior patterns that are very similar throughout a population of animals. The class includes both displays and many other acts, all of which can be analyzed in much the same ways as anatomical and physiological properties have long been studied (see Lorenz 1950). By treating units of behavior as equipment analogous to organs, Lorenz thus legitimized for biology the study of behavior that had been begun by Darwin, J. Huxley, Heinroth, and others (Tinbergen 1963). He won ethology a place among biological disciplines.

The means of making ethology a legitimate part of biology also set the principal directions in which the field would develop, but set them too narrowly. Tinbergen, referring to ethology's debt to comparative anatomy, has recently said that "much of what Ethology has done is no more than the application of already practised biological procedures to behaviour rather than to structure . . . not doing anything really new" (1972:24). His implication was that novelty was not needed, that the procedures ethologists have borrowed and built on have been useful and abundantly productive. They have. But it is unfortunate that further procedures, more appropriate to the study of interactional behavior, have not usually been devised. As useful as the traditional perspective is, it derives limitations as well as advantages from its origins in disciplines

that seek to explain what governs the structure and processes of individual organisms.

The application of this traditional perspective led Tinbergen to set as the "central" problem in analyzing the behavior of a communicator: "What urges the actor to signal?" in his influential book, *Social Behavior in Animals* (1953a:73). His topic, display behavior, is indeed a social phenomenon, but the problem stated is not. Rather, it reduces the behavior patterns observed in interactions to explanations correlating the apparent internal states of individual participants with their responses to specified stimuli. Nonetheless, this approach became characteristic of most ethological research on display behavior.

Tinbergen's objective was developed logically within the perspective that Lorenz established. Dealing entirely with mechanisms controlling individual behavior, Tinbergen wished to show that an animal displays not because it has the foresight to see a need to display, but because it has an innate propensity to display when in certain states. (The assumption of innateness can tend to oversimplify interpretation of developmental mechanisms, but the instinct-learning controversy is peripheral to the subject of this book.) Tinbergen's objective derives from the traditional view that among the things to be studied about behavior, the most important is its proximal "causation," that is, the physiological mechanisms underlying its use.

Physiological mechanisms are not directly accessible through the study of behavior, and hypotheses about motivation have customarily been introduced to represent them. To Darwin (1872) and Huxley (1914), displays were the results of particular "emotional" states, to Tinbergen (1953a:73–74) displaying was "due to relatively rigid and immediate responses to internal and external stimuli . . . the 'language' of animals is of the level of our 'emotional language.' " There is a considerable gap, however, between postulating the existence of a motivational state and understanding the physiological mechanisms that postulate represents. Observations of behavior alone generally have not provided very precise clues as to the complexly variable physiological nature of these mechanisms. Indeed, there are circumstances in which observation cannot be adequate (see McFarland and Sibly 1972).

Motivational systems remain difficult for ethologists to describe. It is not clear, for instance, how many systems are needed to account for behavior and how they are interrelated. In practice, although by no means always in theory, much research has lumped the postulated causation of observed behavior into three "unitary" systems: aggression, fear, and what has been called sexual motivation (see, for example, Tinbergen 1959b). Andrew (1972) correctly points out that this traditional trio still continues to dominate and oversimplify causal interpretations of display behavior, and it is unsatisfactory in several respects.

For one thing, as Hinde has argued since the 1950's (his criticism is summarized in 1970), these gross and basically functionally defined motivational categories do not represent unitary causal systems. Each is said to control such variables as the persistence, directiveness, and temporal clustering of behavior patterns, which do not always operate in concert. Further, each category is based on observations of a diversity of behavior patterns that may depend on different motivational systems. The kinds of activities called "sexual" are so heterogeneous (Moynihan 1962b) that the term is more obfuscating than useful. At one extreme the term refers to acts of inseminating and at the other to a tentative associating without contact, but the single term implies some sort of continuity in motivational systems. By one of Tinbergen's definitions (1959b), " 'sexual' may refer to mating motivation, or to an inclination to be near a mate, which with a shorter or a longer interval develops into a tendency to mate"; the linkage of motivational systems involved in this "development" is not clarified. He recognizes the problem, however, and has also proposed that "a distinction must be made, at any rate in pair-forming species, between the sexual urge and the urge to pair formation" (1959c). Using behavioral data alone, it is difficult to decide where or how to split functionally defined categories—that is, to know the limits of the behavior that results from a single motivational mechanism. In fact, Hinde (1970, 1974) has argued that the problem is wrongly stated: causal categories are numerous and overlap extensively, and in their influences on behavior are better viewed as networks of factors.

Departing from the traditional trio of aggression, fear, and sexual motives, and from efforts to predict the nature of internal mechanisms from observed behavior, Hinde (initially in 1955, and 1956a) suggested that descriptions be made of the variables involved when an animal shows a tendency to behave in any particular way. Thus we can say that there is a tendency to drink a certain amount of water after a certain period of deprivation. That is, measurable amounts of drinking behavior (a dependent variable) are elicited by situations defined by a measurable period without water (an independent variable) through the action of an internal mechanism (the intervening variable), which Hinde calls a "tendency." This mechanism is able to relate the dependent and independent variables to each other, but no other properties of the processes or structures it involves are postulated. For example, the behavior observed when there is a tendency to drink water need not result from unitary physiological processes that are all controlled in the same way by the same stimuli. Indeed, there is no single cause for drinking: it can result from several physiological processes and diverse stimuli (see Epstein 1967).

One of the most important and fundamental contributions of causal research in analyses both of behavioral tendencies and of more traditional motives is the concept of conflict. Most displays are performed by animals that are behaving as if unable to make a clear choice between different kinds of behavior—as if two or more motivational systems are in conflict for control of the individual's behavior. Much of the formalized behavior used when individuals encounter one another during courtship, for example, has been analyzed in terms of such a conflict, usually among motivational systems of sexual attraction, aggression, and fear (see for examples, Hinde and Tinbergen 1958:256; Tinbergen 1959b:39). In his work with gulls, Moynihan (for example, 1955a, 1962b) developed several observational techniques for revealing the different absolute and relative strengths of these motivational components as they underlie the performance of different displays, and Blurton Jones (1968) experimentally verified such differences among the threat displays of the great tit.

COMPARING MOTIVATIONAL AND MESSAGE INTERPRETATIONS   Many such displays that are traditionally interpreted as caused by a conflict of sexual and agonistic motivations could probably be analyzed into different message assortments comprising the behavioral selections of copulation, association, interaction, escape, and attack, the last often with a depressed probability of occurring. Some such displays may provide identifying information indicating that the communicator recognizes a bonded relationship with a particular recipient individual. The gross motivational categories of traditional analyses thus obscure much of the information that is made available by such displays.

Despite the different results, there is little fundamental difference in the sorts of behavioral observations that are used to interpret motivation and messages. The observations used in determining motivational causation have been discussed by Moynihan (1955a) and summarized by Tinbergen (1959b).

First, temporal correlation: displays are often performed simultaneously or in quick succession with acts such as approaching, pivoting, or turning away, and must therefore share causes with such acts. (These acts, or those which occur consistently with a display, are also related to the behavioral selection categories referred to by the display's messages, and, in fact, are the only ideal source of information about behavioral selection messages.)

Second, circumstances: the circumstances in which a display is performed can be compared with those in which directly functional acts occur without display. For example, a male bird will attack another who intrudes into his territory and flee from such an attack if he is an

intruder, but he will display against the other male, occasionally approaching or retreating, if they encounter each other at a mutual territorial border. The displays are thus assumed to be caused by a motivational conflict of aggression and fear, the two causal factors tending to come into balance with each other (that is, they make approximately equal demands of the communicator) when the birds are facing over an accepted boundary. (The circumstances in which a display occurs help an investigator to begin to see the limits of the behavior with which it correlates, and sometimes provide the bulk of the evidence for inferring a behavioral selection message. For example, in discussing the singing behavior of birds in chapter 5 it was argued that messages of interactional and seeking behavior were encoded because singing birds are alone, they patrol and take high conspicuous stations, and they stop singing and interact if an appropriate participant becomes available to them.)

Third, physical form: display postures or movements that incorporate actions recognizably related to motor patterns with direct functions are assumed to share motivational causes with these. For example, an act like striking incorporated into a display would implicate an aggressive motivational state. (In both motivational and message analyses this kind of evidence is accepted only as suggestive, but it may guide a researcher to seek kinds of events in which the postulated behavior can be observed.)

As indicated previously, however, although there are extensive similarities in the types of observations that are used in motivational and message analyses, the products of the two procedures differ. In fact, they often differ more than would be expected on the basis of just their different goals.

One reason is that descriptions of motivational states often are not sufficiently exhaustive. Ethologists have often sought primarily to interpret the behavior of displaying animals in terms of a limited number of motivational mechanisms (such as the traditional trio of aggression, fear, and sex), although there are now many exceptions to this approach such as in the experiments of Blurton Jones (1968) and the work of Andrew (discussed in this chapter in the section about alternative ethological approaches). A priori sets of motivational postulates are often inadequate. They may largely account for prominent features of a communicator's behavior in many circumstances, but not for all the behavior that a message analysis finds to be consistently correlated with the performance of a display from event to event.*

* The description "consistently correlated" must be clarified here. To say that a given class of acts consistently correlates with a display does not mean that it always occurs if the display does. That is one useful criterion. Two alternative cri-

We cannot specify the amount of behavior that either approach should leave uninterpreted. There must be some, because inevitably some observations are incomplete, and others involve changing events in which it would not be appropriate for a communicator to follow through with the behavioral selections usually correlated with a display (that is, to remain in the same motivational state he was in when he displayed). The point is that although message interpretations seek to reduce the number of uninterpreted observations as far as possible, motivational interpretations have often stopped short. This practice can lead, for example, to recognizing that aggressive or sexual motivation *can* under-lie use of a display, while failing to recognize that the display is consistently correlated with some form of *interactional* behavior, even in the minority of events in which this is not attack or copulation.

A second important reason why motivational and message analyses often yield markedly different results is that the latter often focus more narrowly. A message analysis asks what information a *display* makes available. The most closely related motivational question would be to ask what internal states cause that display to be performed. But the question often asked has been what motivates the communicator's be-havior, not only its displaying, but its other behavior in events in which it displays.

When communicators are observed in a variety of circumstances, their motivational states are found to differ in some ways among the circumstances. These differences are usually described as part of the motivation of the displaying animals. For instance, Tinbergen (1959b), in a major review paper, describes many differences in the motivational states of gulls performing such displays as the mew or the head toss at different times. These displays are important in, among others, interac-tions in which the communicators are sexually motivated, but sexual motivation is not necessary for their performance. Because either dis-play is caused by only a part of the motivational state of a communi-cator in any particular event and the rest of the state varies among events, a recipient gull cannot fully understand a communicator's state just from perceiving either a mew or a head toss. (In fact, there may be little reason why he should get so much information from a display: many aspects of the communicator's motivational state may be of little interest to him or may be abundantly evident from sources of informa-tion other than the display.)

Moynihan (1962a) recognized that the motivation of displays them-

---

teria are that one of a limited set of acts—for example, attack or escape—always occurs in correlation with the display if indecisive behavior ceases, and that a speci-fic kind of act—for example, interaction—follows if seeking behavior yields an opportunity.

selves needs to be made clear in analyses and proposed a distinction between their "direct" and "indirect" causation. Direct causes are always involved when a display is performed, whereas indirect need not be. Moynihan's distinction deals very nicely with the kind of problem raised in the preceding paragraph, as he illustrates in describing the greeting ceremonies of the Chilean great skua. Greetings are particularly common when unmated birds approach one another, and Moynihan reasons that such approaches are sexually motivated. However, the approaches lead to situations in which aggression and fear are aroused. The displays that are performed in greeting also occur in purely agonistic encounters in all larids (skuas, gulls, and terns). Thus aggression and fear are the direct causes of the displays (and of the agonistic behavior consistently correlated with the displays; that is, the direct causes are related to the behavioral selection messages of the displays). Sexual motivation is an indirect cause when the displays are given in greetings—by being the cause of these encounters—but is not causal in any sense in many events in which the displays occur. In a message interpretation, indirect causes are *not* interpreted as part of the information provided by a display, because they cannot be known to a recipient who has perceived insufficient detail about the particular circumstances of the display's occurrence. If indirect motivational causes can be known at all, it must be from contextual sources of information.

Moynihan's direct-indirect distinction is rarely made in causal analyses, and thus the different informative contributions of a display and of information sources contextual to it are rarely obvious. The traditional lack of concern over how consistently any motivational state is operating when a display is used is not surprising. This concern comes with an interactional point of view, in which it is necessary to know what information is made available to a recipient just by a display.

Examples of the ways in which motivational and message interpretations can be compared and contrasted are readily made by referring to Tinbergen's review paper (1959b). Tinbergen's work has had a major formative influence on the tradition of motivational interpretation (and on much of ethology). Thanks to him, his students, and other workers, the displays of larids have been among the most intensively studied of any group of animals. (And this is an interesting group of birds, many species of which are ecologically plastic to an unusual degree, and are social year round; general accounts of the life histories of three species are available in Tinbergen 1953b and 1958).

Tinbergen's review of his group's work on gulls provides an excellent paradigm of the motivation approach (1959b). He took the view (p. 29) that displays are caused by conflicts of attack, escape, and sexual components of motivation. Of the nine displays he analyzed, the most

convincing interpretations involve conflicts of aggression and fear. For instance, in the aggressive upright posture the communicator extends its neck upward, angles its bill downward, and lifts its carpals (figure 8.1*A*); it may approach and then attack an opponent, or stop approaching and then withdraw. In two closely related postures the carpals are not lifted and the bill angle differs: it is horizontal in the intimidated upright and upward in the anxiety upright. Tinbergen interprets these postures as a result of relatively less aggression and more fear, but these two motives do not appear to account for all the behavior correlated with each posutre. In fact, gulls adopting these two postures seem to be in a state of conflict between fear and motivation to do virtually any other behavior from foraging (for example, when intruding on a shoreline territory), to association (for example, when forming a pair bond), to attack (see Tinbergen 1965, in which he presents some evidence from his student Manley 1960b). This range of alternatives to escape is not adequately characterized by any of his motivational terms. If a motivational concept were designed to underlie the general set behavioral selection message then it might be adequate, at least by indicating that some motivational component is always in conflict with escape when these display postures are used, and that this component can be very different at different times. This same concept must be used to represent the motivational alternatives to aggression in the case of the oblique-cum-long call displays, with which the communicator may attack, approach and associate, or forage.

Tinbergen interpreted the larid choking display (figure 8.1*B*) as the result of a very high level of conflict between defensive aggression (initiating an attack rarely correlates with choking, but the communicator will fight if attacked) and some motivation to remain with the nest site or territorial boundary. However, aggression may not always be part of the communicator's motivation, because there is both hostile choking with neighbors and friendly choking with mates, and the display forms seem to be the same. Howell, Araya, and Millie (1974) correctly interpret these different uses of the display as being functional differences dependent on contextual sources of information. The choking display occurs with a wide range of behavior, and may be caused by *any* strong motive to remain with a site (that is, more than one motivational cause can lead to staying). Aggression, fear, and sexual categories of motivation alone are inadequate to explain it. Viewed from the perspective of a message analysis, choking must provide some information about the likelihood of interacting, as well as indicating that the communicator will remain at a site; it does not specify the kind of interacting.

Tinbergen recognized something special in the causation of three formalizations, the hunched posture (like the uprights, performed with

A

B

Figure 8.1. Display postures of the black-headed gull. *A*. The upright display posture. Both birds have their necks stretched upward and are holding their wings slightly out from their bodies, with the carpals showing. The bird on the left is in the aggressive upright posture, with its head and bill pointed strongly downward. The bird on the right is in an intimidated upright, with its bill horizontal, neck relatively thinner than the other bird's, and its carpal on the side toward that bird extended much less far than the carpal that extends away. (It is also doing the head flagging display: its head and upper neck are bent away from its opponent.) *B*. The choking display posture. The birds have tilted their bodies so far forward and downward that their breasts are on or close to the ground while their tails point up. Their necks are moderately extended and arched so that their heads and bills point downward. Their carpals are held away from their bodies (sometimes black-headed gulls even raise their wings when choking). They are rapidly and rhythmically repeating a low-pitched "kruh," "kro," or "krohr" vocalization. (From Moynihan 1955a; the intimidated upright and anxiety upright postures had not yet been distinguished by gull researchers, and in keeping with the current conventions Moynihan used the term "anxiety upright" to describe the bird on the right in part *A* of this figure.)

variable bill angles and carpal positions) and the facing away and the head tossing displays: a "tendency to stay [that] is not specifically and compulsory [sic] linked with one particular major motivation" (1959b:37). He also described this as a tendency to approach, and it may be more an expression of motivations to interact than to remain with site, or it may differ among these three formalizations. Tinbergen has not speculated in detail on the motivational mechanisms involved, however. One of the formalizations, the head tossing display, certainly carries a message of receptive behavior. Howell, Araya, and Millie (1974) point out that the head tossing of the gray gull (which does not appear basically different from that of other gulls, although the display varies somewhat in form even within species) always correlates with solicitation, whether of copulation, courtship feeding, or parental feeding. Some motivational basis for this predictable behavior should be recognized in any analysis of the internal causation of this display.

A common display of gulls that Tinbergen found particularly refractory to analysis in terms of the traditional motivational trio is the mew call. He felt that fear always contributes to its performance, but that other motivation is less clear. In fact, if we examine what is known about this display in a species such as the herring gull, fear (and aggression) may *sometimes* be part of the motivational state of a mew-calling individual, for example, as a male turns from a hostile encounter with a neighbor and walks in parallel with his mate, both mew calling. Attraction between a male and a female is evident then and when two individuals approach each other during pair formation: as they meet they turn and walk in parallel, mew calling. This situation probably evokes aggression and fear, and the parallel walking suggests a balanced conflict between approach and withdrawal. But attack and escape are not seen, and hence aggression and fear are perhaps absent or insignificant in many situations in which one individual approaches the other on the nest, such as when a male brings nest material to his mate, or a male or a female comes to relieve its mate from incubation or to feed the small chicks. Mew calls in these cases usually just precede alighting in the territory, then are uttered again while walking toward the mate or chicks. Attack, escape, and sexual behavior are all conspicuously absent when a parent returns to a territory and its chicks are away from the nest, hidden in the vegetation. It usually stands and mew calls; the chicks emerge, run to it, and are then fed.

What seems characteristic of all these performances of the mew display by herring gulls is, first, that the communicator is ready to interact, whether agonistically or not. Second, it is likely to show some indecisive behavior concerned with locomotion: it may be alighting, may approach part way to some individual, may walk in parallel, or

may stop walking and call. What is indecisive about a recently returned parent standing and mew calling? It is trying to feed its chicks but cannot approach them further, not knowing where they have hidden. Or it may be with one chick, unwilling to leave it to approach or find the other chicks. Although all this may add up to many different motivational states in different circumstances and thus complicate a traditional motivational analysis, at least a partial message analysis of the herring gull's mew call would appear straightforward: the communicator is prepared for interaction and for locomotion, but its behavior is likely to be indecisive.

MESSAGE AND TENDENCY   Although using only the three traditional motivational postulates is obviously inadequate to account for the causation of the mew call displays of gulls, there is no reason why Hinde's proposal of behavioral tendencies could not provide an adequate description. Tinbergen in fact speaks of a tendency to stay or to approach.

Insofar as they describe the behavior of communicators that consistently correlates with the performance of displays, studies of messages and of tendencies should give the same results—it is the subsequent analyses, not these descriptions, that are complementary (as explained above). In fact, the terms used to describe behavioral tendencies and behavioral selection messages can be homonymous: tendencies to attack, escape, copulate, associate, interact, locomote, and so forth have all been proposed. The match between descriptions of tendencies and descriptions of messages is, however, not always exact. For instance, the tendencies that have been described include staying, flying, approaching, and being gregarious or with others (see, for example, Moynihan 1960; Andrew 1961a; Stokes 1962; Delius 1963; Crook 1963, 1964; Tinbergen 1964; Fischer 1965; and a partial review by Hinde 1970). Discrepancies result from naming a tendency after studying the performance of a display only in a narrow range of circumstances and hence underestimating it, from defining terms differently (for example, some authors may use "being gregarious" and "associating" as synonyms), or from the current imprecise state of the art of both approaches.

The indecisive behavior found in message analyses to correlate with the performance of many displays is usually a result of conflicting tendencies. Seeking behavior is often the result of an active tendency to behave in a certain way at a time when no opportunity is at hand. Motivationally, indecisive and seeking behavioral selections have been linked through the postulate that both result from the "thwarting" of tendencies (D. Morris 1956b), in the former case through internal and in the latter through external agencies.

The probability and intensity of behavioral selections are both related to what has been loosely termed simply the "intensity" of tendencies. The relationship is not simple, however, because highly intense tendencies to attack and to escape may lead to a low probability of either and to a high probability of indecisive behavior (see chapter 6 and Blurton Jones 1968). Further, the relative probabilities of two incompatible selections and of indecisive behavior may remain the same over wide ranges of intensity.

RELEASER THEORY   Ethology's preoccupation with motivational causation has also led to assumptions about how displays operate as social stimuli. The traditional model of interaction came to represent each display as a stimulus capable of releasing recipient behavior, as a key that unlocks particular motivational states in recipient individuals (see, for example, Lorenz 1937, 1950; Tinbergen 1953a, 1959b). Despite caveats (such as by Tinbergen 1939b) problems then arose through the incautious but too common assumption that the "releasing stimulus" by itself was often a sufficient cause for a response.

Certainly the stimuli claimed to be releasers or "sign stimuli" (a broader term embracing stimuli with releasing properties, even though they have not evolved specializations for providing information) are usually treated as very significant by their recipients. The releaser concept is in fact based on two well-substantiated phenomena. First, animals respond selectively to any stimuli available to them, in part because of the processes of perception that organize stimuli according to various criteria of salience: some become of central concern, others become contextual to them. In visual perception, for instance, this leads to the figure/ground distinction. Selective responsiveness is also caused in part by the fact that animals must act to satisfy different needs when in different physiological states (Hinde 1974); the states bias the selection toward specific classes of stimuli. Second, where evolution (genetic or cultural) has affected the development of some signals and responses to them, it has produced characteristics that an appropriate recipient will automatically classify as salient; it has biased the organisms' response tendencies or fit the signal forms to them.

That there are especially salient stimuli does not imply that stimuli of lesser rank are simply discarded. It implies mechanisms that order perceptual processing, assuring that a given stimulus or set of stimuli is selected and is treated as important by a recipient for whom it should be relevant. Thus, a releaser seems to be a characteristic specialized to be a particularly noticeable stimulus; recipients have become specialized to accept it as a stimulus to which other stimuli have a contextual relationship within a particular frame of reference.

The ability of a releaser to elicit responses in any event depends, at least for vertebrate animals, on information from contextual sources. Many traditional descriptions of the use of releasers, however, tend to underestimate the ability of recipients to vary their responses to a fixed stimulus under different circumstances. Such descriptions misrepresent the contributions made by displays to the complexly adjustable and often very subtle interplay between the participants in a social inter-action.

In experiments with which ethologists have investigated the releaser effect, displays have been used less often as stimuli than have such specialized features of physical appearance as markings or color. Color, of course, is much more easily controlled by an experimenter than is behavior. Although not behavioral, colors are otherwise comparable to displays, and they are often brought into play by either display acts or other forms of behavior that have limited patterns of occurrence. As an example of early experiments on releasers, work by Tinbergen and Perdeck (1950) appeared to show that the presentation by an adult herring gull of the red spot on its bill both released and directed a pecking response from its chicks, which it would then feed. Because models with spots elsewhere (such as the forehead) were much less effective, the chick was thought to have a very precise preconception of the configuration of the pecking releaser. A model with a highly contrasting spot in the right place, even if only two-dimensional, was sufficient: the releaser involved the spot and its location, and other contextual details appeared to matter relatively little in eliciting the full response (see figure 8.2).

Hailman (1967, 1969) redesigned the experimental procedure to control for the rate at which the spot moved back and forth above the chick as the model of the adult's head was moved like a pendulum above it. Using laughing gulls, he found that newly hatched chicks responded just as well to a model with a forehead spot as to one with a bill spot if their spots traveled at the same rate. Yet, although these naive birds had even less complex configurational expectations of the releaser than Tinbergen and Perdeck had concluded (and used a different source of information contextual to the spot), older chicks learned other characteristics of parental heads and came to demand much *more* detail in the appearance of the model. That is, with experience they saw the simple "releaser" (including its motion) as one among several relevant sources of information. It was important, but its effectiveness came to depend on the presence of rich and appropriate contextual information.

Traditional ethological studies of releasers recognized that animals rarely respond to a single stimulus but respond to multiple stimuli. For instance, in a careful set of experiments Tinbergen and Kuenen (1939)

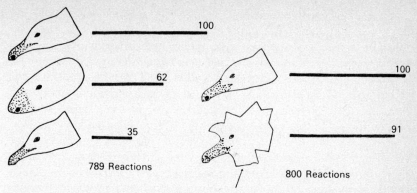

Figure 8.2. Examples of experiments with a releaser. Elicitation and direction of a pecking response used in parent-offspring feeding interactions were studied by showing herring gull chicks flat cardboard models of the heads of adult herring gulls. Numbers of pecks were scored during various presentations, each 30 seconds long. A "standard" model was painted in natural colors (white head, yellow bill, red spot on bill), and provided the basis of comparision for the relative numbers of pecks given to other models testing different stimulus features.

Results of initial tests of head shape are shown on left. An egg-shaped "head" received fewer pecks than did the standard model illustrated above it, but many more than an otherwise standard model that lacked a red bill spot. (Previous tests had established the importance of a highly contrasting bill spot.) The egg-shaped model, however, distorts the shapes of both the head and bill, so another set of tests (on right) was run comparing a "cock's head" model and a standard model. In spite of its bizarre head shape, the cock's head model was nearly as effective as the standard, indicating that the shape of the parental bill is far more important to the chicks than is the shape of the parental head. (From Tinbergen and Perdeck 1950.)

showed that slight jarring of the nest, a gentle puff of air, and the size of an object, its motion, and its relative height are all useful in eliciting food-begging behavior from nestling thrushes. Further, the orientation of the begging actions is selected in response to additional stimuli: at first to gravity (upward), and then later steered toward the parents' heads, which can be distinguished from their bodies by a few visible clues. The Tinbergen and Kuenen experiments revealed both stimuli that do elicit or direct responses and stimuli that do not: for example, suddenly decreasing illumination, as if a parent were casting a shadow on a nestling, does not elicit gaping.

Despite the number of potential stimuli investigated, however, such experiments test each stimulus primarily in isolation from the others and from some or many of the other stimuli that would ordinarily accom-

pany each. Although something of the relationships among stimuli may be shown, the range and complexity of pertinent relationships is difficult to explore and too often is underemphasized in discussing response mechanisms. When animals interact it is *not* simply by providing each other with appropriate sets of keys. Rather, each participant assesses the significance of the information provided by each important stimulus in terms of its relationships to information from various other sources. Responses are chosen on the basis of the relationships among stimuli.

Hinde (1974) points to other common problems in applying the releaser theory. For instance, the theory has often been thought to imply that effects will be immediate, but many effects of displays are cumulative, the gradual result of information received over prolonged periods (see chapter 10). In cumulative effects, as in more immediate responses, much of the necessary information must come from sources other than displays. Further, some information may be provided by very inconspicuously formalized acts such as the different postures assumed by individuals of different status in standing and sitting. The problems presented to observers and theoreticians by such acts are discussed in chapter 13.

Alternative Ethological Approaches

Though most ethologists have been preoccupied with Tinbergen's central problem of "what urges the actor to signal," more recently some have developed a few approaches either less concerned with motivation and more explicitly concerned with communication, or concerned with both. The message-oriented framework for this book thus far is not concerned with motivation at all; rather, it is strictly interactional. Other approaches acknowledge that the task of displays is to make information available about the communicator and they attempt to provide causal analysis based on the stimuli to which communicators are responding when they display; yet others concentrate on the quantitative features of the information provided by different sources.

Mixed Approaches

Andrew has developed a fundamentally causal approach concerned primarily with the mechanisms controlling individual behavior, but it also recognizes the need to describe the kinds of information made available by displays (1964, 1972). He describes the "immediate" causes of displays as properties of those stimuli that correlate with acts of displaying. He chose this procedure to avoid explicit use of "major intervening variables" such as behavioral tendencies (1972:182), because the evidence from behavior alone too often fails to permit an investigator to

know precisely which intervening variable applies in a case, and because the traditional variables are too grossly defined. The important stimuli in any situation provide a display's "immediate causation" and have features that remain similar from event to event, even though the traditionally described motivational states vary. Thus Andrew suggests that the immediate causes of a mammal flattening its ears, as is done in the display behavior of many species, are stimuli that correlate with possible danger of physical injury to its facial area (1972:183). This explanation is more consistent than the aggression-fear conflict often proposed to account for ear-flattening in fights and threatening behavior; it remains valid for the same movement of the ears in such other circumstances as boisterous play with an infant.

Andrew's analyses are phrased in terms of responses made by communicators to stimulus situations. Thus motivational conflict is handled as a conflict among incompatible response patterns of the communicator, much as it is described in analyzing the messages of displays. (It is, of course, entirely appropriate to speak of displays as "responses." I have largely avoided calling them responses made by communicators, using instead terms like communicator "acts" or "behavior," but simply because it is also necessary to speak of responses *to* displays: the responses made by recipients.) Stimulus properties are said to cause displays if they elicit the behavior of attending, leading to alert, protective, or exertion responses. Alert responses span the time it takes an animal to recognize a stimulus, or to classify it as novel, and protective responses or preparation for exertion may occur simultaneously. After a period of "recognition comparison," more precise information about the communicator's response to a situation is made available to potential recipients by vocalizations, locomotion, exertion, immobility, autonomic changes, urination, defecation, scent-marking, attack, escape, threat, copulation, or other physical contacts, grooming, shaking, or stretching. Only some of these response patterns are displays, of course.

The elicitation of attending behavior seems to be a primary criterion for recognizing that a stimulus situation has properties sufficiently salient to be the immediate causes of display behavior. One important such property is the extent of the "contrast" between a stimulus and a neural model the animal has in storage; this contrast can determine whether the response will be approach or withdrawal. Somehow, Andrew recognizes this and other stimulus properties largely from the communicator's behavior, accepting the risk of circular reasoning (van Hooff 1967). His procedure is very flexible, and in practice Andrew sometimes appears to interpret immediate causation less in terms of stimulus properties than in terms of motivational states (as pointed out by Blurton Jones 1972b) that differ from those traditional "major intervening variables" he criticizes primarily in being more diverse, less

inclusive, and more consistent from event to event in which a given act occurs (all important differences).

Andrew's departure from conventional causal models of display behavior has the potential advantage of not requiring that we understand the nature of physiological mechanisms. Further, the approach can describe particular events with considerable accuracy (Hinde 1972:205), and its potential consistency is a very important improvement. However, its unsystematic listing of communicator behavior patterns has as yet provided no structured concept of the "kinds" of information made available by displays, and its emphasis on causation is appropriate primarily to research on mechanisms that control individual behavior. Andrew examines the communicator as an individual responding to stimuli, not as an individual providing various sources of information for other participants in real or potential interactions.

Approaches in which the communicator's behavior is interpreted primarily in terms of eliciting stimuli can be readily reinterpreted for analyses of communication if the behavioral detail is sufficiently extensive, and if not reported primarily in terms of abstract causal mechanisms. For example, the black-throated green warbler *Dendroica virens* has been shown by Morse (1970) and Lein (1972) to have two forms of song vocalizations that correlate differently with other behavior of a singer. Lein showed that one is uttered by males living in populations of relatively low densities and by males alone when well within their territorial borders; this song form becomes less frequent as the season progresses. The other song predominates at high population densities, when a male is in a border region or in the immediate presence of another male, and during twilight bouts of singing. Lein's causal interpretation was in terms of sensory stimuli (such as the presence or absence of other males nearby) and hormonal levels (the first song requires a relatively high androgen level). His description of the occurrences of the two song types, although highly condensed, suggests that the difference in the behavioral information they supply may be similar to that between an eastern phoebe's two song forms (see chapter 3), in that one song correlates with readiness for more intense interaction (perhaps in this species with a relatively high probability of attack behavior).

The possibility that the songs of the black-throated green warbler differ in the messages they make available led Lein to study their relative frequencies of occurrence in different circumstances. He has subsequently demonstrated in an even more detailed, quantified study that males of a related species, the chestnut-sided warbler *Dendroica pensylvanica,* have at least five and perhaps seven basic song forms (figure 8.3), each of which can also be shortened or muted. Each song

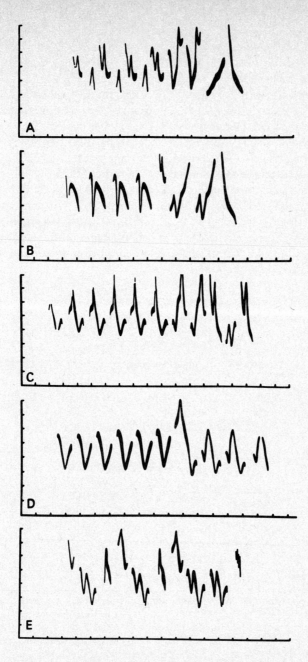

Figure 8.3. Five song types of male chestnut-sided warblers. *A*. Accented ending type 1 song. *B*. Accented ending type 2 song. *C*. Unaccented ending type 2 song. *D*. Unaccented ending type 1 song. *E*. Jumbled song. (From Lein 1973.)

has a very well-defined range of employment (Lein 1973). One is sung primarily during territorial encounters. Two are sung primarily at or near territorial borders, and two almost exclusively within the central areas of a territory; the distance from a neighboring male also influences the choice of song (table 8.1). An abrupt change in a male's situation, for example, as he flies from his border to the center of his territory, correlates with an abrupt shift in song forms from effectively one extreme to the other; a gradual change, as may happen during the dissolution of a territorial encounter, leads to a more gradual change in song forms and a male may then shift progressively through all five. Finally, as the probability of a territorial encounter increases or decreases while a male is singing, he switches song forms in a very predictable fashion. Lein has used these observations to support both a differential message analysis of the song forms and a neurophysiological–external stimulus model of individual causation.

Table 8.1. Song types sung by male chestnut-sided warblers.

Song type	Position (percentage of cases analyzed)			Cases analyzed
	Central	Intermediate	Edge	
AE-1	95.3	1.1	3.5	86
AE-2	89.4	7.6	3.0	66
UE-2	28.8	22.0	49.0	153
UE-1	2.0	20.0	78.0	50
JS	3.8	7.7	88.5	26
	Location of other male (percentage of cases analyzed)			
	Distant	Intermediate	Near	
AE-1	92.2	7.8	0.0	51
AE-2	90.0	10.0	0.0	40
UE-2	39.3	35.5	25.2	107
UE-1	18.2	27.3	54.5	44
JS	0.0	15.0	85.0	20

The upper portion of the table shows that different songs are favored when the singer is in different parts of his territory (regardless of the presence or absence of neighboring males), and the lower portion that the sight or sound of neighboring males at various distances also influences his singing behavior. (*Distant* indicates a neighbor singing over 100 feet away, *Intermediate* between 30 and 100 feet, and *Near* within 30 feet.) Most accented ending songs (AE-1, AE-2) are sung well within the territory and not in encounters with other males, whereas jumbled songs (JS) have a reversed pattern of employment and the two unaccented ending songs (UE-2, UE-1) are to different extents intermediate. These five song types also differ in the extent to which each correlates with yet other kinds of activities in which the singer engages. (From Lein 1973.)

Studies of the Quantity of Information

An approach very different from any discussed thus far is to quantify the amount of information made available by displays. Although very explicitly oriented to the study of communication rather than the causation of individual behavior, such an approach compares only the quantities of information encoded by displays and is indifferent to the *kinds* of information that are encoded (MacKay 1972). In this lack of concern for referents, such measures differ from our commonly held concepts of information, which are semantic. These, instead, are syntactic-level analyses, concerned with a property of signal acts as vehicles for carrying information—their *capacity*.

The mathematical theory proposed by Shannon and Weaver (1949) measures the information supplied by a signal in terms of units called "bits." (A bit is the amount of information needed to make a dichotomous choice.) For example, a honeybee that has obtained one bit of information about the direction of a foraging source from the hive could divide the 360° range of selections available for her next flight into a sector subtending 180° that was indicated, and a 180° sector that was not. With a second bit of information her choice would be narrowed to one of four sectors subtending 90°, with a third bit to a single 45° sector, and so forth. This procedure for measuring the quantity of information in a signal has been applied to the dance of honeybees, first by Haldane and Spurway (1954) and then by E. O. Wilson (1962a, with a minor correction in 1971a:255), in considering the information encoded about direction and distance (see chapter 6). For comparison with a very different display procedure, Wilson also analyzed the efficiency of the functionally similar fire ant odor trail, although in this case estimating the amount of information actually used by recipients based on measurements of their errors. (The difference in employing measures based on the display versus measures based on responses is important, although it has usually been ignored. For example, although forager ants appear to use information about distance from their nest in deciding how far to follow a trail, they do not obtain this information from the odor trails—they must know their position in space when they encounter a trail. Honeybees, on the other hand, *do* encode distance information in their dances.)

Comparing the amounts of information in the display behavior of different species has at least two goals: to reflect the precision of the signaling in each species, and to compare quantities and rates of informing for vastly different species. Despite the considerable differences between the means by which honeybees and fire ants inform one another, Wilson's analyses appear to indicate similar amounts of information.

Quantities of information involved in signaling have been measured in a few other kinds of animals. For instance, to analyze the amount of information transferred per display in each of eight species of hermit crabs, Hazlett and Bossert (1965) observed 1,000 aggressive encounters of crabs that met while wandering about in large aquaria. The individuals could approach, retreat, or perform various displays and sometimes employed over 50 acts in a single encounter. The average amount of information transferred per display in each species was 0.41 bits, and it was transmitted at a rate of from 0.4 to 4.4 bits per second (these calculations are revised upward slightly by Hazlett and Estabrook 1974). This figure is slightly higher than but fairly similar to the rates calculated for the honeybee and fire ant. Aggressive encounters of mantis shrimp, predatory crustaceans of the order Stomatopoda, were analyzed by Dingle (1969) who found that these animals transmit more information than do hermit crabs. Further, the upper limit of the rate at which they transmit is about one and a half to two times that of the upper limit of hermit crabs, and as much as six times that of the fire ant (Dingle 1972). The reasons for this are not clear. Dingle has suggested that they could relate to the use of visible versus olfactory display systems, to the kind of information being carried, or to habitat differences. In addition, most of the ten stereotyped patterns of stomatopod behavior in which he was interested do not appear to be displays and are specialized for more direct functions, which must have had some influence on his findings.

Mantis shrimp, incidentally, are excellent animals with which to develop and apply measurements of the amounts of information being used by recipients. Because they are very active and alert, highly visual, and individuals usually fight readily when they meet, they generate quantities of data very rapidly. Recognizing the opportunities they offer for research, Dingle has not only been able to compare the rates of information transmission in different species (finding differences that correlate with kinds of habitats), but has also been able to show differences in the rate of transmission at different stages of an agonistic encounter (1972, and personal communication). Further, in testing interspecific fighting over territories he found that different species in the genus *Gonodactylus* interacted with little loss in the amount of information transferred, although there was slightly more loss in similar encounters between *G. incipiens* and a member of the different genus *Haptosquilla glyptocercus*. Mantis shrimp of these species regularly hold interspecific territories in nature, suggesting that evolution may have favored efficient communication among the species.

For the ants, crabs, and mantis shrimp a quantity of information was calculated on the basis of responses by recipients and was thus actually a measure of information received and *used* in selecting behavior. The

measures are thus based on pragmatic-level data, and must misestimate the amounts of information actually encoded to the extent that some is lost through the interference of noise, and to the extent that some from sources other than the signaling behavior contributes to a recipient's selection of a response.

Though these comparisons are interesting and useful, the number of observations needed to assess higher-order Markov response chains quickly becomes astronomical in studies of this kind, even when dealing with invertebrates that may attend to many fewer simultaneous stimuli than do, say, birds and mammals. And there is the problem of the assumption in Shannon's measure that the system is statistically "stationary": that frequencies of the behavioral components will not change as interactions proceed (see MacKay 1972; Slater 1973). They do change, which must weaken the power of or in some cases invalidate the computations.

Altmann (1965) studied the probabilistic constraints between events in sequences of actions of rhesus monkeys on Cayo Santiago. He developed a catalog of 123 behavior patterns, only some of which are displays. Any event that affected the behavior of other members of the rhesus social group was considered communicative, and Altmann asked to what extent knowledge of one event in the group enabled an observer to predict the next—to what extent it reduced his uncertainty about what would be done next. As isolated events, the behavior patterns had an uncertainty of 4.8 bits, and approximately 1.9 bits "were picked up by the monkeys from the immediately antecedent behavior in their social group, thus reducing the uncertainty of their behavior to 2.9 bits." However, no contextual information was considered in deriving the measures, for example, whether a monkey was hungry (and thus relatively likely to be aggressive), or whether an infant was with its mother when its behavior was tallied. Altmann's model does enable some prediction of the social behavior of monkeys in this group, but it must be developed much further and incorporate many more contingencies of importance to the monkeys before it can become very practical; such expansion will lead to difficulties in sample size as higher-order Markov processes are investigated. Further, examining the relevance of behavioral sequences in communication deals with only part of the context of any act. Animals are capable of dealing with much more complex relationships than those of sequences alone, and important antecedent events for many responses may not be readily discernable by an investigator using this approach.

A syntactical study, based on signal form rather than the responses of recipients, was made by Rand and Williams (1970). They surveyed the redundancy and information content of the dewlaps of *Anolis* lizards of eight species living together at a mountain site in Hispaniola. (A dewlap

is a morphological structure evolved to convey information, and is thus comparable to a display although not itself behavioral; it belongs in the category of "badges," discussed as a contextual source of information in chapter 9.) Behavioral observations implicate the dewlap and its deployment in displays as serving in at least species recognition and territorial functions. Enumerating differences in just the badge characteristics of color, pattern, and size of dewlaps, Rand and Williams found that all eight lizard species differed, and differed in more than the minimum extent necessary for species discrimination. They determined that the dewlaps of these lizards had more than 6 but fewer than 9 bits of information, all but 3 of which were present as redundancy (exceeded the theoretical minimum needed for a choice).

Rand and Williams further recognized that the dewlaps are not the only sources of species identifying information available to the lizards from physical characteristics of their appearances, and also that the dewlaps were involved in display behavior made up of movements that differed among the species. Altogether, 9 to 15 bits of information appeared to be present, of which between 6 and 12 were redundant for species discrimination. Various functions were considered for the redundancy, and they proposed that the same procedures used to estimate the amount of redundancy in dewlaps could be used in studying quantitative aspects of communication by comparing faunas of different sizes and closeness of phylogenetic relationship, as well as faunas living in environments with different levels of "noise." Among the most interesting predictions made by Rand and Williams was that "the number of species which coexist may be limited by their ability to evolve displays of sufficient redundancy to be distinguishable."

It has been suggested that providing ways to quantify the concept of redundancy, as Rand and Williams did, may be one of the fundamental contributions of Shannon and Weaver's information theory to the biological study of communicating (Pumphrey 1962). Measures of redundancy will be particularly interesting if it should be found that organisms use the minimum amount of redundant information that they need and thus permit as many errors as they can get away with, that is that their need is not to reduce errors as far as possible, but to reduce them only to an acceptable level (a paraphrase of Dancoff's principle, from Quastler 1958).

Traditions in Other Disciplines

Ethologists are by no means the only ones studying the behavior of communicating, although they lay claim to the richest field. As evolutionary biologists they find it pertinent to study the phenomenon in many different kinds of animals, whereas other disciplines involved with

this research tend to limit their attentions or at least their ultimate goals to understanding the behavior of communicating in a single species: man.

Most research on human communication has been devoted to our crowning achievement of language. Surprisingly little of this research has been behavioral. Linguists traditionally study language as a set of abstract principles, something that language users tend to degrade. Even the challenge to our traditional views of grammar by Chomsky (for example, 1965) and his followers is not behaviorally oriented. Others, however, have found the perspective wanting; Labov, for instance, emphasizes the fundamentally creative ways in which people use words, and Hymes and other "sociolinguists" argue that traditional lingusits fail to recognize the diversity of speech patterns available to any member of a language community. Hymes has found that people switch among different forms of language as they change their activities or interact in different relationships, that "rules of speaking" determine word selection, the tone or manner of speaking, and similar considerations, and vary depending on the situation, the kind of interaction, the characteristics of the participants, and the goals (1967). Labov (1973) has begun to study the referents of words by learning how people actually *use* them, a task more customarily performed by lexicographers than by linguists. He finds that the objects referred to by a noun like "cup" are not grouped in precisely defined, invariant categories but have both invariable and variable properties that enable the noun to shift as it is used in the contexts of different words and events, and in shifting to match the novelty its users inevitably encounter. Labov finds this context-dependent shifting adaptive, because we must emply a "finite set of words to describe an unlimited number of objects in the real world around us"; it is what I described as the use of metaphor in chapter 4.

Other linguists have found the behavioral phenomena of "paralinguistics" (Trager 1958). They study not words, but the ways in which words are said: modulations of amplitude, pitch, or voice quality, the use of pauses and the pacing of speech, and the use of sounds other than words, such as moans, groans, laughs, belches, "uh," or "um." Related research concentrates on "kinesic" behavior (Birdwhistell 1970, reviewed and discussed by Kendon 1972b): the visible acts performed along with, or as substitutes for, speech. Many of these acts occur quite apart from language use and, like many utterances studied by paralinguists, are behavior patterns ethologists would call displays. Along with some aspects of spatial behavior, they form the main part of what has become known as nonverbal communication. (A representative collection of papers that covers major research trends in this area, annotated with insightful commentaries, has been published by Weitz 1974.)

Nonverbal modes of human communication are studied by members

of many disciplines, most prominently by social psychologists. In social psychology, as in ethology although for somewhat different reasons, the tradition is to study emotional states and even the personalities of individuals; social phenomena are examined primarily for their effects on the emotions of individuals. That is, mechanisms internal to the individual are really the focus; social circumstances are little more than stimuli.

Observation of naturally occurring events has not been widely used by social psychologists, whose preferences in general run to the experimental testing of a priori hypotheses. Their experiments are often very artificial and restrict the complexity of events in which displays occur to a very small fraction of that which is naturally available. Further, the experimenter usually imposes a predetermined set of responses on the subjects, often giving them a list from which they must choose. For example, in testing for different relationships among preconceived "facial expressions," subjects have usually been shown photographs (a technique actually pioneered by Darwin!), cartoons, or occasionally brief sequences of motion pictures or brief performances by actors or actresses. They are supplied with a limited list of "affect states" such as happy, sad, hopeful, or disgusted, or, in recognition of the problems involved in knowing what constitutes a distinctive affect state, are given affect "dimensions" such as pleasantness/unpleasantness, or attention/ rejection. The subject has only to match between a limited choice and a stimulus that is static or at least largely removed from its usual contextual sources of information.

These procedures have produced some dismal results. On the other hand, the attempt to isolate stimuli and simplify responses is potentially very powerful if its dangers are recognized and minimized, and some very interesting results have also been obtained. Ekman and his coworkers, defining their behavioral units as components of facial expressions that involve movements of particular facial muscles, showed that the same signals are correlated with the same concepts of affect by members of diverse cultures, including the preliterate Fore of New Guinea (Ekman, Sorenson, and Friesen 1969; Ekman 1970). Like ethologists, these workers have postulated that there are universals in human nonverbal signaling behavior, even though they are obscured by cultural differences to some extent. They have found evidence for the existence of "display rules," socially learned techniques for managing facial expression through appropriate de-emphasis, overemphasis, or masking, or through adoption of a neutral poker face.

Working with paralinguistic phenomena of pitch, amplitude, and tempo, Scherer (1974) divested these nonvisible signals entirely of the content clues of speech by producing his stimuli artificially with a tone

synthesizer. Subjects consistently associated variations in features of the tones with various emotional attributes, for example, slow tempo with sadness, disgust, and boredom; fast tempo with interest, fear, happiness, anger, and surprise. Moderate pitch variation produced judgments similar to those made for slow tempo, and extreme pitch variation similar to fast tempo, but fear was associated with both; moderate amplitude variation led to judgments of happiness or disgust, and extreme amplitude variation to fear and anger. Some affect states appeared to be associated with more than one sort of paralinguistic behavior; for example, subjects differentiated "hot" and "cool" anger. Scherer postulated that such features of vocal expression might depend on "respiratory phenomena and physiological correlates of affective state," and if so might be widespread among at least primates. In earlier work with content-free speech based on a technique for randomizing sequences of recordings from prolonged interactions, Scherer (1972) also demonstrated cross-cultural regularities in the ways people attributed personality characteristics to speakers on the basis of audible clues.

Interesting as such results are, they are more comparable to the work of ethologists on the causation of displays than they are to analyses of messages. Affect states such as happiness, sadness, or disgust contribute to a person's mood, as do other kinds of states like contemplation and preoccupation. But "mood" is not behavior, it is a composite of abstract postulates representing internal causes of behavior.

People do attempt to assess the moods of individuals with whom they interact; animals of other species may do the same. However, when asked for free interpretations of a photographed expression, subjects usually describe the situation that could produce it or the behavior they would expect from the communicator (Frijda 1969); for them to recognize an emotional state is "uncommon in real interaction situations" and requires a mediational process. This is in part because moods and their attributes are not easily assessed from single sources of information. A crying person, for instance, may be sad or happy. Such states are in practice probably evaluated by integrating clues accumulated over an appreciable time and from multiple sources, in terms of a store of hypotheses about the factual and emotional implications of the situation and the relationship to this of the communicator's expressions. Frijda (1969) has offered a detailed model of this complex processing of information. The further judgment of a communicator's personality characteristics is not independent of judgments of mood, but accurate assessment of personality probably takes considerable time even if based only on limited data.

Because displays, presumably including all "nonverbal communication," provide probabilistic information about behavior, they must imply

information about the moods and personalities that could contribute to that behavior. That is, it is not wrong to say that displays provide relevant information for these assessments; the weakness of approaches that concentrate exclusively on such information, however, is that they miss so much. If psychologists sought nonverbal communication in naturally occurring events rather than in preconceived experimental circumstances, they would find much else of interest in its immediate relationships with the ongoing behavior of communicators and be less lulled by the siren pull of issues that are of background relevance.

Although the tendency to experiment dominates, it has not fully precluded the use of naturalistic observations by psychologists and others interested in the affect states correlated with nonverbal communication. Ekman and Friesen (1969a), for instance, have drawn considerably on observation in categorizing the repertoire of nonverbal communicative acts. Their scheme recognizes not only affect displays, but also "regulators," used to control the flow and the pacing of the exchange of roles in a conversation, and two classes that are tied to speech itself: "illustrators," which augment, repeat, substitute, or contradict verbal information; and "emblems," which are consciously used substitutes for words. Psychiatrists, of course, have been interested in acts that in any way augment information provided verbally. Such pioneers as Scheflen, Condon, and Ogston (see chapter 13) have studied a wide range of signaling behavior as it appears in therapeutic sessions. Various others have analyzed specific acts with special significance: for example, Scott (1955) has found a "blathering" movement of the tongue, and Loeb (1968) a fistlike movement associated with anger. E. C. Grant (1968) assayed a classification of the nonverbal acts observed during interviews with normal students, acute neurotic patients, and chronic schizophrenics, and found these correlated in each group with four types of "moods": flight, assertive, relaxed, and contact.

Labor and childbirth were observed by Leventhal and Sharp (1965) as a natural circumstance in which they could seek facial indicators of changing levels of "distress." They found regular changes in the creasing of the forehead, movements of the eyebrows, and fluttering, creasing, or closing of the eyelids that correlated with "increases in stress stimulation." Their very useful analysis is restricted by being in terms of an affect variable (level of "discomfort"), and observations were not made in an attempt to correlate behavioral details with different kinds of use of the forehead, eyebrows, and eyelids. In common with all the approaches reported thus far, they took what Wiener et al. (1972) call a "decoding perspective." This perspective tends to interpret what can be learned from a display act in terms of a kind of information that has been preestablished by the observer, and it usually

entails inferences based on information made available by more than one source. Exactly what information is encoded in the act is a question that is not usually asked, even in naturalistic studies.

The relationship of kinesic behavior (popularly called "body language") to naturally occurring speech has been investigated by Birdwhistell (1970) from less of a decoding perspective, although a specific, predetermined structure like the hierarchical organization found in language has been sought. The structure, a grammar of nonspeech signaling, is in fact Birdwhistell's major interest. He is acutely aware that no source of information stands alone, and that signals can be considered in isolation only for purposes of analysis.

In recent years several workers, most notably Goffman, Kendon, and Duncan (whose work is discussed in chapter 14), have begun to make detailed analyses of how naturally occurring human interactions are controlled. Goffman particularly is interested in the widest possible range of interactional behavior and postulates control mechanisms in terms of the sanctioned rights and obligations of participants to maintain "face" and to preserve the flow and structure of encounters. Kendon and Duncan's approaches are more consistently involved with displays, and they do study particular signals in great detail. Their fundamental goal, however, is to understand the rules governing particular kinds of interactions, and they do not attempt to trace single displays through all the diversity of activities in which they occur.

These lines of research are part of neither the experimental nor the affect-oriented traditions of the disciplines studying human behavior, but are based on naturalistic observation. The trend of this work converges with the ethological approach to the study of interacting favored in this book. Just as the study of the messages of displays can contribute to an understanding of interacting, the study of the rules of interactions provides a framework within which to view the uses and functions of display behavior.

9

Contextual Sources of Information

If you heard someone say "help me" and turned to see him lifting one end of a table, you might respond by lifting the other end and then carrying it with him where he led. If, on the other hand, you saw that he was up to his armpits in quicksand your response might be to throw him a rope, hold a branch out toward him, or check the security of your own footing. Many responses to the signal "help me" are relevant, depending on what else a recipient knows and can learn about the immediate circumstance. And this is characteristic of all human speech. Even supposedly definitive words such as "yes" and "no" must be interpreted in context. There are questions, or so tradition has it, to which a lady will never say "yes"—but she may still indicate that she is willing. Diplomats are in the awkward position of not being able to say "no," even when otherwise indicating that they are unwilling. To interpret a lady diplomat one must presumably have a very clear idea of the frame of reference that is momentarily appropriate. Similar skills must be used whatever the signals: displays are as dependent on contextual sources of information as are words, and no point is more fundamental to understanding the behavior of communicating.

Thus, although displays and the kinds of information they make available have been the central concern of previous chapters, the information provided by a display alone can rarely if ever permit a recipient to choose an appropriate response in a particular event. The information can be used adaptively only when fit within a perspective derived from yet other sources of information. These other sources are contextual to displays.

That sources of information contextual to displays are crucial in communicating was realized independently by several ethologists in the early 1960s, including at least Manley (1960a, in an abstract based on

his unpublished D. Phil. thesis 1960b), J. T. Emlen (1960), Marler (1961a), E. O. Wilson (1962b), and W. John Smith (1963). Their sense of the term "context" is what Ogden and Richards (1936) called that in ordinary use, and it refers to events or entities that accompany or surround whatever is said to have the context. Another sense, the "determinative context" of Ogden and Richards, includes all the sources of information in an event, *even* the display under consideration.

Unfortunately, some ethologists have adopted the term in the latter sense. The problem with using it that way is that it leads to thinking of a list of the circumstances in which a display is performed as a "contextual analysis." Such a listing, though necessary, falls far short of analyzing the sources of information contextual to a display and their informative contributions in different circumstances. At least when discussing the behavior of communicating, it is better to use the term "context" to refer only to sources of information that are available to a recipient *in addition to* some particular display source. The term "circumstance" can be used for what is called the context of a display in the determinative sense. Thus, to say that a display is performed in territorial patrolling or greeting is merely to take the first step of naming circumstances, whereas to say that it is performed as the communicator travels from station to station during patrolling, or occurs with a meeting of eyes during the approach phase of a greeting, is to take the second necessary step and name other acts that occur contextually to the display, acts that can themselves serve as sources of information. It may well be that other displays are done only while perched on stations during patrolling behavior, or with eye aversion in the terminating phases of greetings; many sources of information contextual to these displays would be very different from those contextual to the previous ones.

Sources of Information

Students of human communication have long recognized that a recipient's interpretation of a signal is influenced by information from various sources contextual to it. In his lucid review (1966), Cherry mentioned memory of past experiences, the states of mind and present circumstances of recipients, and "all those aspects which serve to distinguish one communication *event* from any other" in which the same kinds of signals are involved.

The sources of information when nonhuman animals communicate are fully as diverse; anything that can be recalled or perceived is informative in principle. When a display is performed a recipient's response may depend largely on where he is (such as in his territory or

not, near his nest or not, near cover or in the open), on whether he sees the communicator approaching, withdrawing, or remaining still, on what he has seen the communicator do in past events, on what actions he sees being taken by individuals other than the communicator, and even on inherited predispositions to respond. Indeed, with enough such current and historical information the performance of display behavior is often unnecessary.

For instance, Menzel showed that often groups of captive young chimpanzees can learn what one individual is doing and act in concert with him even though he apparently performs no displays, or none specific to the problem at hand (1971). Menzel hid a treasured resource or a feared object and then revealed it to one chimpanzee. This "leader" would then be returned to the group and shortly afterward they would all be released in the large (30.5 m by 122 m) test area. The companion chimpanzees, having been through the experiment repeatedly, could discern whether or not something had been hidden from such acts as the leader's pace, style of locomotion, and glancing patterns. These behavior patterns also told them the direction and probable location of the hiding site, whether the object was desirable or aversive, and even approximately how desirable it was.

The abilities that Menzel's chimpanzees demonstrated are probably widespread among birds and mammals, especially in the more continuously social species. They are, in fact, to be expected of species whose sensory systems are acute, environments variable, and life-styles flexible. Such animals regularly process and evaluate a great deal of information and become accustomed to a great many kinds of correlations between events of importance to them. This means, of course, that to focus the ethological study of communicating exclusively on display behavior would be folly. Displays are only part of the story, and many more sources of useful information are not formalized than are. Because displays are specialized to be informative they are likely to be unusually pertinent among most of the other information sources with which they occur, but the richness of these other sources must be understood.

It is useful to visualize the many sources of information available to an individual as having three main origins: the communicator, the recipient, and the other features of the situation. First, the communicator itself provides information from many sources in addition to any display it performs, such as other displays done simultaneously or sequentially, as well as many sources that are not displays. Some of the latter sources have in common with displays a history of formalization for signal function—for example, the color of a bird's plumage, the pattern of the dewlap on the throat of an *Anolis* lizard, and even some of the subtle behavioral adjustments two interacting individuals consistently make once familiar with each other's expectations. Many addi-

tional informative features of the communicator, both behavioral and structural, are not specialized to be sources of information. For instance, an animal feeds when it is hungry and has the opportunity, and the activity of feeding is informative: it can lead other animals to seek food at the same site. A predisposition to join actively feeding individuals may provide an adaptive basis for flocking behavior in many bird species whose food is patchily distributed in time and space (discussed in chapter 10).

For some general analyses of information sources, ethologists do not separate acts which are specialized to inform from those which are not. For instance, Altmann (1965:492) measures the interdependence of individuals as each makes successive behavioral selections; as mentioned in chapter 8 he treats as "communicative" any behavior pattern whose occurrence changes the probability of behavior patterns of other individuals in a social group. In the appropriate circumstances, literally any act can exert such an influence. By concentrating on the minority of acts and other attributes that have become specialized to be especially informative, we have approached the topic of social interdependence from a narrower and more manageable perspective.

Second, some sources of information are brought to any event by the recipient of a display. One such source is its experience. It will see any communicator as a stranger or a familiar individual with whom it may even share a social bond, and as an individual with a particular temperament. It will regard any display as expected or unexpected, any kind of event as familiar or unusual, any local environment as well known or foreign, as safe or dangerous, and so on. In addition, the recipient has a store of genetic information that predisposes it to particular response patterns and that equips it to "read" or decode the display. The recipient's experience and evolved predispositions act in concert to enable it to select appropriate responses. They also render the recipient a unique individual with its own expectations and temperament, able to respond uniquely to any signal.

Third, even apart from a communicator and a recipient themselves, a great deal of information can be made available by other events that are roughly concurrent with the performance of a display. Such events may immediately precede the use of the display, be simultaneous with it, or follow so soon after that they precede any detailed response to it. The number of such sources of information is potentially very large, and their diversity restricted only by what is relevant—often an enormous range of things. Systematic study of this class of information sources may be practical only after responses themselves have been determined and the investigator can ask what features of a situation are reasonable candidates to have contributed to the selection of a response.

Most of these information sources are certainly not specialized to be

informative. They must, however, be meaningful to a recipient in the sense that they provide information he can use in selecting appropriate courses of action. Most should thus be familiar as continuous or predictably recurrent features of his environments. Animals tend to avoid the unexpected or unpredictable; if they must deal with much novelty they have elaborate behavioral adaptations for doing so (see, for example, Rozin's 1968 work on neophobia for the two issues of "specific hungers" and the avoidance of poisons in rats; see also Moynihan's suggestion in 1968a that limited neophilia is advantageous in the foraging behavior of birds that use "indirect" clues to find food). In normal circumstances, most animals seek or remain in environments with predictable features, unimportant vagaries, and where they can thus expect to be able to cope with what they encounter. Further, they more or less continuously monitor these environments or at least sample some aspects of the commonplace events in them so that they have a sustained background of pertinent information against which to judge momentary deviations.

This chapter outlines some common classes of information sources that often serve contextually to displays, beginning with behavioral and proceeding to nonbehavioral sources. Although a few act only with displays, many make their information available whether displays are used or not; some representatives of these classes may never occur with displays; for example, if they act only in the communicator's absence. To restrict the chapter to sources of information that are contextual in events when displays occur would provide an artificial limitation; thus the topic is actually nondisplay sources of information, with a bias toward classes that can be important contextually to displays.

Most sources discussed are provided by communicators, the first of the three classes of origin mentioned previously. This is partly because the focus of the book is largely upon communicators, but mostly because there are few other kinds of information sources for which we have as yet much systematic knowledge.

Behavioral Sources

Behavioral Sources That Are Not Formalized

An important portion of the events that form a familiar background for the lives of individuals of relatively social species is often generated by their conspecific associates. Many of these events include no formalized sources of information; that is, no acts that are specialized to be informative.

McCullough (1969:71) has described an excellent example of some of these background sources in what he calls the "involuntary sounds"

made by a herd of foraging tule elk: "when one gets close to the herd, it is discovered that there is a continuous array of sounds—foot bone creaking, stomach rumbling, teeth grinding, and others" that include a "continuous snapping and crunching of vegetation associated with feeding." McCullough found that if he himself made a sound similar to only one of these, say by accidentally breaking a twig, the herd would immediately be alerted. From this he concluded that the elk must monitor all these sounds continuously, using them to keep aware of the positions of other members of their herd. At least two sounds frequently confirm that the others are being made by elk: the creaking of foot bones that can be heard with each step, and the nonbehavioral noises produced by gas in the digestive tract. Most species living in herds or other closely knit groups must provide functionally similar information sources while foraging. These sources may be quite comparable even in very different species. For instance, Eisenberg, McKay, and Jainudeen (1971) have described an array of sounds made by Asiatic elephants, both sounds that are similar in kind to those of elk (such as sounds made by the motility of the gut) and that are very different (such as sounds made by flapping the ears in thermoregulatory behavior). For many African ungulates and other kinds of herd animals, the information from companions may come more from more or less continuously visible sources than from intermittently audible ones.

In persistently social species, behavioral adaptations provide an ordered framework for their interactions by simultaneously bonding together and differentiating the members of a group. The stable web of relationships among individuals that results forms the structure of what we recognize as an organized social group. Behavioral specializations of bonding or of differentiation are not displays, yet individuals' adherence to the modes of interaction they permit is informative and implies that behavior generally will follow expected courses.

1. Bonding   Bonding patterns link each individual in dependent, supportive, or interdependent relationships to each other member of its group and, in territorial species, in reciprocal relationships with its neighbors. Within a stable group, the bonding patterns can be quite diverse. For example, at least nine functionally different kinds of bonds have been detected among the individuals of a wild baboon troop (Ransom and Ransom 1971): the bonds linking all members of the troop; bonds in major, cliquelike subgroups within the troop; recurrent consort preferences of sexual partners; long-term pair bonds; maternity bonds between adult females whose infants are born at about the same time; mother-infant bonds; sibling bonds; peer bonds; and supportive relationships evident when aid is enlisted in aggressive situations. As ethologists begin to devote more attention to detecting bonding patterns,

comparable structures will be revealed in the groups of many kinds of social species.

Through extended familiarity, individuals sharing bonds should learn much that enables each to understand the other and to interact in a relatively efficient, orderly fashion. That is, they should be well informed about each other's usual behavior patterns and capable of responding appropriately to slight and informal actions. Several studies of pair formation in birds have actually shown that well-bonded individuals do interact with less disorder than do newly met individuals. Captive ring doves *Streptopelia risoria,* for instance, can take at least two breeding cycles and much displaying to complete the development of an exclusive pair bond (Erickson 1973), but established mates will repair without premating display even after a separation of seven months (Morris and Erickson 1971). In nature, kittiwake gulls *Rissa tridactyla* who breed with the same mates they had in previous years lay larger clutches earlier and have a higher breeding success than those who change mates or are nesting for the first time (Coulson 1966). Also in nature, sandpipers known as greenshanks *Tringa nebularia* are always noisier when forming new pair bonds than when reforming previously established ones each spring (Nethersole-Thompson 1951). In the related stilt sandpiper *Micropalma himantopus,* when a female rejoins her mate of the previous year there may be virtually no formal courtship, whereas new pairs are involved for days in very conspicuous and time-consuming aerial displays and get a later start at nesting (Jehl 1973). These sandpipers have so little time to nest in the Arctic summer that their overriding concern each year is to seek to reform old pair bonds, and a male may fail to breed at all if his former mate does not return to their previous territory.

2. Differentiation The individuals bonded together with a group play different parts as they interact. They are differentiated by sex, age, kinship, relative statuses, the roles they assume, and their individual temperaments among other things.

The terminology with which to describe this differentiation is confusing, however, because ethologists have yet to sort out clearly the kinds of patterns that govern interindividual relationships. Indeed, ethology lacks a sufficiently elaborate conceptual structure for social relationships to be able to deal with the implications of a term like "role": ethologists have used it either simply to describe a persisting position, part, or slot that one individual fills within a group or to denote any of the various functions that one individual can perform for other group members at different times (see Crook 1970; Bernstein and Sharpe 1966; Gartlan 1968; Rowell 1966, 1972; and Hinde 1972). Even in anthropology, where social relationships have been studied for much longer than in

ethology, the necessary concepts and terms to deal with social differentiation have not been fully clarified. Some anthropologists do seem to have sorted out the basic issues, however. In particular, Goodenough (1965) has argued that statuses, and the rights and duties they imply, are always relative. They are not the properties of individuals but of relationships between individuals. Goodenough describes each individual as having a number of "social identities," some combination of which it must assume in any interaction. Some social identities, such as age and sex, are usually relevant; others, such as occupation, are relevant in many fewer kinds of interactions. Some identities combine appropriately with each other; others do not. Goodenough's concepts need to be extended to the social relations of nonhuman species, and social identities, rights, and the analogues of duties must be cataloged.

In the meantime, however, what matters from the standpoint of communication is that social differentiation appears to be highly patterned and thus permits prediction of behavior. Adherence to patterns of social differentiation must be at least as important as adherence to bonding patterns in providing sources of information to participants in interactions.

In what ethologists have called "dominance" relationships, for instance, an individual may either take precedence as a dominant or defer as a subordinant in encounters with other recognized individuals. Status differences between individuals apparently become established either as the animals mature together or later, during specific agonistic encounters. Although establishing a dominance-subordinance relationship often requires much displaying, the relationship is maintained with little further use of displays, largely because subordinant animals readily defer (Rowell 1966, 1972) and are continuously attentive to the dominants (Chance 1967). The Altmanns noted in their studies of rhesus monkeys and baboons that dominant males "can communicate with greater efficiency, either by a less intense variant of a display or even a different display—one that apparently involves a smaller expenditure of energy" (Altmann 1967). Altmann found this comparable to Jay's descriptions of a dominant male langur as being vocal less frequently than other males, and relying instead on gestures and other subtle clues. Their observations show that in events in which status order is relevant, information from any source that simply identifies a particular individual enables recall of learned information about its behavior and the ways in which its displays should be interpreted.

This finding can likely be generalized, although ethologists have yet to study the effects of other sources of social differentiation with anything like the detailed attention lavished on dominance status and the related spatial concept of territoriality. Whatever the kinds of stable interindi-

vidual relationships, the course of most interactions within a group is broadly predetermined as long as each participant continues to adhere to the behavior expected of it by the other members of its group. This adherence necessarily provides information that creates a setting within which the performance of displays must be viewed. Displays, for their part, contribute to the management of interactions by providing information that is somewhat more attuned to the immediate issues of particular events.

Formalized Behavioral Sources

Displays can be performed as some of the contextual sources of information for other displays, or, by being repeated, displays of one kind can become sources contextual to each other. Either procedure gives a communicator the opportunity to structure some of the sources of information with which a given display will be perceived by its recipients, a form of control that is potentially very useful. Displays can also be performed within formal behavioral programs that govern the moves of two or more participants in what are called "formalized interactions." Adherence to these programs provides information that is contextual to that provided by displays.

First, repeating a display offers perhaps the simplest way of structuring part of the context that is available with it. Repetition is informative, and can affect the behavior of recipient individuals as Shalter (ms.) has shown by playing back recorded "alarm calls" to isolated birds. Hearing the recorded calls always aroused the subjects (either red domestic fowl or their wild counterpart, red Burmese jungle fowl), and playback bouts 30 seconds long elicited many more vocal responses than did bouts that were only 10 seconds long. In other tests, using captive laughing gull chicks, Beer (1973) discovered that repetition of a display was a crucial contextual feature in eliciting an alarmed response but that repetition of a single communicator's calls was not enough. The chicks gave no overt responses to repeated playback of a single recording of the species' alarm call, but fled and hid when they heard a recording of a sustained chorus of alarm calls uttered by many gulls. This confirmed Beer's observations that chicks "ignore isolated alarm calls in the field, even when they are uttered by their own parents. It is only when several gulls have been put to flight, and hover, alarm calling overhead, that chicks break off what they have been doing and flee and hide."

Repetition of a display can be irregular or it can be organized into a regular (that is, formalized) pattern. When repetition can be expected, the displaying is "tonic" (Schleidt 1973), and variations of the rate and regularity of repetition should be usefully informative. Each recurrence

of a display in a tonic pattern becomes a special event to which preceding units are contextual. For instance, black-tailed prairie dogs have a bark vocalization that may be uttered singly by an individual that is likely to avoid interacting, may be uttered intermittently by an individual recurrently facing decisions between interaction and other behavior, or may be uttered as a continuous barking display in regularly paced strings by an individual that has become ineffectually balanced between selecting to interact or to engage in some alternative behavior, and is closely watching the animal that makes its choice difficult (W. John Smith et al. in press 2). In the last case, the relative stability of the communicator's behavior is assessable from the pattern of repetition: for example, the likelihood of the communicator beginning or resuming its alternative to interaction increases and the regularity of its attending behavior decreases as the interval between successive barks is lengthened.

In prairie dog barking, the communicator's control over context extends primarily to its choice to repeat the bark vocalization and to the duration it assigns to the interval between successive utterances. It can also vary the form of barks to some degree, so that it need not repeat what is, strictly speaking, the same kind of display. In fact, animals often perform more than one distinctly different kind of display in an event, either sequentially or in more or less simultaneous clusters. The different acts may almost invariably accompany one another, as the long call vocalization of gulls is done with an oblique posture and series of movements, and so be largely redundant. Or the displays may combine and recombine more freely, as do many of the component displays that make up the complex "facial expressions" of primates (see chapter 13).

When many displays are combined and recombined contextually, the resultant message combinations probably provide very detailed information. Recipients may need procedures for assigning relative weights to the component displays, that is, for recognizing one or two of preeminent significance in a given circumstance and viewing the others as contributing contextual information. The processes that contribute to the performance by both communicators and the use by recipients of complex display combinations present some of the most interesting challenges in studying the behavior of communicating.

In addition to the use of detailed contextual relationships among displays, it is also possible that cumulative effects can result from the tonic characteristics of continuous displaying activity. Quantity—the number of repetitions of a display, the number of diverse displays, or the duration of a bout of displays—may be significant in some circumstances. For example, it could help a female to select among rival males.

Wolf and Hainsworth (1971) have argued that a female insectivorous or seed-eating bird seeking to form a pair bond with a male on a rich territory might get a clue about the relative qualities of the territories of different males by assessing how much free time each male has to sit and sing. Males who can obtain all the food they need relatively quickly should produce a greater quantity of song, if singing and foraging are mutually exclusive activities. Another possibility is that a male who displays copiously may sustain a female's interest while he is otherwise occupied. For example, when ducks are courting the diversity of a male's displays may be determined largely by his responses to the activities of other males: Dane and van der Kloot (1964) have shown that male goldeneye ducks attempting to copulate are often interrupted by other males and distracted by the need to display to these competitors. While a female goldeneye lies prone on the water waiting for him to mount, her mate may be directing displays less to her than to individuals with whom she has no direct dealings. She continues to lie prone for many minutes, however, suggesting that her mate's continuous persistence in displaying may be a sufficient clue to her that he is still in a precopulatory state. Comparable observations of eiders have led McKinney (1961:56) to propose that this sustained continuity of courtship displaying, not the detailed diversity of male displays, is significant to a female eider.

It is difficult to know to what extent displays figure among the sources of information contextual to other displays, in part because some acts in the repertoires of most species may or may not be formalized (a problem considered in chapter 13). Many are probably acts for which evolutionary or cultural shaping processes have not been carried very far; others appear to be displays that are falling into disuse and becoming less formalized (Moynihan 1970b). (Like the words of a language, displays appear to have "lifespans" in which they become formalized, flourish, and then disappear—a phenomenon that is most obvious in the high turnover rate of popular slang.) Still other acts have retained most of their direct functions while becoming formalized in very limited characteristics, sometimes making it hard to recognize that they are displays. Formalization may simply restrict an act's directly functional range of variation and perhaps the precise timing with which it correlates with other acts, as Kendon has shown in distinguishing between the informative and monitoring functions of different patterns of gazing during human conversations (1967; also discussed in chapter 13).

Some other students of human behavior have, like Kendon, evaluated the formal characteristics of subtle behavior by very detailed analyses of small segments of interactions, usually through frame-by-frame dissection of filmed encounters. Ethologists have not often worked in the same

exhaustive detail, although they have repeatedly made more casual observations of subtle signals. For instance, the difference between the postural orientations of directly facing or obliquely facing toward an opponent have often been reported as significant sources of information in bird displaying (see, for example, Hinde 1956a; Tinbergen 1959b; Moynihan 1962b; and W. John Smith 1966). Marler (1965) has summarized the use of various body positions and modes of carriage that appear to be subtly specialized in Old World nonhuman primates, and Jay (1965) described several clues that characterize the behavior of dominant male langur monkeys: slight postural differences from other individuals, the use of slight pauses and hesitations, physical nearness to other individuals, and the "duration of visual focus" with which they will attend other individuals. Man is clearly not unique in having an array of subtle formalizations, but no repertoire of any species has been adequately investigated. For the present there is little alternative to treating these acts, be they fully displays or not, as sources of information contextual to what we recognize traditionally as display behavior.

Finally, in addition to acts that can be performed by a single individual formalization can occur at another level of complexity: such formalized patterns of interacting as greetings, leave-takings, reaffirmations of relationships, and appeasing-reassuring exchanges that two individuals produce by their mutual efforts. Erving Goffman (for instance 1963, 1967, 1971) describes the cooperative use of many "interaction rituals" by persons engaged either in "interchanges" or in avoiding encounters. His concept of an interchange involves two or more participants in an interaction of two or more predetermined kinds of moves, and he sees interchanges as being abundantly used and fundamental tools of interacting (Goffman 1967:20). Humans as well as many species employ this sort of formalized behavior, which is discussed more thoroughly in chapter 14.

The formalization of interactions provides each participant with a repertoire of parts or roles (in the theatrical sense of these terms) to be assumed in different kinds of events—to at least this extent, the world *is* a stage. A particular interactional formalization can be available for use by many or even all individuals in a population (although used by each to quite different extents), because the roles are a property of the kind of event, not of the individuals themselves. That is, a set of standardized parts to be played is peculiar to each type of formal interchange. Each participant temporarily adopts one as a suitable tactic to employ within a formalized sequential pattern that provides all participants with an established means of managing their interaction. For instance, in a common sort of human interaction described by Goffman (1971) as a "remedial interchange," one participant has the role of transgressor, the

other that of having had his rights violated. In the formal interchange the first apologizes, the second accepts, the first then thanks the second for accepting, and the second may then say "you're welcome." In comparable circumstances, nonhuman primates use formalized interactions that are analogous to at least the first two steps of this human procedure.

The behavior that evidences the adherence of a participant to a formalized interactional pattern, coming from both informal acts contextual to displays and from displays themselves, is highly informative. It thus provides a major part of the contextual framework within which the use of displays must be viewed.

METACOMMUNICATION   The term "metacommunication" denotes a postulate concerned with the contextual relationships among formalized acts or comparable specializations. The postulate, basically that the function of some formalized sources of information is to provide information about other formalizations, is attractive. However, to set this phenomenon apart with so special a name seems more misleading than helpful, at least with respect to the nonlinguistic behavior that so far has been called metacommunicative.

The concept of metacommunication was first applied to the study of nonhuman behavior by Bateson (1955), who proposed that social play behavior would not be possible without signals that could distinguish it from similar nonplay behavior (for example, that could distinguish play fighting from attack). He initially defined metacommunicative acts as those providing information about the relationship between participants in an encounter, but went on to define as metacommunicative any message that provided information about the frame of reference within which another message could be viewed. This led many ethologists like Altmann (1962, 1967) and Bekoff (1972) to adopt the working definition that metacommunication is "communication about communication," although Lyons (see editorial comments by Hinde 1972:124) has objected that this definition is not in line with philosophers' use of the comparable term "metalanguage." Altmann's discussions introduced the term to ethology and have established its accepted uses. To Altmann, metacommunicative signals accompany play, correlate with differences in status, or provide what I have termed the behavioral supplemental message of direction.

Altmann's metacommunicative phenomena are treated as two separate issues in this book. First, information about direction is treated as behavioral information provided by many displays in a supplemental relationship to the information they provide about behavioral selections. If the provision of directional information is metacommunicative be-

cause of this relationship, so too is the provision of all other behavioral supplemental messages (probability, intensity, and relative stability). By this reasoning, *all* displays incorporate metacommunicative messages because all provide information about probability. In this usage the concept of metacommunication is superfluous to the concept of behavioral supplemental messages and has the disadvantage that its restricted application to only a part of the information covered by the latter concept seems arbitrary.

The view that messages about behavior can be described as referring both to behavioral selections and to attributes of performances (the behavioral supplemental messages) is economical; it makes it possible to encompass chronic displaying within the same descriptive scheme used to deal with displays performed primarily in more acute situations or in immediate response to changing circumstances. The chronic postural signaling of dominant and deferent animals, called "metacommunicative status indicators" by Altmann (1967:338), is treated in chapter 6 as a way of providing information about behavioral selections that have persisting probabilities of occurring. The behavioral selections may be realized frequently (for instance, the avoidance behavior shown by deferent individuals) or infrequently (precipitous attacks by high-status individuals). The latter may rarely attack even if chronically prepared to do so, because the avoidance of the former individuals obviates their selection of attack behavior. But for either frequent or infrequent use of the selections, no conceptual baggage needs to be added to describe the communicative characteristics of the persistent signaling.

Second, that a communicator should make formalized sources of information available contextually to other formalized sources which it provides is treated as a common adaptation engendered by the pervasive importance of context (see preceding section of this chapter). Granted that so much communication depends upon contextual interpretation of signals, widespread evolution of displays that are performed along with other displays, or with badges and the like, is inevitable. Indeed, Altmann's proposal (1967:356–357) that "metacontingencies between messages and the responses that they elicit are the essence of metacommunication"—that metacommunication occurs when the contextual relations between two or more display acts influence a recipient's selection of responses—reveals that metacommunication is nothing more than this common contextual phenomenon. (Altmann's proposal specified sequential metacontingencies, but his examples include the use of directional indicators that are made available simultaneously with the messages they supplement.)

Social play behavior is the original and most commonly cited kind of

event in which ethologists claim metacommunication. The claim is usually based upon the performance of displays in the initiation or reinitiation of social play. Some of these displays appear to correlate only with play activities and thus to provide a message (not a meta-message) having social play as its behavioral selection referent (see chapter 6). These displays provide both the information that species' typical behavioral components of social play (such as biting acts in canids, felids, and many primates) are highly probable, that other such acts (cuffing, chasing, fleeing, and so on) are nearly equally probable, with little overall stability in the choice of any particular selection, and that the acts will have certain intensity characteristics (for example, biting will be gentle and fleeing less than all-out). Yet there is a sense in which such displays communicate about communicating. Much social play appears to be rule-bound *cooperative* activity, a member of the class termed "formalized interactions" in the paragraphs that immediately precede this discussion of metacommunication; displays referring to play thus provide information about formalized behavior, about activities that are specialized in part to be communicative. But by this reasoning, *any* display having all or part of any formalized interaction as a referent would be metacommunicative, not just "play signals." The result would be that a considerable amount of behavior would be called metacommunication, although we should know no more about it for having labeled it.

Metacommunication, or whatever communication about communication is called, seems to be a reasonable conceptual category. But at present, at least as applied to nonlinguistic communication, it is a category that suffers from arbitrarily restricted use and that seems to explain no more than we already understand, even if it were much more broadly applied. Perhaps it is an idea whose time will come, but its time has not come yet.

Nonbehavioral Information Sources

Most animals provide information through various formalized sources in addition to their display behavior. Among the categories distinguished below are "badges" (W. John Smith 1974), which are effectively orna-ments worn by individuals; "constructions," which animals build and which in some cases can function in the absence of their makers; and "tokens" of substances given by a communicator to a recipient, which either provide information about the thing sampled or create a contex-tual framework for specific kinds of interactions. In addition, some chemical products that are released by an individual (called "phero-mones"; whether specialized to be informative or not, see chapter 2) are used similarly to badges and constructions.

These sources of information tend to be more persistent than displays, because displays last only as long as the behavior is sustained. Depending on its molecular weight, a pheromone may be detectable for seconds or weeks after it has been released or deposited. A badge may exist unchanged through a breeding season, an ontogenetic phase of an individual's life span, or even throughout life.

Because they are capable of making information available for relatively prolonged periods, such sources can function in ways that displays cannot. However, these sources are not important only because they persist, nor do they always need to be continuously available to recipients. Some badges are customarily kept concealed or inconspicuous until revealed by a display act; pheromones may be released only at appropriate times; and constructions may be left largely functionless except when their builders visit and actively use them. Though the information sources themselves are nonbehavioral, their effects are often dependent on the behavior of communicating.

Badges

Badges are characteristics of an animal's appearance that have been modified to be informative adornments. Many badges are colors and color patterns, or specially shaped feathers, fur, scales, horns, or other outgrowths, or even modified appendages such as the enormous claw waved by a male fiddler crab. Badges may be kept prominent or may usually be hidden, may be lifelong or seasonal, stable or behaviorally modifiable through changes in chromatophores, vascular dilation, inflation of air sacs, and so forth.

Badges operate only in the context of other sources of information, but under some circumstances a badge may be remarkably effective and an animal seeing it may respond even if this entails ignoring many other potential sources (the observational basis for the releaser theory discussed in chapter 7). In a now classic set of experiments with the European robin, Lack (1940) showed that the patch of red feathers from an adult's breast could elicit display postures of the sorts used in threatening if placed within the territory of a breeding male. The responses were elicited even when the red feathers were presented in stuffed mounts from which Lack had removed the head and tail or even all the remainder of the robin (that is, the sources of information that should inevitably be present along with the breast badge). Although they responded vigorously, the tested birds did not behave as if interacting with a living rival. Lack showed that in more natural circumstances robins do take into account information from other sources in addition to this badge. For example, adult females wear a conspicuous red patch of feathers identical to that of an adult male and occupy the same

breeding territories, yet they are not usually threatened by their mates. The test situation isolated the badge in an unnatural way for which the robins were unprepared, and although the results indicate that the badge is important, they do not imply that robins usually respond solely to it. A similar finding emerges from an essentially reversed procedure employed by Peek (1972). He removed only a badge, the bright red epaulets from seventeen otherwise normal male red-winged blackbirds (these epaulets are shown in figure 9.1). Peek discovered that although these males displayed vigorously and were normal in behavior and appearance apart from the missing badge, they could not maintain their territories. Likewise, when Crews (1975) removed the hyoid cartilage from captive male *Anolis carolinensis* lizards, he obtained individuals who could not raise the dewlap flap of colored skin beneath their chins; receptive females would not respond sexually to the otherwise normal courtship display behavior of these males.

BADGES PROVIDE IDENTIFYING INFORMATION   Like the robin's red breast feathers, badges are often important in identifying a species. In fact, they are often the principal means of assuring that animals form pair bonds only with conspecific partners. Hinde (1956b, 1959) reported that it is common for the badge patterns that animals display to differ more among species than do movements and postures, and badges have been implicated as what are called "species-isolating mechanisms" in innumerable cases. Among the most convincing demonstrations has been N. G. Smith's study (1966a) of four species of arctic gulls. Each is very similar in appearance and yet they nest near each other without interbreeding. The displays of all four species are extremely similar and probably not useful as species-isolating mechanisms. Differences in plumage colors, particularly in wing-tip patterns exist, but are slight; Smith showed them to play only limited roles in interspecific discriminations. The degree of contrast between the appearance of the eye plus the bare, fleshy ring that surrounds it and the white of the head feathers is the crucial source of information used by the gulls in identifying mates of their own species (figure 9.2). Two species have light eye rings (in one yellow, in the other orange) and light yellow irides. The other two have dark reddish purple eye rings, one with a dark iris and the other with iris colors that vary among individuals from dark to light. The case of the last species is complicated, as it overlaps in part of its breeding range with a light-iris species and in the other part with a dark-iris species. Its females apparently select as mates males that closely resemble their own parents, and this permits the geographical variation to be sustained. (That a learned preference for parental coloration can be a mechanism that affects mate selection has been demonstrated

Figure 9.1. The epaulet badge of male red-winged blackbirds. Territorial male red-winged blackbirds are shown in their high-intensity song spread and flight-song displays (*A, A'*, and *C, C'*, respectively). Normal flight is shown in *B*. The red epaulet extends from shoulder to carpal (the bend of the wing) and is always visible, but it is erected and made more conspicuous while singing. (In the population of red-winged blackbirds studied for preparing these drawings, the epaulet is red only and lacks the yellow border that is characteristic of most other populations of this species.) (From Orians and Christman 1968. Courtesy of the Museum of Vertebrate Zoology, University of California.)

experimentally in another species, the dimorphic snow goose *Anser caerulescens;* see Cooke and McNally 1975.)

Smith developed techniques for capturing individual gulls of known breeding status, and further techniques for painting their eye rings to mimic those of other species. He found that females of three species would form pair bonds only with males having an eye-head contrast like their own (the criterion used by the fourth species may be based on matching to the parental pattern). Further, after pairing a male gull would copulate with its mate only if her eye-head contrast was appropriate; copulation ceased and pairs broke up when a female's eye ring color was painted to alter the contrast. Thus both males and females discriminated, although at different stages of the breeding cycle. Yet Smith's experimental alterations became ineffectual once the eggs had been laid; if species discriminations had not been accomplished by then, it was too late. The gulls were using the badge as a source of information for what Mayr (1963) calls a "prezygotic" isolating mechanism: a procedure or structure that prevents attempted fertilization between members of different populations.

Markings such as the eye rings of these gulls are only one of many kinds of species-identifying badges. Even the outline shape of an animal can also be specialized to this end. Estes and Estes (in press) shows that among African antelopes, even species that have very nondescript coat patterns specialized for concealment instead of for displaying are entirely distinctive in silhouette. They propose that silhouettes of antelopes' antlers and facial configurations, which can be used to distinguish among all species, have evolved as species-identifying badges.

Many kinds of animals have badges that identify particular classes of individuals within a species. For instance, in a North American woodpecker, the yellow-shafted flicker, males have a black "moustache" streak that females lack. Noble (1936) tested to see if this sex-specific plumage badge elicits appropriate responses. He captured a female, painted such a mark on her and released her; she was then attacked by her mate. In similar experiments, Tinbergen (1953a) showed that males of a small freshwater fish, the three-spined stickleback, determined the sex of a territorial intruder by the presence or absence of a bright red belly. Chaffinches may also tell the sex of an individual by its breast color; when Marler (1955a) dyed the breasts of females red to resemble those of males even formerly low-ranking females were avoided in aggressive encounters and were able to become dominant over other females.

Among African ungulates, various prominent specializations of size, pelage badges, and horns may distinguish males from females. Still, many of the least and the most gregarious species are monomorphic;

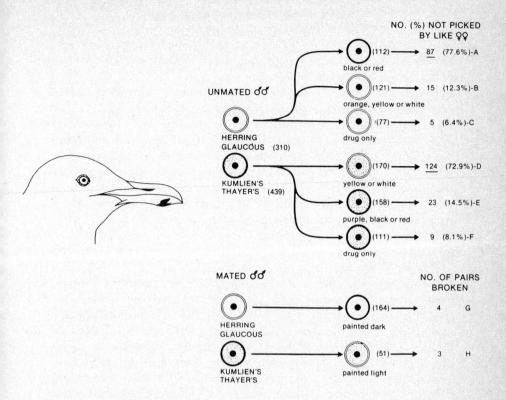

Figure 9.2. Experiments with facial badges of gulls. On the left is an outline of a herring gull's head, showing lack of contrast between white feathers and the yellow of the fleshy ring surrounding the pale yellow iris. The eye ring is represented by the outer, narrow circle in the schematic eyes at right; it surrounds the wider iris and the black central pupil. The diagram summarizes the results of artificially altering the contrast between the eye region and the head of male gulls at two different stages of the breeding cycle.

With herring and glaucous gulls, increasing the contrast between the eye region and the white head (by painting the fleshy eye ring dark) makes unpaired males unacceptable to prospective mates (A), but has no effect once a pair bond has been formed (G). Conversely, decreasing the contrast for unmated male Kumlien's and Thayer's gulls interferes with pair bonding (D), but, as for the previous species, has no effect after a pair bond is formed (H).

For both groups, control painting of eye rings without altering contrast has little or no effect on pair bonding (B and E) as compared with the bonding success of males who were drugged and captured, but not painted at all (C and F). (From N. G. Smith 1966a.)

this is apparently either because dimorphism never evolved, in the former, or because females have evolved to resemble males, in the latter. Estes (ms. 2) suggests that a lack of sex-identifying badges in ungulates may be a result of selection for unmistakable species-recognition characteristics, and that in the highly gregarious species it may facilitate integration of both sexes into a herd.

Infraspecific classes determined by the abilities of individuals to take particular positions in a social hierarchy are sometimes identified by badges. For instance, in two species of mountain sheep, in which males fight by ramming head-to-head with their horns, individuals can distinguish different horn sizes at a distance. Rams prefer to encounter other rams with similar-sized horns (Geist 1966, 1968). Deer apparently employ similar visible clues, and a dominance hierarchy may shift if an alpha male sheds his antlers in the annual cycle before a beta male—with reversion to the normal status relations once both males are without antlers (see Portmann 1961).

Antlers and horns are usually adapted both as contact weapons and as badges. They can be important for fighting and, along with color patterns, frills, and other cranial ornamentation, for providing information relevant to agonistic and sexual encounters—and not just in ungulates, but also in various lizards (Rand 1961). Their dual use was quite possibly very significant in the evolution of such remarkable animals as *Triceratops* and other ceratopsian dinosaurs (Farlow and Dodson 1975). However, there probably are cases in which the evolution of such cranial structures has become dominated by their informative functions. For example, it has been argued that the sole use of the antlers of the extinct Irish elk *Megaloceros giganteus* was as a badge (S. J. Gould 1974). With a spread of nearly 4 meters in a large stag, these antlers were visually impressive. As 90-pound appendages to a 5-pound skull, however, they must have been poor weapons.

Gross features of an animal's appearance, horns included, incorporate various functional adaptations, and it can sometimes be difficult to know if any formalization of their appearance has occurred. Often none would seem necessary. For instance, the relative sizes of individuals are effective sources of information about how they can be expected to interact, but size as such often differs as a result of age or differential roles of the sexes even in the absence of formalization to be informative. In the hermit crab *Clibanarius vittatus* relative size is highly informative, although often a result just of age. Hazlett (1968) found that in chance encounters one individual usually retreats. Larger individuals are more aggressive than are smaller, and in his experiments the greater the size difference the more likely that two individuals would ignore each other. Further, the larger crab was less likely to display and the smaller more

likely to retreat without seeing a display by the larger. A similar relationship between relative size and the likelihood of avoidance by the smaller individual was found for two nonterritorial grapsid crabs by Warner (1970), although in one species the size clue was complicated by the existence of different cheliped (large claw) sizes in males and females, perhaps a formalization for sexual differentiation.

SOME BADGES PROVIDE BEHAVIORAL INFORMATION   Badges that simply identify their wearers provide a sustained background of information about individual communicators that is a necessary part of the frame of reference within which recipients perceive their displays. However, other badges both identify and provide information about specific kinds of behavior patterns that a communicator may be expected to select from its repertoire. These badges must be adjustable to operate over more limited periods of time; they must have more than one state of availability. Their durations are correlated with differences in the transitoriness of the behavior about which they provide information: some endure for months, some for minutes, others achieve evanescence through being incorporated into display acts based on supporting structures.

The colors of animals, for instance, may differ when the animals are in different physiological states, primed for different kinds of behavior. Many bird species have breeding and nonbreeding plumages: a male scarlet tanager is scarlet only during the spring and summer, and in autumn and winter is green like the female of the species. Some birds change the colors of their bills on an annual cycle: the American goldfinch has a yellow-orange bill when breeding, then a melanin deposit makes the bill dark for the remainder of the year (Mundinger 1972). With adulthood, all males of the superb blue wren *Malurus cyaneus* of Australia assume a darkened bill color, but the timing of the annual molt from dull nonbreeding to brilliant breeding plumage is determined by each individual's "social position" (Rowley 1965). Males maintaining territories make the change fully a month before those who live within other males' territories. The "supernumerary" males of a group are adults who do not pair or inseminate females, but who later assist pairs in feeding offspring.

Physiological changes relevant to interactional behavior occur abundantly on much shorter time scales than an annual cycle, of course. Many species of fish, squid (Moynihan 1975), or octopus (Wells and Wells 1972), and other animals can change color as a form of display behavior. Even humans of some skin colors can blush and pale. The Asian teleost fish *Badis badis* can assume eleven different color patterns in different circumstances. *Badis* usually changes its color more slowly than its behavior, so that the color changes have limited short-term

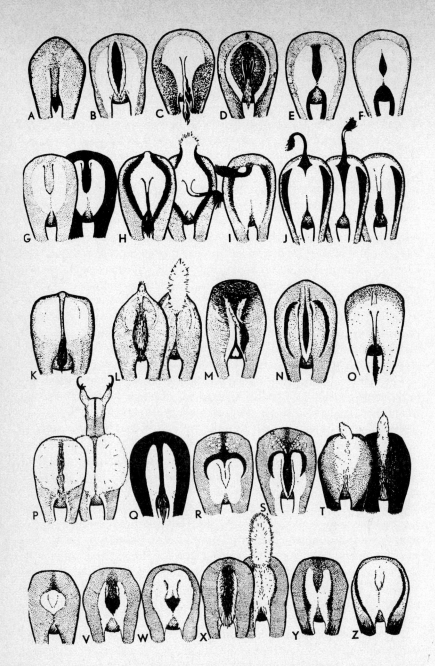

Figure 9.3. Some ungulate rump patch badges. *A.* Uganda kob (*Adenota kob*). *B.* Defassa waterbuck (*Kobus defassa*). *C.* Bontebok (*Damaliscus pygargus*). *D.* Common waterbuck (*Kobus ellipsiprymus*). *E.* Stone sheep (*Ovis dalli stoni*). *F.* Bighorn sheep (*Ovis canadensis*). *G.* Backbuck (*Antilope cervicaptra*), female and male. *H.* Springbok (*Antidorcas marsupialis*), showing erected crest. *I.* Thompson's gazelle (*Gazella thomasoni*). *J.* Grant's

predictive utility as compared to most other displays. Nevertheless, in the absence of more definitive overt behavior they are a useful source of information (Barlow 1963). Baerends et al. (1955) showed that in another fish, the guppy, different color patterns provide a relatively sustained background against which other displays are employed, and in some fish species such color changes are more finely tuned and continuously variable information sources than are the display movements they accompany (for example, in the courtship of *Tilapia melanotheron;* see Barlow and Green 1969).

Badges employed in display behavior that provides information of immediate and usually very transitory relevance may be hidden or visible primarily as relatively inconspicuous identifying marks between periods of active display. That is, although badges are structural and hence enduring they need not necessarily be continuously available as sources of information nor persistently fully conspicuous.

Some badges, for instance, are made conspicuous by displays that provide information about agonistic behavior. A species-specific white or black and white patch adorns the rump of the North American pronghorn, many Old World antelopes, mountain sheep, deer, wapiti, rabbits, and many other mammals of different phylogenetic lineages. On becoming very alert or fleeing an individual usually makes the rump patch conspicuous by erecting its hairs or, in some species, by lifting the tail (figure 9.3). For herd-living animals such a display serves as an indication of danger. It would be a mistake, however, to characterize the displaying of all mammalian rump patches as simply conveying messages of escape probability. Guthrie (1971) has suggested that some rump patches may also function in status interactions because in some species the patches change size and shape seasonally and in other species vary geographically in correlation with horn or antler size. In some primates, colored rump patches are involved in both status encounters and sexual behavior. Some evolution of rump patches may even converge through quite unrelated selection pressures. Rabbits, for

gazelle (*Gazella granti*), male and female with tail raised. *K*. Banteng (*Bos sondaicus*). *L*. Greater kudu (*Tragelaphus strepsiceros*), showing erected tail. *M*. Gerunk (*Litocranius walleri*). *N*. Impala (*Aepyceros melanupus*). *O*. Coke's heartbeast (*Alcelaphus cokei*). *P*. Pronghorn (*Antilocapra americana*), with erected rump patch. *Q*. Sable antelope (*Hippotragus niger*). *R*. Sika deer (*Cervus nippon*). *S*. Fallow deer (*Dama dama*). *T*. Caribou (*Rangifer tarandus*), winter and summer. *U*. Roe deer (*Capreolus capreolus*). *V*. Black-tailed deer (*Odocoileus hemionus columbianus*). *W*. Mule deer (*Odocoileus hemionus*). *X*. White-tailed deer (*Odocoileus virginianus*), showing rump patch under tail. *Y*. Red deer (*Cervus elaphus scoticus*). *Z*. Wapiti (*Cervus elaphus canadensis*). (From Guthrie 1971.)

instance, often forage alone and do not usually spread alarm visually to other rabbits. Smythe (1970) proposed that their conspicuous rump patches might induce predators to attack before coming within the distance at which a successful pursuit can usually be begun, effectively communicating interspecifically to goad predators into tactical errors.

Other badges are made conspicuous in displays indicating some likelihood that the communicator will attack. The expandible dewlaps under the chins of *Anolis* lizards that carry such redundant identifying information (see the discussion of measurements by Rand and Williams 1970 in chapter 8, this book) are normally collapsed and scarcely visible; when expanded and employed in correlation with some display movements they can threaten. Many species of tyrannid flycatchers (including those in the genus *Tyrannus,* W. John Smith 1966) have a brilliant patch of orange-red to yellowish crown feathers concealed atop their heads that is rapidly uncovered and even erected when there is a high probability that the communicator will attack. This patch effectively mimics the color and size of the gaping mouth that is opened immediately below it in an attack. Many other species of birds have less elaborate crests of elongate feathers that are raised when either attacking or escaping is at all probable.

Information about locomotory behavior is commonly provided by displaying badges, at least in gregarious species. The Canada goose, for instance, has two badges that are emphasized or revealed through display actions that correlate with an increasing probability of flight. One is a white patch under the chin that contrasts with the species' black head and neck. Although always visible, this patch is made much more conspicuous by its head-tossing display. Raveling (1969) found that head-tossing precedes flight, being done especially often by the members of a family that first become ready to fly. Only if the head-tossing is done by the gander, which leads flights, will the family fly soon after this display begins; in fact, the badge's conspicuousness is marred on immature individuals by black flecking in the white area. Just before a Canada goose takes flight it extends its wings, suddenly revealing another, more fully hidden badge: a white patch at the upper base of the tail.

The Canada goose's white rump patch is an example of a "flash pattern" (Moynihan 1960, 1962c). Flash patterns are conspicuous patches of contrasting color on the wings, tail, or lower back that are revealed as an individual either takes flight or makes intention movements of taking flight. Many birds that flock, whether intra- or interspecifically, have flash patterns; further, the plumages of many gregarious species are largely black, white, or pied. Bold plumages and attention-attracting flash patterns make their wearers conspicuous in

motion, and thus facilitate the formation and coherence of flocks by helping each member be more aware of the others' comings and goings. (Despite the obvious correlation of flash patterns with locomotory behavior, the class of badges was at times misinterpreted in the earlier literature—for example, by Hingston 1933, whose simplistic notion of "colour-conflict" admitted interpretation only in terms of the expression of aggression and fear.)

DEFINITIONAL PROBLEMS   The term "badges" should perhaps not be extended to all the structures of organisms that are specialized to be informative. Some such structures are properties of communicators that are plants, not animals; because these communicators can only elicit behavioral responses and cannot enter into behavioral interactions, their special structures might better be called "badge analogues." Colors, color patterns, flower shapes, and scents have all evolved to take advantage of the responsiveness of the insects, birds, and bats that are the major pollinators of most flowering plants (reviewed by Faegri and van der Pijl 1966). This responsiveness permitted the evolution of flowering plants, which began about 140 million years ago. In any one group of plants these badgelike or chemical adaptations are usually specialized to be attractive to specific kinds of pollinators. For example, some flowers employ pigments such as flavonols that absorb ultraviolet light and so produce patterns that are visible to insects but not to vertebrates (Thompson et al. 1972). Or flowers may mimic the appearance of insects, like the various orchids that provide no nectar but that visually attract male bees and wasps who attempt to copulate with them (see Wickler 1968). Fragrant flowers and the nectar they promise attract pollinators, and tasty fruits are taken by animals who then disperse the seeds they harbor. Some flower scents are adapted in very special ways to the pheromonal communication (see chapter 2, and below) of some insects: Dodson et al. (1969) have suggested that the attractiveness of the scents of some New World orchids for male euglossine bees may lie in their incorporation of chemical compounds that males can use as precursors in the production of sex attractant pheromones. Plants evolve such intricately specialized badge analogues because providing information is adaptive if recipients can, in making use of it, provide a service for the organism that is the information source.

A second minor definitional problem is created by specialized features that are intermediate between the concepts of "badge" and "construction" (considered next). These objects are badges that are manufactured; they are behavioral rather than physiological products. Humans, of course, manufacture ornamental clothing and even such ornamental items as jewelry which may glitter and flash, providing information

about the wearer's movements. Self-adornment may not be confined entirely to human practice, although it may more often function less for intraspecific signaling than for camouflage against discovery by predators, for example, in crabs that affix algae to their shells (Wickler 1968). Yet dominant laboratory mice are known to trim the vibrissae of subordinates with whom they are caged (Long 1972). Because they do this after the fights in which status is established, it may be a matter of constructing badges that identify the subordinant animals. On the other hand, the trimming might also function by restricting the sensory capabilities of subordinant animals, or may be an anomalous result of the restrictions of social life in a cage; it is by no means understood.

Constructions

Objects are constructed by diverse species and employed as stages or as elaborations of sites used for display behavior. Such objects may function only to provide information or may have additional, perhaps crucial, direct functions.

Nests, for instance, are often used in courting behavior. In the three-spined stickleback fish each territorial male builds a nest before trying to court females (Tinbergen 1953a). He must lead a female to this nest before she will lay eggs; it then holds the eggs while they develop. Courting males of many other fish species prepare similar structures to receive females and their eggs. Males of weaverbirds that nest in dense colonies on the African savannahs build nests and entice females to them by display performances (Crook 1962, 1964). In some species the males go out and seek females, returning with them for further ceremonies at the nest much as do stickleback fish. Males of other species rely entirely on mass demonstrations made at the nests, and may not leave their nests even to follow a female that is leaving after having been attracted to the nest tree. Some of the differences among these weaver species correlate with the conspicuousness of the nest and with differential ecological pressures to form pair bonds rapidly (Crook 1963, 1964; see chapter 12).

The form of nests is necessarily adapted to the problems of rearing young. However, in many cases aspects of nest structure have probably been specialized to facilitate communication, for example, to attract a mate. Some nests have even been specialized for delayed communication: females of some solitary wasp species build nests in which they deposit eggs and supply food as provision for the young that will develop later. These nests have a single exit, and the developing larvae must align themselves to face it before they grow too large to turn around in the chamber. Cooper (1956) has shown that a mother wasp

marks the appropriate nest walls by their geometry and texture, thus achieving communication with the offspring she will never meet. Yet other nests may simply mimic those built for the eggs and young: grebes and loons (divers), for instance, cannot copulate while swimming and build rudimentary nests upon which this interaction takes place (see Huxley 1914, 1923). Because any solid substrate would support the pair out of the water, these nests are informative constructions rather than just mechanical necessities.

Not just nests, but also more arbitrary constructions are built for courting. Males of the Panamanian fiddler crab *Uca musica terpsichores* defend circular territories that center around the entrance of a burrow used for displaying. Some sexually mature males build crescent-shaped sand structures called "shelters" or "hoods" beside their burrows (figure 9.4). Long a subject of debate, these shelters have been thought to modify the local environment, perhaps by shading the burrow. Naida Zucker (1974) has shown, however, that the open side of the shelters is not oriented with respect to the sun or any particular feature of the environment. Zucker noted that shelter building occurs much more in dense populations than in sparse ones, and that shelter-owning males display only in the 180° sector in front of the shelter opening. She mapped the distribution of neighboring crab burrows around the shelters of 261 males. Displaying male neighbors were much more common within 10 to 20 centimeters from a sheltered burrow in the sector behind than in the sector in front, although individuals of all other classes were equally distributed all around. She therefore proposed that the shelters

Figure 9.4. The shelter constructed by a male fiddler crab. The crescent-shaped sand shelter of this displaying male fiddler crab indicates that intruders approaching from in front will elicit territorial defense more readily than those approaching from behind. Without the shelter he would defend all surrounding sectors equally. (From Zucker 1974.)

may help to keep agonistic behavior from increasing in direct proportion to population density. Her hypothesis was substantiated with the following experiments.

Display "shakes" were elicited from test males by pulling a captured individual along a 120-cm track of a pulley system to 30, 20, and then 10 cm from a test burrow in successive steps. The closer the stimulus animal came, the more shakes it elicited whether approaching from in front or from behind a shelter, but at each distance male stimulus crabs elicited more displaying when in front than when behind. In her tests, males with shelters displayed less to individuals approaching at all distances within the 180° sector behind the shelter than did males without shelters. However, the shelters did not block their view, because males did respond to rear approaches. Instead, males with shelters had apparently redefined their territories and were much less prepared to defend the hind sector. The shelters serve effectively as sign posts providing this information. (This need not be their only function or even their only informative function; the shelters may, like weaverbird or stickleback nests, help the males to attract females. Such a function is compatible with the one Zucker discovered.)

Animals that come to special sites and use these only for displaying may alter the appearance of the sites. In most species of New World tropical forest birds called manakins, males gather into small groups called "leks" and display communally, apparently stimulating each other to more continuous performance than might be done by solitary individuals. Males of some manakins modify display sites within the general lek area, for example, by clearing an "arena" on the forest floor (Chapman 1938; Sick 1959, 1967; Snow 1963). In the more diffuse leks formed by the bowerbirds of New Guinea, each male actually builds an enormous, elaborate bower, stage, or tower where he displays and waits for females to be attracted. He may incorporate thousands of sticks and grassblades in his structure and ornament it with several kilograms of pebbles and snail shells, bringing fresh flowers to it daily. Gilliard (1956, 1963) found that in at least two genera the males of the species with the least elaborate plumages (these males' plumages may be indistinguishable from those of the drab brown females) build the most elaborate bowers. As elaboration of plumage is typical of lek-forming bird species, and extreme in the relatives of bowerbirds known as "birds of paradise," Gilliard saw this trend as a secondary evolutionary development that reduces males' vulnerability to predators through an evolutionary trade of elaborate badges for elaborate constructions. The constructions are separable from the birds, so that the males do not have to bear continuously the increased conspicuousness they obtain. Still useful for attracting females, constructions should be less disadvantageous than elaborate badges if predators are also attracted.

Humans also alter sites in ways that make display behavior more conspicuous by building stages, ornamented podiums, and the like. Some of these devices directly augment the dissemination of displays—loudspeakers and television, for example. Other devices carry information in the absence of the communicator, some by recording and others by mimicking displays and perhaps substituting for them: for example, the sculptures that Bali natives are said to use as threat signals guarding their houses and fields (Eibl-Eibesfeldt and Wickler 1968), or arrows and similar signposts on city streets. Other devices substitute for linguistic behavior rather than displays.

Scents and Scent-Marking

Badges and constructs are visible or sometimes tactual sources of information, but both have their chemical counterparts in pheromones. Pheromones, informative chemicals that animals make available either incidentally to metabolic processes or through display behavior, can provide several sorts of information about communicators. To begin with, they usually identify a communicator. They may also indicate the state of some of its physiological processes, for example by providing a measure of its estrogen or androgen levels. In both rhesus monkeys (Michael, Keverne, and Bonsall 1971) and humans (Michael, Bonsall, and Warner 1974) vaginal secretions of volatile aliphatic acids increase near the midpoint of the estrus or menstrual cycle; these secretions have been shown to be attractive to male monkeys, although comparable studies of human recipients have not yet provided a conclusive test of attractiveness.

Pheromones also usually provide information about an individual's propensity to perform certain kinds of acts such as seeking a mate or attacking. In some kinds of interactions, mammals may busily surround themselves with their own odors. For instance, Mykytowycz (1974) describes how two male wild rabbits threaten one another. Each rubs pheromones onto the ground from its chin glands, defecates, and erects its tail to expose the inguinal glands. The male who more persistently displays the white undersurface of its tail wins the encounter, the loser pressing its tail down tightly, apparently thus suppressing the dissemination of the inguinal pheromone.

Some pheromones as they decay indicate the recency with which the act of depositing them was performed. Rasa (1973) has shown that the anal gland pheromone of the African dwarf mongoose identifies the communicator, and that the extent of decay of the cheek gland pheromone which is deposited with it enables recipient individuals to assess the immediateness of the threat that the latter scent implies.

In many cases pheromonal molecules have evolved to have fading

times appropriate to the functions they support, which was mentioned in chapter 2 in discussing chemical communication. Alarm pheromones are relatively small molecules and diffuse faster than do the larger molecules of trail pheromones and the pheromones used for what E. O. Wilson (1971a) called "group identification" and "simple assembly" functions—that is, for bringing individuals together. Fading time has been further adjusted evolutionarily by differences among pheromones in the ratio of the number of their molecules customarily released to the response threshold in molecules per cubic centimeter of recipient sense organs (the Q/K ratio; see E. O. Wilson 1968, 1971a). Short fading times are achieved by lowering the emission rate Q or raising the threshold concentration K, or both.

Most pheromones are disseminated by diffusing into the air or water in which the communicator lives, many as they are released from the animal's body. Others remain on the animal as "surface pheromones" and are probably tasted during antennal inspections (E. O. Wilson 1971a) or allogrooming (Steiner 1974). These are effectively chemical badges. Some, like the "colony odors" of many social insects, may be environmental odors that are absorbed directly into an individual's cuticle, but others are special products which the communicator may spread over its body with elaborate "self-anointing" behavior (figure 9.5) or may rub onto another individual (reviewed by Steiner 1974). Dominant male flying phalangers mark the other members of their group (Ralls 1971), as do dominant male Columbian ground squirrels (Steiner 1974). Dominant male African dwarf mongooses do most of the "allomarking" in their groups, but are also marked by other members, particularly by the juveniles who seem to be very dependent on scent badges (Rasa 1973). Like this social mongoose, rabbits mark each other and their offspring with two different pheromones (Mykytowycz 1965), and an individual may also mark itself. In anointing its body with pheromones a rabbit may first mark a dung heap and then roll in the scented dung (Mykytowycz and Gambale 1969). Although more fastidious, humans also manufacture scents to wear as odorous badges.

If another individual is marked it is usually a mate or an offspring or shares some other bond with the marking individual. The individual who does the marking often dominates the other, but may not. For instance, a male mara *Dolichotis patagonum*, a large caviid rodent, will at times stand up on his hind legs and direct a powerful jet of urine onto the rump of his mate. This is particularly common when he has just driven a rival male away from the female. In part, what he is doing seems to be marking her from a distance—she does not usually tolerate his close presence. In response, however, the female projects a jet of urine back into the male's face, apparently not so much marking him as rejecting

Figure 9.5. Self-anointing with pheromones. This male arctic ground squirrel *Spermophilus undulatus* is rubbing its upper back, shoulders, and the back of its head against a rock on which it has just deposited a pheromone from the apocrine glands located at the corners of its mouth. (From Steiner 1974.)

his further approach (she directs urine toward any approaching individual; see Genest and Dubost 1974).

Other pheromones are deposited by the communicator in special marking behavior at sites, or on specific objects at important sites, and may attract recipients. A male bumblebee, for instance, marks special places along his flight routes with a mandibular gland pheromone, and females are attracted to these marked sites for mating (Free and Butler 1959). Males usually ignore females elsewhere, but interact with those who are in the vicinity of their marks (Free 1971). Honeybees mark rich foraging sites with secretions from their Nasanov gland, attracting other bees to gather nectar at the same flowers. Indeed, foraging honeybees use olfactory information both from the Nasanov scenting and from the odors of the flowers themselves (see von Frisch 1950; Ribbands 1954); they use it so extensively that some experimenters have proposed that the remarkable dance behavior described in chapters 2 and 6 has no social functions (for example Wenner 1967; D. L. Johnson 1967; Wenner, Wells, and Johnson 1969).

Although these authors have shown very convincingly that olfactory information alone can be sufficient to guide bees to good foraging sites in some circumstances, their postulate about the dances does not follow. It would be extremely surprising if the considerable information being made available by such elaborately stereotyped performances were not used by the individuals who so carefully attend the dancers, because it must be the fruits of their attentiveness that have provided the selecting

advantages leading to the evolution of the dancing performances. In fact, experiments have been done that amply support the claim that the followers act on the basis of information that can be obtained *only* from the dance (see, for example, von Frisch 1967, 1974; J. L. Gould et al. 1970). In one set of experiments, for instance, dance-followers within a hive were made to orient to an artificial light source while they attended dancers who were orienting to gravity (the dancers were rendered unresponsive to the light by painting over their ocelli). Misdirection resulted: the recruits did not go to the foraging sites used by the dancers but rather to the sites specified by their dance performances when interpreted with reference to light instead of gravity (J. L. Gould 1974, 1975).

In other species, pheromonal marking of sites provides information that may lead recipients to avoid the immediate area or, if they continue to be attracted, to become cautious in their behavior within the marked region. Such site-marking can indicate the existence of a territory, or at least of a space within which some rights and privileges are claimed by the marking animal. Bourlière (1955) reviewed literature indicating that some larger species of mammals such as bison mark territories by rubbing against specific kinds of objects such as a tree from which they may remove the bark (a visible modification) before annointing it with pheromones. Cats may urinate near their scratching trees, and some canids conspicuously scratch up surrounding earth after defecating and urinating (Kleiman and Eisenberg 1973). Other mammals deposit pheromones at regular but less conspicuous scenting places that are not visibly marked, or build up large heaps of dung marked with scent. Even a bird, the Japanese quail *Coturnix c. japonica,* marks its droppings with a special chemical product, although in its case one that foams and is visibly striking (scents being of little use to most birds); the significance of this foam-marking is not yet understood (Schleidt and Shalter 1972).

The function of site-marking by mammals is not necessarily to indicate that the communicator will behave territorially and exclude other individuals; individuals who mark frequently are usually able to win fights, and they will either dominate or exclude others encountered in the marked region (Ralls 1971). The territorial function that scent-marking sites may sometimes have appears to result more through predisposing invading animals to be alert and ready to withdraw on sighting a resident than through eliciting avoidance responses themselves. Invasions into marked areas are not unusual. In fact, in many species particular marking sites are sought out and used by several individuals; they may act as places where a general exchange of information is facilitated, helping these individuals to organize their social and perhaps spatial relationships (R. P. Johnson 1973).

Many pheromones have an immediate effect on a recipient's behavior, but others act more as "primers" (E. O. Wilson 1971a). That is, they influence a recipient's behavior by first altering its hormonal state. For instance, primer pheromones have been implicated in the control of sexual behavior in many mammals, including primates. In fact, a good case has been made for the possible pheromonal control of menstrual synchrony within small groups of girls living in dormitories of a women's college (McClintock 1971), although phenomena of this sort are better understood where they affect the estrus cycles of female mice (see reviews by Parkes and Bruce 1961; Bronson 1971). When females of laboratory mice (strains of *Mus musculus*) are caged in large groups some individuals become anestrus (the Lee-Boot effect); when kept in small groups, some develop pseudopregnancies through continued secretion by the corpus luteum in the absence of fertilization. Anestrus can also be induced by removing the females' olfactory bulbs. Adding a male or males to a group of females causes estrus cycling to become more regular, and the cycles are shorter than in the absence of males (the Whitten effect). In McClintock's study, college girls who saw men more than three times a week had menstrual cycles that closely approximated the national norm of about 28 days and were significantly shorter than those of girls who saw men less often. At least in laboratory mice, the Whitten effect is pheromonal: for example, Marsden and Bronson (1964) showed that male urine added to female nares produced the effect, and Whitten, Bronson, and Greenstein (1968) showed that female mice caged below or downwind from male mice developed synchronized estrus cycles, whereas females caged upwind did not and, in fact, tended to become anestrus.

Mice show one further effect of pheromones on the physiology of reproduction. Pregnancy is blocked if a newly impregnated female laboratory mouse is exposed to a male other than the one with whom she mated, and estrus usually resumes within a week (the Bruce effect; see reviews by Parkes and Bruce 1961; Bronson 1971). A similar effect can be obtained by exposing her to just the urine of an alien male. Female prairie voles *Microtus ochrogaster* have been shown to abort during most (at least the first 15 days) of their period of gestation, but whether pheromonal cues alone are sufficient throughout this period is not yet known because the new males introduced in the experiments were left in the females' cages (Stehn and Richmond 1975).

Urine figures in many of the above experiments, although it is not usually considered to be a pheromone. However, it can contain pheromones, including metabolites that are important chemical sources of information. These probably include sexual hormones and their breakdown products by which males may learn the estrus status of females

(Ewer 1968; Michael and Keverne 1968; Estes 1972), or females the identity and reproductive status of males (Parkes and Bruce 1961; Whitten 1966; Estes 1972); even males may learn about the current state of other males (for territorial wildebeest, see Estes 1969, 1972). Where urine carries such chemicals recipient animals usually approach urinating individuals to sniff, in many cases apparently employing the special vomeronasal accessory olfactory organ (see chapter 13; and Estes 1972). Other cases in which urine acts as a solution containing odoriferous chemicals include marking behavior (Ralls 1971) in which the information relates to the readiness of the communicator to dominate other individuals. Rottman and Snowdon (1972), however, have shown that for at least an alarm-inducing pheromone of laboratory mice which *can* be carried by urine, aerial dissemination is quite possible without the pheromone being in this solution.

Display behavior is not necessarily involved in the Lee-Boot, Whitten, or Bruce effects. It is certainly not part of the use of *Schreckstoff,* which generates alarm responses in some fish (von Frisch 1941; see also critique by Williams 1964); this pheromone is not released behaviorally, but only as it escapes from injured tissue. Schreckstoff and many other pheromones may not be in any way modified to be informative and, modified or not, are in many cases made available without the aid of any formalized behavior. These issues are very often overlooked in research on pheromones, which is very much response-oriented and which seeks primarily to identify or at least to detect the presence of chemicals that are produced by animals and that affect recipient behavior.

Tokens

The remarkable dance performed by a honeybee forager on returning to her hive provides information about the general location of a food source (see chapter 6) but fails to tell her recipients what kind of flowers are worth seeking there. To accomplish the latter, the dancer stops periodically and regurgitates from her honey stomach a sample drop of scented nectar collected from the flowers she has visited (von Frisch 1950), a token that supplies the necessary information. Many social insect species are known to engage in such "trophallaxis," the passing of food from one individual to another within the colony, and it is not always tokenism. E. O. Wilson (1971a) shows that in many species it both enables the sharing of food and may also inform each worker of the nutritional state of the colony as a whole, permitting individual decisions about when to forage and when to engage in other tasks. In such cases trophallaxis contributes to the phenomenon that Wilson has called "mass communication," discussed in chapter 12.

Making tokens available in display behavior seems to have developed in three different ways among diverse animals. First, as just illustrated, a token can be a source of information about a substance or object that has intrinsic relevance: a kind of flower at which to forage, or the amount of food available to a colony of social insects.

Second, in other cases it can be a token of the communicator's probable behavioral selections. The display act of providing the token may yield information about kinds of behavior that are not closely connected with the token itself. In many bird species a male may present his mate with samples of the kind of material used in building nests, although he may take no part in the actual construction of the nest and she may not use the sticks, plant down, and other substances that he offers. In different species a male may present the material when he is seeking to associate closely with his mate, or when he is about to attempt to copulate. In other display employing tokens one bird may provide another with food, an act usually called "courtship feeding" although it is not universally restricted either to courtship phases of behavior or even used only by members of a pair. Although often having only token nutritional function, courtship feeding is in some bird species necessary to augment a female's own ability to collect food while she is making eggs (for titmice, genus *Parus,* see Royama 1966). It may even take on, very indirectly, a function similar to that described for trophallaxis in some ant species by E. O. Wilson (1971a): the amount of food she obtains may enable a female to respond physiologically to the prevailing availability of food in the environment, helping to time her initiation of egg laying (see Perrins 1970 for a detailed discussion of the ecological functions implicated in this behavior).

Something like courtship feeding has also been described for some insects, predatory flies in *Empis* and related genera. In some species of this group a male captures insect prey and gives it to the female before attempting to copulate; she devours the prey while he mounts. In other species the male envelopes the prey in a frothy, baloonlike construction and gives the female a present that is effectively wrapped up. In yet others, the "present" is depleted—the male has sucked the prey dry himself, but the female need only see its remains in the wrappings. In a bizarre derivative of this wrapped token-giving behavior, a male of the fly species *Hilara sartor* captures no prey, but provides the wrappings without content (Kessel 1955; see also Portmann 1961, based on a survey by Meisenheimer). Kessel sees the evolutionary origins of this token-giving behavior in a very practical male tactic: in the genus *Tachydromia* males bear no gifts, and occasionally one is eaten by a female with whom he seeks to mate.

Third, tokens may sometimes be substituted for badges, not for environmental objects. The bowerbirds mentioned above not only build

bowers, they decorate them with fruit, flowers, pebbles, moss, and other colorful objects (Gilliard 1956, 1963). In at least one drab-colored species the male holds fruit in his bill while twisting his head back and forth before a female. The behavior may derive evolutionarily from courtship feeding, but in this case the berries are now token substitutes for a badge—for the brilliant crest feathers found in males of phylogenetically related species.

Tokens, constructions, badges, pheromones, and formal and informal acts of a communicator are together representatives of only one of the three origins for information sources outlined at the beginning of this chapter: sources originating with communicators. To understand more fully how individuals engage in the behavior of communicating we must eventually study all relevant sources of information, including those brought to an event by individuals who are recipients of information from communicators, and those from all other sources related to the event. Devising procedures with which such a large diversity of information sources can be studied systematically is a formidable task and one that cannot conceivably be handled by amassing all the data that might be relevant. Our choices will have to be guided by models of the process of communicating and by an understanding of the adaptive results generated in functionally different kinds of events.

10

Meanings and Functions

Most of us, on considering an event in which signals permit information to be shared, promptly become interested in the uses to which this information is put. We are oriented to considerations of responses and their functional implications. Perhaps this is because we depend so much on communication in our own lives—it is an abundantly used and literally pragmatic tool. On the other hand, the topic stressed in much of this book is more abstract: the information itself, the property of displays and things and events contextual to them upon which recipients base their responses.

Information is central to the process of communicating, but analyzing it alone provides us with a grasp of neither the meanings nor the functions of displays. Although meanings and functions depend on information, their study requires concepts and procedures specific to the pragmatic level of analysis. Only when we begin to deal with meaning, with the responses to signals, can we begin to understand the process of communicating and its potential as a tool of interactional behavior. And only when we have also considered function (defined by Hinde 1956c as the "biologically significant" consequences of acts) can we understand why particular meanings are important results of sharing information— that is, how these responses enable the evolution of interdependence between participants engaged in the behavior of communicating.

Because communication does not occur unless there is a response, many ethologists appear to take the view that responses are *the* crucial components of communicating (for example, Marler 1961a, 1967a; Altmann 1965; Struhsaker 1967). This view has tended to give ethology a preoccupation almost solely with responses and their functional implications, with a consequent lack of research on the information contents of displays. The emphasis is regrettable for at least two reasons. First,

communicating has no *single* crucial component. Each of its three basic units—a communicator, a signal (formalized or not) with its information content, and a recipient—is essential. Communicating does not occur in any event unless information is both made available and received. Second, the study of responses and functions is very complicated, and the topics are enormously diverse. Despite insightful field observations, clever experiments, and many attempts to classify, the study of the responses and functions that depend on displays is still fragmented.

Concepts of Meaning

What is meant by "meaning"? This is a longstanding question; the term is loosely and widely used. Cherry (1966) called it a "harlot among words." He made, however, a very practical suggestion: observers can infer what some stimulus means to a recipient at a given time by seeing what response he makes to it. On the basis of being informed, recipients select responses; their choices are the best evidence we have as to what the information meant to them in the incident.

Because ethologists study behavior, they can take the position that a display has meaning to a recipient in some event to the extent that they see him base his selection of a response on the display and on information from other sources contextual to it (W. John Smith 1965, 1968). But this covers only one of two main ways in which we use the word "meaning." The other way of comparable importance in common speech is to say: "what I mean is. . . ." We are accustomed to signals having meaning for communicators as well as for recipients; a little reflection will show that this meaning, which involves the communicator's intentions, very often differs from the information that a signal actually makes available—that is, from its message. That is why we may have to explain what we mean from time to time.

Cherry's solution to this is consistent: for a communicator, the meaning of a display in a particular event is the response he expects or intends to elicit by using it (Cherry 1966), even though the information provided by the display may sometimes fail to have the foreseen effect. Many theorists working with human communication attach enough importance to the communicator's reasons for signaling to argue that communication does not occur without such intention (for example, Ekman and Friesen 1969a; Goffman 1969; Lyons 1972—but not Kendon, Bateson, Birdwhistell, or Scheflen). A criterion of communicator intent is too restrictive to be practical for an ethological definition of communication, but the issue must be examined.

Do communicators intend to elicit responses when they perform

displays? Certainly humans often have such intentions when communicating, but sometimes we seem not to ("I couldn't help smiling"; "I tried not to blush"). Sometimes our intentions lead us to misuse displays: for example, we mimic the use of a smile to be misleading. In so doing, of course, we run the risk of our manipulative behavior being discovered when our subsequent acts reveal our intentions, and this can endanger our ability to have stable and productive social relationships.

To what extent do animals of species other than humans foresee possible results of signaling and communicate intentionally? Captive chimpanzees can be induced to communicate the location of hidden food sources or aversive objects to each other if an experimenter shows one individual the hiding site, then releases it in the area with its companions (Menzel 1971). As mentioned in chapter 9, the leaders appear to direct the others not with special display behavior, but primarily by the way in which they set out to go to a site: their pace, style of locomotion, and glancing patterns. All the chimpanzees in Menzel's group know the game, so to speak, having been through the procedure many times. Intention becomes clearly evident in events when no other chimpanzee spontaneously follows the leader, especially if the leader is one of the smaller chimpanzees in the group because these individuals are unwilling to move out on their own. The leader will then glance from one to another of his companions, beckon with his hand or his head, walk backward toward the goal while orienting toward the group, and if all else fails try to pull or drag one of the others with him. Intention in these cases can be assessed from the animals' attempts to manipulate each other's behavior in terms of their own known goals, the communicator continuously adjusting its performance to the kinds of responses it elicits. As anyone who has kept dogs or cats as pets is aware, this sort of behavior is not restricted to primates. Nonetheless, scientists and philosophers have long tended to sweep such evidence aside so that man may be seen as unique. Because of this intellectual climate, we know little about the phylogenetic roots of intentionality, though the story may be fascinating (Griffin in press).

If animals do attempt to maximize the effectiveness with which their display behavior serves their goals, however, it is not usually obvious in the field. Laboratory experiments using reinforcement techniques have shown that at least the frequency with which displays are used can be altered when individuals learn the characteristics of particular circumstances: ducklings, for instance, can learn to emit distress calls to elicit a brief presentation of a stimulus to which they have been imprinted (Hoffman et al. 1966). But in natural circumstances it often appears as if most animals perform their displays whenever an appropriate set of circumstances arises, whether or not the displays appear to be necessary

or effective—as if the displays were more or less automatic responses. Further, in natural circumstances each participant does not usually appear to perceive itself as the target of another's signaling, as conversing humans do, although it might be very difficult for an observer to demonstrate this (see Hinde 1974:89–90).

One disturbing possibility is that nonhuman animals might sometimes use displays to manipulate their recipients, intentionally misleading them. If this occurs (apart from the special interspecific cases discussed in chapters 12 and 13, which need not be intentional), it is at least possible that ethological observers will detect it through contradictory patterns in the behavior that correlates with displaying. But lying is usually socially risky, and hard to do convincingly—or hard at least for most people, as Ekman and Friesen (1969b) have demonstrated in their studies of nonverbal "leakage" and "deception clues." Often many kinds of signals are performed more or less simultaneously, and it is difficult, when lying with some signals, to keep performance of all the others completely in line (human liars often "give the game away" through inappropriate movements of their legs and feet, even when their facial expressions agree with their verbal lies). Ethologists have not yet caught nonhuman animals in intentional acts of lying with display behavior.

The possibility that nonhuman animals might sometimes use displays metaphorically is less disturbing and much more exciting. In a situation in which no display in the repertoire has a set of referents that fits, does an animal ever try to communicate by pressing into service a display whose referents have something in common with the messages he needs to convey? For the attempt to be successful, a recipient individual would have to be able to recognize that the circumstance is one in which the established referents of the display apply only in part. Such a recipient would respond on the basis of only some of the information made available by the display—that portion which fits appropriately with the pertinent information available from sources contextual to the display. Humans do this regularly, and it can be a powerful procedure when the recipient grasps the relevance of the metaphor. It is at the same time a risky procedure because the common knowledge of a display's referents held by both communicator and recipient must be partially set aside, both individuals making the same unprogrammed transformation.

In human use, the communicator often intends to foster a particular meaning by using a metaphor. Although it is not clear that the use of metaphor need be intentional, the capacity to try, consciously, to communicate might readily encourage increased dependence on the device and may increase the likelihood that a communicator will monitor his recipient more closely for indications that the metaphor did its job. Thus the issue of communicator intent may be unusually important in con-

sidering this aspect of communicating. Unfortunately, we still lack adequate definitions of intent and operations for measuring it.

Intentional use of displays, which would show that we must develop ways to assess meanings to communicators, is thus difficult to establish as a form of goal-directed behavior among nonhuman animals. Hinde (1974:89–90) has proposed that the question of its presence or absence is misleading because there is a gradation in the complexity of goal-directed behavior and hence perhaps of intentionality. For the moment, this is where the matter rests. We need not assume that nonhuman animals are always consciously aware of their goals—even much of our own behavior is automatic—but neither need we assume that they never are. This point is not addressed here. Goal-directedness is seen simply as being inherent in "adaptive" behavior—behavior with consequences that are, in an evolutionary sense, useful.

(Philosophers and linguists may see a somewhat related problem on the recipients' side of the issue when displays are described as providing information or, in particular, as informing recipients. This description should not be taken as implying that the method by which displays function is necessarily propositional, that they offer information simply for consideration by recipients as if a communicator were proposing courses of action in a reasoned discussion or debate. No restrictions on how recipients process information from displays or other sources are implied, and certainly displays can function to bluff, coerce, appease, and the like. Whether they ever operate simply as propositions in the sense we use that term in discussing human behavior is a question for research on cognitive processes and not a problem for this book.)

If the use of displays is, on the average, adaptive, then we need not be overly concerned with the presence or absence of internal mechanisms such as "intentionality." If the intentions of nonhuman communicators ever lead them to attempt deception, study of the behavior correlated with the displaying should reveal it. Until we find such cases, the only meaning with which comparative ethologists must always contend is that evidenced by a recipient, for responses by recipients generate the functions of displays.

This narrows the scope of investigation, but not to a simple task. At least three major kinds of difficulties arise in studying the meanings of displays to recipients that make the job formidable indeed.

First, a display need not elicit an immediate, overt response from a recipient individual. In fact, as anyone who has ever studied displaying in, say, songbirds can attest, most displays appear to have no effect at all in most events seen by an observer. A bird may sing repeatedly and perhaps utter a variety of other vocalizations, and other members of its species in the area will continue unhurriedly with their business. How-

ever, the behavioral evidence available in such an incident should not be interpreted as indicating that they are not responding. Some may not be, of course; they may have excluded perception of the display through sensory filtering. For other recipients, the information supplied by the displays may have a latent effect by providing them with a changed perspective, setting the stage for responses or the inhibition of responses to later stimuli (Thorpe 1967a; J. M. Cullen 1972). The information may have a "primer" effect (Wilson and Bossert 1963; E. O. Wilson 1971a:234) by altering the long-term responsiveness of recipients through changes in their physiological states (see, for example, Lehrman 1965; Hinde 1965). Huxley postulated such a mode of functioning in 1923, calling it "stimulative" as opposed to "informative." Although his terms are somewhat ambiguous he clearly saw the need for displays to function over shorter and longer time-spans, both with respect to regulating current activities and to initiating and regulating underlying "psycho-neural" states.

Some displays are repeated at more or less regular intervals or predictable times, apparently having a continual effect on individuals that have come to expect them ("tonic communication," Schleidt 1973): prominent examples are the songs of mated territorial birds and the howls of howler monkeys. Other displays may be repeated in sustained bouts (such as in fighting fish *Betta splendens*) or at quite irregular intervals (for example by male chimpanzees) and produce cumulative effects on their recipients, perhaps leading to changes in a group's social structure (Simpson 1968, 1973a).

The production of cumulative effects can be complicated by varous factors. A display can elicit from a recipient both very short-term responsiveness that decays in a few seconds, and somewhat longer responsiveness that decays over a period of minutes (reviewed by Heiligenberg 1974). Successive repetitions of the display within the longer period can cause recipient responsiveness to build up in a stepwise fashion, whereas successive repetitions at still longer intervals do not have this effect. Further, during the long intervals over which repeated displaying produces effects, more and more diverse sources of information can contribute to the effects. It becomes difficult to determine what sources do contribute, particularly because some of the additional stimuli will be formalized and readily observed, but others may be so inconspicuously specialized that they escape a human observer's detection (Simpson 1973a; Hinde 1974).

Second, there may be more than one response when a display is used. Thus, even if an observer does detect an immediate response to the use of a display he cannot conclude that it is the only response. Most signals that are sufficiently significant to elicit prompt responses must also alter

the subsequent responsiveness of their recipients (Cherry 1966:250; Marler 1967a), setting in motion sequences of behavior or at least becoming important as contextual sources of information to be used in evaluating subsequent stimuli.

Third, although a "response" is the behavior a recipient performs subsequent to a display, what the recipient is responding to is not necessarily clear. An observer often cannot distinguish between the effects of the display on one hand and those of all other potentially relevant sources of information on the other. The responding animal somehow assesses the relationship between information received from different sources and makes decisions based more on some sources than on others. At the extreme, an observer has to ask if the same responses to a situation would have occurred even in the absence of a display, if all other things were equal. Opportunities to make such comparisons, at least in any detail, are improbable. Nature rarely repeats exactly; the rough approximations to identical situations that it provides may sometimes be good enough for the test, but only if they recur frequently enough to permit large samples. The temptation is thus to reduce natural complexity by eliminating seemingly peripheral variables in the laboratory. This procedure grossly changes the situation and may usually have pervasive disruptive effects on the social behavior of at least vertebrate animals (see Klopfer and Hatch 1968). The difficulty is further compounded by the fact that attempts to control or even to catalog available sources of information should include variables that have affected the histories of each individual animal. Relevant historical variables may not only be very numerous, they are usually largely inaccessible to an observer.

Ethological Concepts of Function

Displays achieve their functions indirectly. That is, their functions depend on their meanings to recipients—on responses made to them by individuals other than their performers. In the absence of any response, a display does nothing for either communicator or recipient. (Or it does nothing via social communication—displaying may be in some ways emotionally satisfying, like whistling in the dark; but that issue belongs at the individual, not the interactional, level of analysis. Any emotional consequences of displaying are probably secondary derivatives based on the use of displays as signals.)

When a recipient does respond, various consequences follow. Some provide the adaptive advantages of the display; others are perhaps only advantageous by-products or are neutral or disadvantageous. It is in the first class, the consequences that are on the average adaptive (that is,

those used by natural selection to keep a display in the behavioral repertoire of a population of animals) that evolutionary ethologists recognize the display's functions (Hinde 1969, 1975a). That is, a display's functions are those consequences that enhance the fitness of individual animals or the bearers of particular genotypes to survive long enough and well enough to be effective in passing their genes on.

Increased fitness is an ultimate advantage, a long-term result that may not be achieved in every incident. Some risks and costs can be entailed in interacting, but they must balance out to be less than the benefits. Further, the benefits obtained by individuals in each incident need not be equal or symmetrical, as long as interacting eventually contributes more to the fitness of each than would a more solitary life-style (barring those evolutionarily unstable cases in which the value of displaying lies in being misleading; see chapters 12 and 13). Dissimilar and unequal advantages result, for instance, when an animal submits to the threats of another and obtains a smaller share of the immediately available resources. Even the relatively disadvantaged animal may find its responses are more to its benefit than either fighting for its share or leaving the group to fend for itself.

The ultimate function of increasing fitness is very difficult to measure and test; usually an ethologist must simply retreat to the assumption that elaborate patterns would not evolve unless they did increase fitness. The more immediate advantages of performing and responding to displays can sometimes be dealt with more easily, however, and these shed light on how fitness is augmented. For instance, if an individual responds to an "alarm call" from an associate by seeking cover, his response may save him from being killed by a predator. The adaptive value of an appropriate response for the recipient in such a case is obvious. Yet even in so simple an event, the adaptive function for the communicator is much less obvious. Instead of helping him to escape, the act of displaying may make him temporarily more vulnerable.

Analyzing even this simple example is complex. First, assume that the communicator giving the alarm call or similar display is an adult. The display of a young juvenile animal may elicit less extreme responses, limited perhaps to orienting and investigating behavior; it may even be ignored. Suppose the communicator is a songbird, and performs the display while in a stable flock engaged in foraging behavior. The display provides the information that there is a very high probability that the communicator is escaping. (The displays traditionally termed "alarm calls" are very different from the "screams" given by animals after they are captured.) In response, most recipients will immediately flee or freeze, perhaps depending on how close they are to cover.

The advantage each recipient obtains is an increased likelihood of

escaping the predator. The advantages the communicator obtains (at some risk) are (1) the preservation of the flock, the members of which may act in the future to provide him with similar warnings, (2) the protection of his investment in any members of the flock who may be carrying his genes (his offspring or other kin), and sometimes (3) the confusing and distracting of the predator, if the response of the flock members is to scatter explosively (see, for example, Charnov and Krebs 1975). These advantages do not all apply in every flock, of course. The first works best with relatively small, closed flocks; in larger flocks advantages may come instead through limiting the predator's success and thus preventing it from specializing on the communicator's species or locality (Trivers 1971); the protection of kin works only in flocks in which competition over resources has not favored kin dispersal (W. D. Hamilton 1964).

Now suppose a different circumstance: one of two individuals who are fighting performs the same display, with the same message content. Its opponent may respond by chasing it, or by simply ceasing to fight and turning its attention to the victor's spoils: space, access to food, or some other resource. In at least the latter case, the function is to save the recipient the physical risk and energy expenditure of pressing an unnecessary attack; the function for the communicator is to decrease the likelihood of being attacked further.

Finally, suppose a third circumstance. The communicator is a bird that has sighted a predator near its nest and nestlings. In response to the display the nestlings will usually huddle silently; if they are old enough they may leave the nest and scatter. The bird's mate may initially flee or freeze, but is then likely to approach the predator and perhaps mob (pester) it—like the communicator itself. Neighboring small birds of other species may also respond in the same ways. The function of the responses of the various recipients is, for the communicator, an increased likelihood that its offspring will survive the predator's approach. The function for the nestlings is survival, and the function for the communicator's mate is first to make her aware of (and hence safer from) the predator, and second to enable her to try to save her nestlings. The consequences for birds of other species that respond are again personal survival and perhaps increased safety for their own nearby offspring, if the predator can be driven out of the area.

In the first and third examples the predator may also be a recipient of the display, although it is effectively "tapping into" the system and providing different selection pressures on its evolution. The predator may respond by using whatever directional information it can get from the display to help it locate the communicator, with the possible consequence of obtaining the latter as prey. Or it may simply pass through the

area, saving time and energy that would be lost trying to catch prey that can no longer be taken by surprise. The consequences *for the predator* of responding to the display are not among its functions—it does not evolve to make life easier for predators.

Note that the functions for the communicator and for the recipients in this set of examples can be somewhat different, and that functions differ among different classes of recipient. Greater differences than these examples show are common, even for other displays with rather precise messages—for example, the postural display with which a rhesus monkey solicits mounting. Even as performed by a single class of communicators, such as adult females, the display elicits different responses in different circumstances, with diverse functions. A male might respond by mounting and copulating. The female's infant may respond by climbing on her back to be carried (see Hansen 1966), perhaps to a new foraging site, or perhaps when the troop is fleeing from a predator. Yet another response is for an adult individual of either sex to mount, and not for copulation (see Hanby 1974). Soliciting mounting in nonsexual interactions between adults sometimes appears to reaffirm the status relationships of the two participants. At other times it may be simply a means of forestalling or terminating agonistic behavior permitting the animals to meet, to rest together or allogroom, or to settle down after excitement. (The effectiveness of a presenting display that solicits mounting in reducing the probability of being attacked can be seen in some data of Chalmers 1968, for another species of monkey, the black mangabey *Cercocebus albigena*. He reports that in 53 cases in which one mangabey approached an individual that was subordinate to it and the latter presented to be mounted, the former never attacked, whereas it did attack in 9 of 30 cases in which the latter did not present.)

Few events in which vertebrate animals display are as simple and few kinds of displays are performed in as few kinds of situations as are "alarm calls." Few displays have messages as narrowly predictive of the communicator's behavior as these examples and as displays by which an individual indicates that it is receptive to being mounted. Yet recipients find that information from sources contextual to the displays is crucial for their selection of responses even to alarm calls and invitations to mount; they respond in diverse ways, and the functions generated are diverse.

Very much more diverse kinds of functions are generated by broader messages such as locomotion or interaction. Because these messages can be relevant in many more kinds of circumstances, they therefore combine with many more kinds of contextual information sources. Further, unlike a message of escape behavior with a high probability, a more broadly predictive behavioral selection message must always be con-

sidered as it combines with other behavioral selection messages in the many displays that encode it. Functions depend on the whole set of messages in a display: a display combining the messages of locomotion and association will have fewer functions than a display combining locomotion and general set messages, and its functions will be somewhat different.

A further problem in assessing the functions to which a display can contribute derives from what Hinde (1975a) calls the "ramifying consequences" of displaying. Responses to a single use or period of use of the display can extend over prolonged intervals, generating consequences that become interlinked with the consequences of other acts. Suppose, for instance, that an unmated female bird approaches a singing, unmated male. The immediate function of his song is to attract her. It probably also prepares her to be receptive to his attempts to establish a pair bond, which may require a few hours or days. Ultimately, the song and her response function to increase the fitness of both individuals by facilitating breeding. A complete description of "the" function of song must encompass this whole nested set of adaptive consequences, as well as adaptive consequences of responses by other recipients such as neighboring males. However, the farther analysis proceeds along such a chain of consequences, the greater become the contributions of other events than the display. Often, the practical thing to do is to consider primarily the relatively immediate consequences.

To summarize, ethologists interpret the functions of display behavior in terms of the adaptive advantages it confers on the individuals that perform it and respond to it. But functions differ for different users in any event, differ in different kinds of events, and their extensions and ramifications over short and long periods must be taken into account. No display that is used in more than one kind of circumstance has a single function, any more than it has a single meaning. Rather, it figures in the generation of many adaptive and inadaptive consequences, and functions insofar as it supports the former.

Procedures

Despite the difficulties in dealing with delayed responses and in sorting out the effects of numerous sources of information, most of what has been postulated about the meanings and functions of displays has been based on field observations of naturally occurring events. No display occurs randomly, and the kinds of circumstances in which it is found usually suggest to an observer the kinds of responses to it that would be adaptive. Often the apparent outcome of such a circumstance, or at least some changes that occur, support the observer's view.

For instance, a male bird may move about almost continuously on his territory, taking conspicuous perches and singing. His song is loud, and enables him to be found readily, at least by an observer. A key feature of this kind of circumstance is usually that the male is unmated; sooner or later he is usually joined by a female, and they form a pair bond. If she arrives ready to be courted in response to the song, then attracting females with whom to mate is a function of this singing. This might lead us to predict that males would sing much less frequently (if at all) once mated, but might resume frequent singing if they lost their mates— and this does describe the typical pattern of song use in many bird species. However, song does not cease when a pair bond is formed, but remains quite predictable in some circumstances that have little to do with interactions between mates, which suggests further functions. Other classes of recipients appear to respond. When a male sings he may be accosted, avoided, or engaged in countersinging by a neighboring male (see, for example, observations of Orians and Christman 1968 on red-winged blackbirds), suggesting that song has functions in territorial as well as pair-bonding relationships. The details of timing of singing with respect to interactions before, after, and even during some kinds of territorial boundary encounters strongly support this interpretation.

Quantified Observations

When situations are sufficiently simple and discretely defined, quantified observations can help to reveal differences between encounters with and without particular displays. For example, observing a winter flock of European chickadees called blue tits at a feeder, Stokes (1962) counted the number of times individual communicators adopted particular postures (including feather postures) in the presence of one other individual; tallied their subsequent behavior as attack, escape, or staying; and similarly tallied the behavior of the recipient individual. Some of his most interesting results are shown in table 10.1.

In this kind of circumstance the first two displays shown in the table correlate with a greatly increased probability that the communicator will escape, the next three with an increased probability that it will attack, and the sixth perhaps with a slightly increased probability of staying. (Because these analyses are limited to a single kind of circumstance, they do not yield an understanding of messages although the form of observations is appropriate. Had behavior also been tallied in other kinds of circumstances, attack, escape, and staying might have been found to be inadequate descriptions of the ranges of behavior with which some of the displays correlate.)

It is also apparent that recipient individuals behaved appropriately,

Table 10.1. Communicator behavior and responses correlated with blue tit display behavior.

Display behavior	Condition	Probability of subsequent communicator behavior			Probability of recipient behavior after the display		
		Attack	Escape	Stay	Attack	Escape	Stay
Crest erect	+	0.00	0.90	0.10	0.21	0.03	0.76
	−	.22	.26	.52	.18	.37	.45
Body fluffed	+	.00	.91	.09	.20	.05	.75
	−	.20	.31	.49	.19	.34	.47
Nape erect	+	.38	.15	.47	.05	.49	.46
	−	.15	.42	.43	.21	.27	.52
Wings raised	+	.33	.19	.48	.13	.48	.39
	−	.15	.42	.43	.21	.27	.52
Tail fanned	+	.39	.15	.46	.16	.51	.33
	−	.15	.41	.44	.20	.27	.53
Beak open	+	.17	.31	.52	.10	.20	.70
	−	.17	.40	.43	.21	.30	.49

The study is based on observations of acts performed (+) or not (−) by individual blue tits during their visits to a bird feeder. The interval for scoring presence or absence was defined by a visit, which usually lasted several seconds. In this time, the displaying individual might threaten for up to 5 seconds and perform more than one display. Recipient behavior was scored only for those instances in which the communicator did not actually attack, although those instances in which the communicator fled were not deleted from the calculations. (From Stokes 1962.)

which suggests that they were responding to the displays or to other sources providing similar information, or both. Although the effect is clearest for the first two displays this cannot be safely inferred to imply a response to them, because the events in which the communicator fled after displaying (an act that might well increase the likelihood of a recipient staying) were considered along with those in which it remained after displaying. However, events in which the communicator attacked were left out, so that the results for the next three displays quite strongly implicate those in eliciting the virtually doubled probability of the recipient fleeing, and the final beak open display seems to reduce the likelihood of any agonistic response.

Comparable results were obtained by Andersson (ms.) in a study of dyadic encounters between great skuas in the territories of a prebreeding social group or "club." He assayed the behavior of the communicators that could be predicted from their formalized acts and, like Stokes, found that recipients behaved appropriately: for example they stayed if the communicator's display indicated a high probability of escape be-

havior, and stayed even in those relatively few cases in which a communicator, having performed such a display, also stayed. Andersson also examined two kinds of information sources that occurred contextually to the displays to see if they affected a recipient's responses: the rate at which a communicator approached, and the communicator's ownership or lack of ownership of the territory that was the site of the encounter. His results reveal that whether a communicator approached and stopped, walked forward, or ran toward a recipient had a great influence on the latter's behavior, and that territorial ownership played some role.

In comparing communicator and recipient behavior, Stokes treated whole incidents of visiting the bird feeder as if each were a "single, instantaneous event"; he thus largely ignored sequences within each visit. Some studies have examined behavioral sequences and tested whether the transition frequencies between two acts (usually the first by a communicator, the second by a recipient) differed from random expectation according to various models (see, for example, Altmann 1965; Hazlett and Bossert 1965; Dingle 1969, 1972; Maurus and Pruscha 1973). Although interesting, the findings of these procedures have as yet been very simple and limited by the enormous number of observations required for statistical testing (see chapter 8). The techniques are further limited in that they entail some rather arbitrary classification of behavioral units and because the correlations they demonstrate are not necessarily cause and effect; further, they can deal only with immediately evident behavior, not with delayed effects (see J. M. Cullen 1972; Hinde 1974:99–100). Yet these approaches do have the advantage of demonstrating and testing sequential correlations by quantified, very objectively described means.

Natural Experiments

More can be gained from observation of natural circumstances when some important variables happen either to be eliminated or to remain constant. These moments provide special insights, giving an observer a situation comparable to that in which an experimenter has exercised control, yet without the disadvantages of interfering and rendering the situation less natural.

Tinbergen (1959b) dubbed such events "natural experiments," and offered examples of three kinds that recur reasonably frequently. In the first, a recipient alters its behavior on hearing a vocalization when it cannot see either the communicator or the stimuli to which the communicator is responding. For instance, Tinbergen reports an instance in which a pair of foraging black-headed gulls became visually separated

on opposite sides of a sandbank. In the next 70 minutes the male responded to approaching gulls sixteen times by long calling in the oblique display posture. In thirteen of the cases his mate immediately ran to him (in the other three she was already engaged in squabbles with yet other gulls). She could only have been approaching in response to his call and a knowledge of the general circumstance.

Tinbergen's second example involves an interaction in which the recipient ceases an activity when a given display is performed, but immediately begins it again when the display stops. His student E. Cullen very often saw pair-forming male kittiwakes peck at females who had joined them on their nesting territories, each a narrow ledge with little room for movement. An attacked female would perform a facing away display and the male would immediately stop attacking. When she would next look back at him he would immediately strike again, and she might flee. Though facing away was often seen to stop or reduce a male's attack, it never caused attack to begin.

The third example is a cirumstance in which an act is performed repeatedly without causing recipients to alter their behavior, and then a largely similar act is performed with a display and elicits recipient changes. A male black-headed gull gathering nest material on his territory repeatedly approached five individuals who were resting just outside the boundary. They simply continued to rest while he was preoccupied. Afterward he preened near his nest, then again approached them, this time in an aggressive upright display posture. When he was within 3 meters they all adopted anxiety uprights and walked away. A comparable type of natural experiment occurs when there are responses to an act both when a display does not accompany it and when one does, but the responses are markedly different. Within pairs of common grackles *Quiscalus quiscula,* for instance, close approaches by a male lead to immediate withdrawal by the female unless he performs the head-down display, in which case the female may repel him or permit him to mount, but never flees (Wiley, in press 1).

Other examples of natural experiments could be added. For instance, when a communicator performs a display in the presence of two or more recipients and they do not all respond in the same way, evidence is available about both the range of effects of the display and the influences of different contextual factors. Some of the latter are readily classified if the recipients differ in age, sex, relative status, or other classes, or if the communicator orients more toward one recipient than another, and so forth.

Natural experiments can be sought. They are more likely in some circumstances than in others, and when the characteristics of a given kind of "experiment" are known an observer can choose a location to

increase the number he or she can hope to see. Circumstances can also be altered in ways that effectively duplicate natural events: for example, a female bird can be caught and removed from a territory, requiring her mate to repeat advertising and pair-forming procedures with another female, or a nest can be destroyed so that a pair repeats a phase of breeding behavior it had already completed. Both these alterations may simply accomplish what a predator might do; other manipulations are more benign. To determine how many honeybees would arrive at feeding sites with different characteristics after dancers had reported, von Frisch (1950, 1967) simplified his task by providing controlled, artificial sources of food.

Of course, as more control is exerted by the observer, the experiment ceases to be natural. More to the point, it soon ceases to duplicate all the characteristics of naturally occurring events. The experimenter begins to deprive his subjects of sources of information that they expect to have and adds sources that do not fit any of their expectations in detail. Interpreting the results of such experiments must consider these effects. Yet, for specific purposes, even confinement to laboratory environments can be worthwhile: a great deal has been learned about the relatively long-term effects of displays on physiological states in the laboratory, for instance in the very detailed studies of ring dove breeding behavior by Lehrman and his group (see, for example, Lehrman 1965); characteristics of the decay of responsiveness to single or repeated display stimuli have been studied (reviewed by Heiligenberg 1974); and the differences between individuals in the displaying tactics they employ when locked into aggressive encounters have been related to the outcomes of the encounters (for example, by Simpson 1968), to cite just a few examples. Our understanding of responses and functions is at present too limited to qualitative description (Andrew 1969). It requires more testing, even with the inevitable difficulties that accrue.

Experimentally Altered Participants

A relatively straightforward way of altering natural events is to remove some individuals from a group, modify their appearances or odors so that they appear to belong to another class, and replace them or put them into a new group. Although such alterations have sometimes been done in nature with small numbers of animals, Marler (1955a) has conducted more extensive experiments in the laboratory, dying the breast feathers of female chaffinches to match the reddish color of males. The results were striking. Normally colored females tended to avoid red-breasted females much as they avoided males, and 38 of 44 altered individuals came to be dominant among members of their own sex. They

even had some success in agonistic encounters with male chaffinches, which usually dominate females, as long as they had not learned to be subordinate to the males before their color was changed. Thus a function of the red breast badge in this species, at least in agonistic encounters, is sexual identification and its concomitant influence on the responses that underlie status relationships.

Artificial Participants

Because modifying the appearance of live animals can be difficult, many experimenters have instead used stuffed animals or more artificial models, as Lack did, for instance, in studying responses to the red breast feathers of European robins (discussed in chapter 9). That models do not behave is at once an advantage and a disadvantage: for example they can be presented in preselected postures, but in most cases must maintain the posture whatever a recipient does. This may not matter in eliciting initial responses, particularly if the model is in posture that would usually be held for a while—such as the precopulation crouch of a female bird. Some models have been made to perform simple or stereotyped movements: Hunsaker (1962), for example, experimented with head-bobbing patterns of lizard models and showed different preferences among recipients of different species. Tinbergen (1953a) learned much about responses to threat displays of stickleback fish by manipulating models and even manipulating live fish constrained within glass tubes which cannot be seen in the water. Wooton (1971), however, found that models are not treated by sticklebacks as full equivalents to live fish. Even a moving model cannot really be made to *interact* appropriately by an experimenter lacking a complete knowledge of interaction processes. Attempts to make models responsive never approach anything like the subtlety shown in most interactions of birds and mammals. (Mirrors, sometimes used instead of models, produce responses that, far from resembling interaction, must baffle most animals when the opponent responds instantaneously by mimicking their every move.) Because of such problems, the use of models to elicit responses has had limited application and the results have often been disappointing (see comments by Tinbergen 1959b and J. M. Cullen 1972 for examples).

The experimental use of models has nonetheless been elaborated in some directions that are potentially very interesting. For instance, Stout and Brass (1969) simultaneously presented recipient gulls in their nesting territories with two different models of territorial intruders and tallied which model was attacked or attacked first. Stuffed gulls or rough wooden outlines of gulls to which real heads or wings were sometimes

Figure 10.1. Models used in experiments eliciting responses from gulls. Responses of glaucous-winged gulls *Larus glaucescens* were examined by presenting pairs of models within the territories of nesting birds. Models were constructed of wooden blocks with or without parts of stuffed gulls attached, and were set up in various display postures to test the relative effects of different body parts and postures in eliciting attack. (From Stout and Brass 1969.)

added were presented in different postures (figure 10.1). Using a real head enhanced the wooden model more than using real wings. Gulls attacked those models in aggressive upright display postures more readily than those in oblique or choking display postures. In natural circumstances a silent, immobile, territorial intruder would be more likely to be in an upright than any other posture, and the recipients may have been most willing to attack models in this posture primarily because of the fit to their expectations. As the authors point out, however, a communicator in the aggressive upright is also less likely to attack than are communicators in the other postures. The full relevance of the results is not easily assessed, although the reliable response differences produced by the technique makes it quite promising.

Models have also been used specifically because they do not interact naturally: in some experiments their limitations have been turned to an

advantage. Simpson (1968) did this in investigating the prolonged aggressive encounters of Siamese fighting fish. Two fighting fish opponents of the same sex tend to alternate with each other in the two orientations used in their displaying: when one turns to face the other will often turn broadside. By using a one-way mirror to separate opponents in some experiments, Simpson had discovered that some individuals escalated certain measures of their responses (such as the frequency of tail-beating) when watching another fish persist in a single orientation (to the latter fish the partition was not a window but a mirror, and it was watching its own reflection). Simpson gained further control over this unusual circumstance by replacing the mirror-watching individual with a model fish whose orientation could be controlled at will. He found that different individual fish respond to overly sustained orientations of their "opponents" by altering the frequency of their tail-beating in one of two ways: some individuals tail-beat more to a persistently facing model than to a persistently broadside one, and others the reverse. Why the individuals in his population of fish should divide into two groups is not yet fully evident from this work, but such tactical dichotomies may be widespread in populations. They could, for instance, reflect a difference between individuals who will attempt to dominate an encounter and those who will more cautiously seek to defer.

Playback

Vocal displays can also be presented experimentally, in this case by playing back tape-recorded samples. The technique has been used frequently to test whether birds can discriminate the song (or sometimes just an unstudied sample of the vocal repertoire) of their species from that of closely related species (see, for example, Dilger 1956; Thielcke 1962; Lanyon 1963; Gill and Lanyon 1964), or can discriminate among songs of subspecific populations (for example Thönen 1962) or local dialects (for example Lemon 1967). By testing for different responsiveness to songs recorded for territorial neighbors and from strangers, playback has even demonstrated the ability of birds to recognize individuals by vocal characteristics (for example Weeden and Falls 1959; Falls 1969; S. T. Emlen 1971; Krebs 1971; Goldman 1973).

The playback tests just cited have all attempted to elicit territorial defense behavior by presenting the stimulus sounds within or near borders. Most studies of this kind had assumed that territorial males can usually be expected to respond to such apparent intrusions or challenges in more or less similar ways, and that their responsiveness to the playback should vary primarily with the effectiveness of the recorded vocalizations for eliciting a readiness to defend. Thus little or no study of the males used as experimental subjects has usually been made.

However, when other factors that might affect a subject's responsiveness have been assessed, they have been found to be numerous and important. For instance, males very quickly habituate to the test situation (Verner and Milligan 1971; Petrinovich and Peeke 1973), although it is not clear that such habituation occurs because playback, like the use of models, cannot begin to simulate the properties of real interactions; the issue is complicated because they also habituate to the songs of their established neighbors. Males also respond differently to playback in the morning and afternoon, differently if they have young than if they do not, and very probably differently on the basis of a host of other variables such as age, paired status, and time since last territorial dispute (Verner and Milligan 1971). The various responses of both male and female white-crowned sparrows depend on which stage of the reproductive cycle they are in, and different responses change in different ways (Petrinovich, Patterson, and Peeke 1976). Even the parts of their territories that at least some species defend most vigorously differ at different stages of the nesting cycle: for example, American redstarts respond more vigorously to playback in defending the centers rather than the peripheries of their territories before obtaining mates, and after pairing defend both equally—although the size of the area may then have decreased (Ickes and Ficken 1970).

Problems with playback experiments arise in choosing which activities of the subjects should be used as measures of response and in calculating the intensity of their responsiveness. Verner and Milligan (1971) found that virtually all the behavior of white-crowned sparrows appeared to be altered by a playback stimulus, but that relatively few response variables were highly consistent and easily measured. For recipients of that species, their flight lengths, numbers of songs, and rapidity of approach to the playback speaker were by far the best measures of responding. For some species in which each male has a repertoire of different song forms, recipients may produce a useful measure of a different sort by matching the selection of song forms they hear (Lemon 1968).

Response is not an all-or-none phenomenon, and during a playback experiment a recipient male often goes through a whole sequence of agonistic acts. Most workers who have tried to compare the relative effectiveness of different playback sounds have dealt with this by proposing arbitrary scales; they have assigned more weight to one kind of response (such as adopting a submissive posture) than to another (a limited approach to the speaker, with peering at it). S. T. Emlen (1971, 1972) has developed a procedure that takes eight different measures of responsiveness and an intensity measure of each, then sums the scores obtained from each category in each test (see table 10.2). Because this approach has the problem of treating different responses as if each kind

Table 10.2. S. T. Emlen's index for scaling the effectiveness of playback of indigo bunting song in eliciting agonistic responses from territorial males.

Behavioral category	Rank score	
Singing rate	0	Singing in 3 minutes immediately post-playback is a normal rate (not exceeding 18 songs per 3-minute period)
	1	Rate increased: $18 < n \leq 21$ songs per 3 minutes
	2	Rate greatly increased: $n > 21$ songs for 5-minute period
Song quality	0	No change
	1	Song lengthened by the addition of at least 3 figures
	2	Song lengthened and delivered at a soft (barely audible) intensity; high-pitched "squeak" notes occasionally added
"Alarm," erect posture (usually accompanied by "chip" call note)	0	Rare
	1	Infrequent; $2 < n \leq 10$ "chip" notes
	2	Frequent; $n > 10$ "chip" notes
Approach (to within 3 meters of loudspeaker)		Never approaches
	1	Approaches, but not during first 60 seconds of playback
	2	Approaches within first minute of playback
Flights over loudspeaker	0	$0 \leq n \leq 2$
	1	$3 \leq n \leq 10$
	2	$n > 10$
"Attack" on speaker itself	0	Not present
	1	$1 \leq n \leq 4$ attacks
	2	$n > 4$
Quiver posture (usually accompanied by soft, repetitive twittering call note)	0	Not present
	1	Infrequent; $2 \leq n \leq 5$ quiver bouts observed
	2	Common; $n > 5$ quiver bouts
"Fluffed" posture (usually accompanied by "tseep" call note)	0	Rare; $n \leq 5$ "tseep" notes
	1	Posture infrequent; $6 \leq n \leq 25$ "tseep" call notes
	2	Posture common; $n > 25$ "tseep" notes

The "squeak," "chip," "tseep," and twittering calls are parts of the species' non-song vocal repertoire. The "fluffed" posture is a display performed by an immobile, submissive recipient, and the "quiver" posture by a recipient behaving indecisively, apparently between attack and escape behavior, and pivoting on its perch. In the rank scores, 1 indicates a moderate and 2 an intense response for each behavior type. The eight measures obtained in a given playback test are summed on a quasi-quantitative 0 to 16 scale that takes into account the broad spectrum of response patterns that is usually seen over the 9 minutes of the tests Emlen ran. (From S. T. Emlen 1972.)

were equally important, he used the technique along with a more conventional scaling procedure that assigns increasing significance to each of the categories (or to seven of them: the two singing categories in the table are combined in his weighted scale).

All such techniques measure differences in the responses elicited by different playback stimuli. They are not necessarily a good indication of how birds respond to these vocalizations in natural circumstances. Their initial responses may usually be natural, but the source of the sounds they hear does not respond in an expectable fashion to their approach and displays, and then the whole event must become quite a novel and problematic experience for them.

The purpose of playback experiments, however, is not usually to study the behavior that is used in territorial interaction, but to use responses only to learn to what extent different kinds of identifying information are made available by particular vocalizations. Because sound recordings are easily modified (for instance by altering intervals between sounds, filtering out the higher or lower frequencies, or rearranging sequences) it is even possible to compare responsiveness in such a way as to ask what physical components of songs carry this identifying information (see, for example, Busnel and Brémond 1961, 1962; Brémond 1968a, 1968b; Falls 1963, 1969; Schubert 1971; S. T. Emlen 1971, 1972; Helb 1973; Gurtler 1973). Falls (1969) and his students have been able to mimic the pure tones of the white-throated sparrow's songs by using audio oscillators. By observing responses they found that pure tones within a particular frequency range, with a minimum length, and separated by intervals of fixed maximum duration carry the information that identifies the species. Information identifying particular individuals was found in the particular frequencies used within the range characteristic of the species, and perhaps in characteristics of the change in frequency between successive notes. Other characteristics such as the number of notes, the amplitude of the song, and the rate of singing are variable and, from observation of their natural occurrence, appear to provide changes in behavioral information. Studies of other species' songs have also revealed a division of the physical characteristics into those carrying identifying information and those carrying other kinds of information (see, for example, S. T. Emlen 1972; Ficken and Ficken 1973).

Playback research has not been limited entirely to the use of responses by territorial males or to studying the physical characteristics of song, although most of it has dealt with responses to identifying information. A few studies have reported on the responses of female birds to appropriate songs, such as songs of the correct species, dialect, or mimicking species (Milligan and Verner 1971; R. B. Payne 1973a), or of an individual mate (Falls 1969). Female grasshoppers have been

presented with playback of songs recorded from lone males and chorusing males of their own species to determine which is more attractive to them (Otte and Loftus-Hill ms.). Both male and female *Engystomops pustulosus* frogs were (separately) provided with playbacks of the "whine" call in versions of different complexity (Rand ms.). These experiments showed, first, that a male increases the complexity of his utterances in response to hearing a nearby call and, second, that females approach the source of the vocalization, and do so more directly the more complex its form (evidence supporting differences in the location information it provides, as discussed in chapter 6).

Nesting gulls have been exposed to the vocalizations of prehatching chicks to learn how such sounds influence their shift from incubating to parental behavior (Impekoven 1973), and playback has been used to study the development and extent of individual recognition of various vocalizations by pairs and their offspring in gulls and other birds (see Beer 1970 and White 1971 for examples). Beer (1973, 1975b) has shown that the vocalizations used by laughing gull chicks to identify their parents when the latter come to the nest or territory to feed them are not the crooning calls uttered in their feeding interactions, but the preceding long calls which come as a parent returns. Further playback experiments revealed that by the time chicks are two weeks old they also distinguish between different forms of these parental long calls, ignoring those with which other adults are addressed, and approaching only chick-directed calls. By playing back recordings in controlled conditions Beer learned that this distinction can be based purely on what the chicks hear, and thus found that they must be attending characteristics of the loudness (amplitude) patterning of the long calls. Amplitude differences are not usually analyzed in studying bird vocalizations, and this discovery might well have been missed without Beer's very careful playback procedures.

Responses to played back "alarm" calls have been investigated in a few cases. R. G. B. Brown (1962) found that oystercatchers approach to about 50 meters from a speaker playing their alarm call and stand watching it; they wandered off when playback ceased. Black-headed gulls fly toward a speaker playing their alarm call, circle over it, gradually gain height, and disperse; they habituated to successive tests. Curio (1971) found that he could attract small birds to alarm calls of conspecific individuals or of species which utter very similar vocalizations on detecting predators, and that they would mob the speaker. There is, of course, a problem in identifying a vocalization by referring to a single function such as "alarm." When Thielcke (1971) played their species' alarm call back to tree creepers it elicited either alarm calls or rivalry; the species utters the same call both in responding to predators and in some interactions with conspecific rivals.

In more complex experiments, Stout et al. (1969) simulated intrusions into the nesting territories of glaucous-winged gulls by erecting a mirror or a full model with a speaker, and then playing back a single kind of vocal display continuously for five minutes. The many variables introduced by these interesting experiments make their results very difficult to interpret, but the five different kinds of vocal displays compared were found to lead to differences in the latency before the recipient attacked and in the durations of its attack. Although the circumstances differ in many ways from natural events, it does seem reasonable that one effect of the species' alarm call display was to decrease the duration of attack whereas choking and long call displays increased this measure. Nonetheless, the authors' contention that the choking and mew calls "stimulate attack" by a territorial defender seems a misinterpretation. In the circumstances of the experiments attack was often elicited as a response, but in more natural interactions it might not have been: most real intruders would have found these displays inappropriate to repeat for five minutes while either maintaining a static posture or mimicking the owner's actions (the mirror effect). Further, for a display to evolve the function of eliciting attack seems a very risky procedure that must be inadaptive in most circumstances.

Pheromonal Experiments

Much invertebrate communication is pheromonal, and it has often proved to be experimentally manipulable with much less violence to the social fabric than occurs with, say, the playback of song to a bird. Thus E. O. Wilson (1962b) has been able to learn about the responses of ants to trail pheromones by removing the Dufour's gland or its chemical product and using it to lay trails himself. Similarly, the responses of ants to the spread of their alarm pheromones can be tested by experimenter-controlled release of the volatile chemicals, which may even be synthesized (E. O. Wilson 1971a). Ethologists studying mammalian pheromones, such as those odors of male mice that control onset and timing of female estrus cycles, have also successfully induced predictable responses, although usually long-term responses mediated through major changes in hormonal physiology. However, response to chemicals is not necessarily response to display behavior. The availability of many pheromones may not depend on formalized behavior (see chapters 2 and 9).

Conclusions

Both observational and experimental procedures can contribute to an understanding of responses and thus, ultimately, of functions. Each

technique, manipulative or observational, has its own contributions and drawbacks. Where the results of different procedures can be compared, the disadvantages introduced by each can be to some extent reduced and their contributions pooled. Probably natural experiments will continue to offer the least distorted results and the insights most faithful to the systems that have evolved, but the advantages of more controlled experiments will have to be sought repeatedly and attempts made to see that their conditions do not grossly violate the expectations that recipient animals bring to the tests.

The careful investigation of responses and functions will require prodigious effort. We shall have to devise theoretical structures and programs that enable us to be very selective about which kinds of events we seek to study—an unplanned attack on the general characteristics of responding would be very wasteful. With judicious choice, though, it should be possible to show how different displays and other information sources contribute in various circumstances to the selection of responses, and to use these demonstrations to construct, modify, and test models of the behavior of responding and the functions that accrue.

Categories of Responses and Functions

The recipient of a display selects responses on the basis of the relevant information available from diverse sources, and his responses are on the average functional for the communicator and (in evolutionarily stable systems) for him as well. Ethologists have tended to give categorization of functions priority over categorization of other aspects of communication. Functional inferences thus mold the perspectives within which most ethological description and categorization of responses or "meanings" has been done.

A number of ethologists have attempted to draw up comprehensive lists of the kinds of functions that appear to be served by displays. Representative of the functions suggested for intraspecific displays of insects and other arthropods is Alexander's list (1967, 1968) of about a dozen categories, which he grouped under a somewhat smaller number of circumstances of use: alarming conspecific individuals (usually about a predator), forming pairs, establishing aggregations, separating rivals, establishing dominance, timing insemination, facilitating insemination, reforming pairs after interruptions in courtship, maintaining pairs, maintaining families (in social and subsocial species), and, in social species, directing other individuals to food or a nest site. E. O. Wilson (1971a:235) listed about nine major response categories that support most of these functions in social insects. Thielcke (1970) has reviewed most of the proposed functions of bird vocalizations and, cautioning that few have been studied with sufficient thoroughness, lists territorial de-

fense, attraction of a mate, maintenance of pair bond, mutual stimulation of males, synchronization of pair activities, facilitating the simultaneous hatching of the eggs in a clutch (see below), familial recognition, group coherence (at various interindividual distances, perhaps including a mutual correcting of orientations during nocturnal migration; see W. J. Hamilton 1962, 1967), assembly of roosting groups, attraction of feeding sites, control of agonistic encounters, maintenance of pair relations and facilitation of the various activities in which mates cooperate, raising of young, and alarming young and other conspecific individuals. Obviously, at this level of categorization the functional categories suggested for insects considerably overlap those for (nonhuman) vertebrates.

Such descriptive lists are helpful in outlining the apparent functional scope of display behavior, but they tend to lack consistent theoretical structures with which to organize their data. The criteria employed for categorization are too variable and too arbitrary to be of much use in guiding research. Loose definition of categories may also conceal relationships among the means by which functions are achieved.

Marler (1961a) has offered a more abstract procedure for categorizing functions, using the ways displays predispose their recipients to act. Roughly speaking, this scheme classifies functions according to where, what, and to, with, or from whom, categories not unlike those into which the messages of displays can be sorted.

Following a procedure of C. W. Morris (1946), Marler attempted to examine the ability of displays to act as "identifiors," "designators," "prescriptors," and "appraisors." Identifiors dispose a recipient to respond within a specified spatio-temporal region (Morris's term unfortunately has different implications from the class of identifying information that I described in chapter 6). Identifior functions can thus control at least initial responses as a recipient alters its spatial relationship to a signaler, or changes its orientation. As an example of the latter, when a sentinel baboon sees a predator and gives a two-phase bark other baboons initially look at the signaler, then look in the direction of its gaze. Designator and prescriptor functions are not well differentiated from each other in most nonlinguistic communication and Marler treats them together. Both involve predisposing a recipient to make particular selections of response patterns. Having oriented itself and perhaps moved in space, what the recipient decides to do next is determined in large part by the identifying, motivational, and "environmental" information that Marler classifies within these combined functional categories. Appraisor functions predispose a recipient to react preferentially to certain objects, for example, by enabling him to distinguish and choose among different communicators.

In addition to classifying the function of displays, Marler is also interested in determining their "information content." Keeping functional considerations central, however, he uses the term "information content" to mean not the information that displays make available, but only that part of the information that their recipients can be shown to use (personal communication). It is possible that not all the information revealed by message analyses is taken into account by recipients as they select their responses to displays. Marler states his position as follows: "information . . . cannot be discussed independently from the occurrence of responses to the signal in other organisms. We thus require a means of inferring information content from the nature of the response given" (1961a:299). His procedures for determining information are thus different from those I describe in chapters 4 through 6, and are based on studying not communicators but recipients.

This procedural restriction does not imply that it is logically necessary to study messages only through recipient responses. Such a restriction would confuse the concept of information as a *property* with the concept of communicating as a *process*. Information is a property of all events and entities, and is available whether or not some potential recipient actually perceives or acts on it. The process of communicating, on the other hand, does not occur unless information is shared, in our social sphere, between a communicator and a recipient. A full understanding of the process *does* require consideration of responses, and only these can reveal what information has been used by the responding animal. But responses alone do not indicate what different sources contribute the information used by a responder. That is, the procedure of studying recipient responses does not make clear what portion of the information acted on was obtained from receipt of a display, and what portion from sources contextual to the display. On the other hand, studying the information made available by a display does reveal both its potential and its limits.

Studying the information content of displays through the responses made to them by recipients is awkward because the responses vary in different circumstances. I have postulated that each display makes a consistent assortment of messages available, and that recipients respond to this set of messages differently in different events because they encounter it in the presence of information from different contextual sources. Marler, although among the first ethologists to recognize the context-dependency of responses, prefers to postulate that displays themselves have a variable information content (1961a:309): "the response evoked by a signal—and therefore the information it conveys —may vary with changes in the circumstances both of the sender and the receiver." As an alternative, he suggests that a display may have a

rich information content from which recipients can elect to respond to different subsets (personal communication). The latter proposal implies what I described earlier in this chapter as the use of metaphor, except in suggesting that selective discarding of messages could be a customary procedure as animals perform and respond to displays. The "metaphor" proposal holds instead that only in unusual circumstances might communicators perform displays that were not fully applicable, and that recipients would need to be able to recognize these circumstances as special.

Marler's approach makes it feasible to study "information content" from responses, but its emphasis on recipient behavior also has the effect of presenting a display as a "releaser," and a releaser that acts differently in different kinds of events. A recipient would have to process each kind of event as a discrete unit. Marler describes communication as if a display were the only available source of information, incorporating information from other sources into its rich information content. On the other hand, I view a display and sources contextual to it as providing information separately—the display depends on contextual information but does not incorporate it. This would require a recipient to assess the relationships of information from a (limited) number of significant sources, rather than processing whole events as discrete units. Like Marler, I can image a display's information content being divisible into subsets, with a recipient choosing to attend some and reject others (the procedure involved in human use of metaphor), but only when the recipient is guided by sources of information contextual to the display. That is, communicator and recipient must share a code, must relate the same referents to the same signals, even if they are to be creative in employing these signals as metaphors. Distorting the code in any way for a particular signal should require a process mediated by other sources of information than the signal.

Viewing a display's information content as variable and described entirely in terms of the responses with which it is correlated enables Marler to say that a display "specifies" responses and the classes of "appropriate recipients" (1967a). This phrasing is also reminiscent of releaser concepts, but seems to derive from his functional orientation of viewing a display as something that acts to direct, sometimes even to command, the actions of a recipient. According to the philosopher Jonathan Bennett (personal communication), displays appear to function as injunctions. In a fundamental sense, this must be correct. That is, whatever adaptive advantages a communicator obtains from displaying must come because the displays yield him, on the average, some measure of control over a recipient's behavior. This view, however, does not depend on the assumption that the information content of a display

varies with the circumstance of its performance. The injunction leaves open many possible mechanisms by which displays might operate.

No currently available scheme seems to provide both a comprehensive categorization of functions or meanings and a strong framework for guiding research. Because responses and functions are far more diverse than the messages of displays, rather than working on further inclusive classifications we might do better first to formulate questions that permit us to develop and test models applicable to particular problems. For instance, we might ask how recipients select among and evaluate information sources when they respond adaptively in agonistic circumstances, and seek general rules for predicting response patterns from available information in such events. Any such attempt must begin with a provisional framework within which to organize what we know or suspect about meanings and functions, but a framework specific to the problem at hand. As these organizational frameworks will tend to limit the directions of research, the structure of each should be continuously questioned and modified as relationships become apparent.

Various potentially useful frameworks can be constructed if the main assumption is that each question a framework engenders should seek to explain how display behavior enables interaction to be initiated, continued, directed or controlled, and terminated. Specific questions could then be phrased for different kinds of interactions (agonistic or courting, to take two overlapping examples), different phases of interactions (such as initiating, maintaining, or altering the flow of an encounter), different kinds of circumstances in which interactions occur (such as whether participants are close together or apart), or perhaps the kinds of tactics selected by a given participant (being deferent or being forceful). These different topics obviously are not mutually exclusive. On the contrary, each makes a necessary contribution to a coherent picture of what the behavior of communicating can accomplish, and how.

Beginnings have been made in studying some of these matters. To study how agonistic encounters are controlled, for instance, Tinbergen (1959b) proposed frameworks dividing responses into relatively offensive and relatively defensive categories, and functions into distance-increasing and distance-decreasing. A representative sampling of such proposals and the more descriptive accounts of meanings and functions that abound in the literature is reviewed in the remainder of this chapter.

This review is not comprehensive but it should indicate the current state of such research. The principal features of its organization are, first, a very gross division of circumstances in which displaying occurs (participants apart or together, without or with established interindividual relationships); and second, an assortment of kinds of interactions within these circumstances. Many of the latter have been chosen because

of their importance in particular circumstances. Description of each kind of interaction is limited largely to a single class of circumstance as an arbitrary convenience; many kinds (such as agonistic) could be treated under each circumstantial category.

Interacting at a Distance

To function at all, displays performed when potential recipients are at a distance from the communicator must identify the communicator and help recipients to find it if its location is not immediately obvious. Such messages are discussed in chapter 6, and responses of orienting toward the sources of displays confirm the use of the information that they provide about location. That use is also made of information identifying communicators is apparent from the playback research previously discussed in this chapter; many studies have shown that recipients discriminate and respond selectively to vocalizations of their own species, dialect, or even neighbors or mates instead of those of other individuals. Under the general function of assuring that participants in subsequent interactions will be relevant to each other, such selective responsiveness ensures the reproductive isolation of different species (see Mayr 1963) and perhaps even the restriction of gene flow among intraspecific populations (for example, Nottebohm 1969; Baker 1975).

Identifying information permits recipients to be selective about which communicator to attend, but attending is only the basis for further responding. Apart from attending, "the first response that the human observer can detect is often merely a change in the respondent's spatial relationship to the signaler" (Marler 1967a:770). Marler suggests (1968, 1972) that the functions displays have in adjusting interindividual spacing fall into four classes: use of a display may result in increased or decreased distance between the communicator and recipient, or in maintaining either distance or proximity between them.

When the recipient is far from a communicator it must decide whether or not to change their spatial relations. If the response is to approach, the display is said to have an attracting function in this circumstance. This facilitates further, more varied interaction. An important distinction must be made between the first response elicited by the display and subsequent responses that depend in any degree on that signal (Marler 1967a). For instance, although two displays may each initially elicit approach, one may predispose a recipient to become defensive, and the other to copulate.

If a recipient withdraws, ceases to approach, or carefully keeps its distance the display is functioning to repel. If the recipient simply orients toward the communicator, or perhaps moves more or less later-

ally or tangentially it is often not clear to an observer what, if any, function was achieved by the display. There are, however, numerous possibilities. For instance, the roaring of troops of howler monkeys or the loud singing of territorial birds may function to reassure neighbors that familiar individuals are in expected places, and hence that they need not rush defensively to a border area. By helping neighbors keep track of each other's locations, both while establishing and maintaining their relationships, singing and howling probably also help them to avoid close confrontations. (In what amounts to a natural experiment in which singing fails to keep neighbors of two closely related species of meadowlarks of the genus *Sturnella* apart, combat and wounding are increased in narrow regions where the otherwise geographically separated populations of these birds overlap. Rohwer, 1972, 1973, found that males of each species are largely unresponsive to the other's songs and thus tend to interact primarily at close quarters where they must fight in response to their nearly identical appearances.) Bird song also has functions that are not immediately observable: for example an autumnal resurgence of song by territorial males in some species functions to teach their young the characteristics of their local dialects, although they may not reveal this until they either sing such dialectal forms or, if females, are shown experimentally to respond preferentially to them (see Thorpe 1961; Marler 1970).

Recipients who are not with a communicator may not receive its information immediately; for considerations of function this temporal separation is at least partly analogous to that of space. Displaying that reaches some or all of its recipients at a later time can do so only through an intermediary, of course, such as a scratched tree trunk or a site where pheromones are deposited. As indicated in chapter 9, the individuals who contribute most to such sites are typically sexually mature animals capable of controlling other individuals. Various attracting and alerting (if not repelling) functions accrue. Further, the delay between displaying and receiving the information makes it possible for a communicator to communicate with itself. Its perception of the act of displaying as it displays does not qualify as social communication, because only one individual is involved (this is not to deny the use of immediate feedback information in regulating its behavior). However, with delayed feedback the recipient individual cannot encounter the display (or, rather, its traces) in exactly the same contextual sources of information that were present initially. It has become, to this degree, a different individual: as the saying goes, you cannot step into the same river twice. The communicator may have performed its marking display as a response to the novelty of entering a new region. Returning later, more familiar with the area, the individual becomes a recipient, detect-

ing its mark and being reassured by the evidence of its earlier behavior (see, for example, Kleiman 1966; Ralls 1971; R. P. Johnson 1973), or, if it is a parasite seeking new areas to hunt for its hosts, it can avoid the area and move on (Price 1970).

Interacting When Participants Are Already Near Each Other

Individuals interact in each other's presence in very different ways depending on whether they have established mutual relationships based on individual recognition.

Individuals who have come into association anonymously may form only fragile groups and tend to respond agonistically to each other's displays. On the other hand, many quite stable groups such as fish schools appear to be anonymous or at least partly anonymous; in partly anonymous groups each individual may know a few, but not most others—as in large flocks of birds—or may know only that other individuals do or do not belong to its own group—as in social insects. Members of most of these groups coordinate their activities and usually do not interfere directly with one another. They may perform little or no display behavior to achieve these functions, often relying upon special badges such as flash patterns (see chapter 9) that make their movements very conspicuous.

In other kinds of anonymous groups the members compete while associating—for example, choruses of calling male frogs or cicadas. Display behavior may enable these individuals to come together and even to achieve synchrony in their advertising (see Alexander 1967). Further, their combined displaying may provide a louder, farther-reaching beacon by which females may be attracted for mating (or, in fireflies, a brighter, more visible beacon; see Lloyd 1971). They cooperate to attract females and, although they do not share them, each individual probably has a better chance of mating when in the group than when not.

As an alternative, associating individuals may establish relationships based on mutual recognition. Such relationships may involve a special convention such as a territorial boundary that limits the further competition of neighbors, or a more positive bond that unites the individuals as members of a pair or larger cooperative group.

Formation of persisting pair bonds has been observed in many species by ethologists interested in "courtship" behavior (see reviews by D. Morris 1956b; Bastock 1967). Often the first step occurs when females are attracted to territorial males who are advertising with displays that provide information about their readiness to interact (see chapter 5). As a female intrudes on such a territorial male he usually directs to her

the same displays and even attack behavior with which he responds to a male opponent. Her response, however, differs from that of a male, and typically includes displays that function to appease the owner (encoding a low probability of attack, and readiness to interact or even just to associate). He is then likely to perform the same displays that she performs, and thus reassures her. Male and female may then enter into mutual bouts of displaying like those described in detail for various gull species (see Tinbergen 1953a, 1959b; Moynihan 1955a, 1962a); these cooperative performances are considered in chapter 13 as formalized interactions. The immediate function of these displays is to control agonistic behavior and permit the participants time in which to get to know each other.

Species that are not territorial at the time of pair formation often have less need to overcome agonistic responses. In many cases, pairs form within larger, partially anonymous flocks that are joined at an earlier season, and the functions of "courting" displays largely appear to lie in forging closer bonds between two particular individuals. For instance, a male duck's displays probably function primarily to convince a female to select him from a group of males, and to bring her hormonal state into readiness for pairing, although they must also be important in helping him compete with other males for her attention (see McKinney 1975). Common grackles gather in groups in which both males and females sing, every bird with an individually specific song. Males follow nearby females as these depart from the groups, and over a period of days each male begins to associate closely with a particular female. She then quits the singing group and begins to frequent the nesting colony with her male in attendance; the two birds continue to sing, although now antiphonally as a pair (Wiley 1976, in press 2).

In many galliform birds, males form pair bonds with females by calling them to food with a tidbitting display (although in one species, this display also appears to help overcome a tendency of the female to escape from a male; see Stokes 1961; Stokes and Williams 1971). In functioning to permit pair bonding, however, displays need not also be functioning to promote breeding. For instance, yearling cocks in the red jungle fowl court only hens who already have chicks, which may function to gain them experience in courting and to sort out status relationships with their peers as they compete for mates (Stokes 1971).

Establishing interindividual relationships is usually accomplished quickly, and then individuals interact on the basis of their mutual arrangement. They periodically have to deal with the terms of the relationship itself, although much more often they accept it tacitly and their interaction has other functions.

First, in interactions that deal with the terms of an existing relation-

ship two individuals may seek to reaffirm it, test and try to alter it, or to add to it (as was just mentioned in the case of birds who belong to the same flock and then become a mated pair for breeding purposes).

Reaffirming relationships, the giving and seeking of what Hinde (1975b) refers to as "social approval," is an important function of various kinds of displays. In many persistently social species of animals, such reaffirming includes formalized interactions in which individuals groom or preen one another (see chapter 13). Their behavior in such interactions sometimes reflects the dominance relationships among individuals (for example, Sade 1972; Sparks 1967), although often in no simple way (Simpson 1973b) if at all (see Bernstein 1970). Nonetheless, considerable leisure time is spent in this activity, which may indicate that it can be a procedure for reaffirming bonds, counteracting the divisive effects of status differentiation. (It is not limited to just this function, as allogrooming is initiated in some agonistic and in precopulatory circumstances and helps individuals manage the situations; see review by Sparks 1967.) Among other means of reaffirming relationships in species that scent-mark is the marking of subordinate animals by a dominant member (see Ralls 1971; R. P. Johnson 1973). Animals may also reaffirm relationships by greeting each other, which often involves repeating, at least briefly, the formalized interactions with which they initially established their bonds.

Interindividual relationships are not always stable. As individuals age, or as their physiological states get out of synchrony with each other, they may periodically test their relationships. Male wildebeest on neighboring territories meet daily and go through a complex challenge ritual that Estes (1969) proposes has the function of checking the status quo. Particularly important may be a step in the performance where one individual urinates and the other tests the urine olfactorily by means of his *Flehmen* response (chapter 13), perhaps assessing the androgen level as an index of the first's "territorial credentials" (Estes 1969:329). Simpson (1973a) has described a two-week period during which a male chimpanzee performed five "charging displays" in the presence of another male who had been able to supplant him, turned the tables, and became able to supplant that male. Cause and effect over that period are not certain, of course, but the displays probably tested and found wanting the second male's readiness to maintain his social position. Testing may require shorter periods when newly met individuals are establishing a basis for the organization of their subsequent interacting, as Simpson indicates.

Second, individuals with established relationships most often tacitly accept the implications of these as they interact. They can then be concerned with immediate issues in dealing with each other, or with

events that arise outside the group. Many perspectives can be used to categorize the meanings and functions of displays in their interactions, depending at present primarily on an investigator's interests. In the literature, four commonly adopted orientations are the control of agonistic events, sexual encounters, group cohesion, and interactions in which parental or comparable services are provided. The following sketchy accounts merely indicate the kinds of topics with which this literature is concerned. They do not purport to be a full review of the literature on the meanings and functions of displays in these or other kinds of interactions.

AGONISTIC EVENTS   Moynihan (1955b) proposed that the most general function of displays performed in agonistic events is to enable the participants to obtain various advantages without having to fight or, more precisely, without being attacked. To risk being attacked even in retaliation is to court injury (see Geist 1974). The strategies employed in agonistic confrontations keep this risk small by relying largely upon formalized moves, or a mixture of formalized acts, probing attacks, and retaliation (see Maynard Smith and Price 1973; they offer models of how benefits accrue to individuals who forego the "hawkish" approach of total war). The assumption that formalized interactions alone can be used to settle contests is investigated further by Maynard Smith (1974). Comparable modeling is done by Parker (1974) for events in which acts provide reliable measures—for example of the strength of the contestants—that enable predictions of the outcome of the escalation. In a further analysis of the kinds of behavior expected to evolve when contests are asymmetric—when one contestant may have more to gain, or may be more able to win fights, or may have some prior claim— Maynard Smith and Parker (1976) conclude that providing the information which completely reveals the asymmetry will enable such contests to be settled without escalation.

Moynihan recognized several different ways in which displays function to gain advantages at low risk of being attacked: threat and appeasement are most commonly observed. In threatening behavior the function of the communicator's displays is to intimidate its opponent and, if necessary, to cause it to flee. In appeasing (or in reassuring) behavior, displays function to forestall an opponent's attack but without inducing its escape.

Such functions are necessary if social groups are to remain coherent and more or less stable. Groups cannot have structures based on status hierarchies, for instance, without display behavior that keeps every even slightly competetive encounter from becoming a confrontation; thus Dilger (1960) found that subordinate individuals in caged redpoll flocks

were able to move about near dominants by maintaining a fluffed submissive display posture, and by avoiding orienting their heads directly at the latter.

Territorial behavior, as usually defined by biologists, depends on defense against intruders; it is facilitated by displays with the function of spacing out would-be competitors (Tinbergen 1959b). Territorial intruders who do not threaten may be treated leniently, in some cases, particularly if they perform appropriate displays: for example, if intruding European blackbirds utter a "seep" vocalization that encodes a high probability of flying, a territory owner will usually approach slowly without posturing or even sit quietly nearby until the intruder leaves (D. W. Snow 1958).

"Deceptive" displaying (Moynihan 1955b) apparently forestalls attacks by the procedure of eliciting some nonagonistic alternative, often courtship feeding, sexual behavior, or allogrooming. It has also been called a procedure for "remotivating" a recipient.

"Exemplary" displaying, Moynihan's fourth category, appears to reduce the probability of attack behavior by stimulating recipients to a communal display performance; it is marked by an infectious spread of the initial display in an event. In greeting behavior, for instance, the head flagging of gulls (Tinbergen and Moynihan 1952; Tinbergen 1959b; also see figure 12.2) or the wing fluttering and calling of kingbirds (W. John Smith 1966), both participants commonly perform the same or very similar displays while apparently becoming less likely to attack or flee. The so-called triumph ceremonies of geese (see chapters 6 and 13, and figure 10.2), Australian magpie geese (Davies 1963), and some ducks (for the flying steamer duck, see Moynihan 1958b) incorporate displays performed simultaneously by whole families when a gander returns from winning an encounter or when the family is disturbed by some outside event. By displaying mutually, participants are apparently prevented from quarreling with one another. Exemplary displaying is a descriptive category, however, and in its functioning probably overlaps appeasement and deceptive procedures, depending in any given case on the messages of the displays that are mutually performed.

Moynihan also recognized a category in which displays have "alerting" functions in agonistic events created by the attack of an outsider, often a predator. Such events are often minimally complex interactions: some response usually follows immediately after an alerting or alarming display, and is selected from a small number of choices. Further, an observer can often know that a recipient of the display could not see the predator or even the nondisplay responses of the communicator, and so must have been responding just to the display and its own immediate

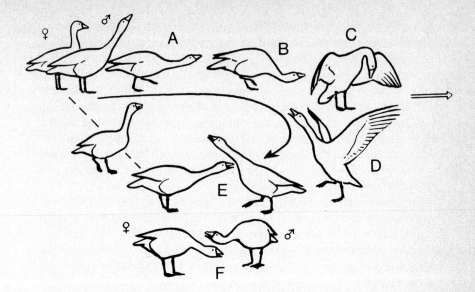

Figure 10.2. Triumph ceremony of graylag geese. *A, B, C:* A gander leaves his family, in this case only a mate, and goes forward to challenge another gander (to the right, not shown). *D, E, F:* After the encounter he returns and both he and his mate adopt display postures resembling those he used in leaving, except that they orient obliquely past each other whereas he had oriented toward his opponent. When a gander's family includes offspring as well as his mate, all members adopt these display postures and oblique orientations on his return. (According to Fischer 1965.)

circumstances. These characteristics make such events very attractive material with which to begin a study of how recipients respond.

The alerting function of a pattern of vocal displaying of the Carolina wren *Thryothorus ludovicianus* were nicely demonstrated in experiments in which Morton and Shalter (ms.) exposed pairs of wrens to a tethered hawk. After an initial alarm response, each pair of wrens settled into a standard procedure for as long as the hawk was visible: the female would monitor the predator, uttering variable bouts of "chirt" vocalizations, and her mate would disattend it and resume foraging. If the hawk moved the female called more rapidly, spaced her vocalizations evenly, and increased their average peak frequency from 2.5 kHz to 3.0 kHz. Her mate would then respond by looking at the hawk. Differences in the pattern of her calling thus determined whether or not he would respond with alertness. Her continous calling provided him with the information that she was remaining alert (see the discussion of messages of attentive behavior with continuing stability in chapters 5 and 6); specific changes

in the pattern may provide information on shifts in the probability that she would flee (not assessed in the experiments). Note that the female's displaying functions differently for each individual. It permits the male to continue foraging without being endangered. A Carolina wren cannot forage while watching a hawk, so the female is allowing him to maintain his reserves of energy in the presence of a predator. By protecting the male even at the expense of not obtaining food for herself, a female increases the likelihood that he will maintain a permanent territory on which she can live. Morton and Shalter have found that females without mates cannot hold territories, are driven away by other pairs, and appear to have little chance of survival.

The contextual information sources that may be important to the recipient of a display with an alerting or alarming function may also be readily assessed and have been measured in some studies. For instance, Hodgdon and Larson (1973) tallied 92 instances in which a beaver slapped its tail on the water (a display that apparently carries an escape message) and in which they knew such contextual variables as whether the recipient was in deep or shallow water, its age and sex, and the age and sex of the communicator. Water depth was an important element of context: "beavers in deep water were ten times less likely to respond to tail-slapping than those in shallow water." However, the responsiveness of any beaver in shallow water also depended on its sex and the sex of the beaver tail-slapping. The recipient's age was another element: adults in shallow water always responded by swimming to deeper water if another adult was the communicator, but yearlings did so less frequently and kits only half the time.

Alarming displays by young animals may be taken more seriously by other young than they are by adults. Among New Forest ponies, for instance, when one alarmed foal raises its tail vertically and joins its mother, other young foals follow suit but adults appear to ignore the signals (Tyler 1972). The adults' lack of response is partly because they have usually seen the source of danger before the foals detect it, and partly because foals are often alarmed by events that do not concern adults.

Some displays have been found that have the opposite of an alarming function. For instance, alarmed gannets, boobies, anhingas, or cormorants take flight without display, which can alarm their neighbors. When one of these birds is about to fly but is not alarmed, it performs an elaborate "pre-take-off" display (figure 10.3) that apparently functions to prevent other individuals from becoming alarmed (van Tets 1965). In potentially dangerous situations (such as approaching an artificial feeding site) a feral male chicken will take a vantage point, inspect, and then crow; his waiting hens then proceed to the site. Apparently one

Figure 10.3. Sky pointing display of the gannet *Sula bassana*. Sky pointing occurs before and during movement, usually walking, and often precedes flight. It is particularly common as a bird leaves its nest site, but at times occurs even while walking along the fringe of the colony. *A*. As one bird is relieved at the nest by its mate it stretches its neck vertically and holds it stiffly, the bill pointing upward. *B*. Same situation; note how the wings are rotated (the posture is thought to be formalized from features of intention movements to fly.) *C*. When walking while sky pointing the communicator uses a marked swaying gait. (From J. B. Nelson 1965.)

function of the crowing is to indicate that the situation is safe (McBride, Parer, and Foenander 1969). When a black-tailed prairie dog that has been fleeing or ready to flee becomes ready to do some alternative it may perform a jump-yip display (W. John Smith et al. in press 1), which can have a similar all-clear function.

SEXUAL ENCOUNTERS The degree to which management of sexual interactions depends on display behavior differs greatly among different sorts of animals. Among different species of birds, for instance, elaborate displays are necessary if copulation occurs without a well-established pair bond, as it does in most grouse, hummingbirds, and other lek species in which males display communally. Such precopulatory displaying always provides very abundant species identifying information, and resembles in part the pair-forming behavior of other birds. Many fish, frogs, and lizards employ displays for similar precopulatory functions. In those species that form pairs long before copulating, on the other hand, virtually no display may be required to facilitate copulation, although the female may solicit by adopting the posture that enables the male to mount. In some species it may be necessary for mates to use the displays that control agonistic responses in such circumstances as greetings in order to come close together, or to perform displays such as the tidbitting of galliforms that attract associates for a variety of functions.

Yet if both individuals are then ready to copulate, displays specific to this behavior may not have to follow.

The main function of precopulatory displays must be to help a male and a female to cooperate in beginning copulation. In various species of gulls, however, males continue to display, and in fact display more conspicuously while copulating than they do before mounting their mates. For instance, a mounted male ring-billed gull *Larus delawarensis* utters a loud copulation call display (Moyihan 1958c), incidentally revealing his bright orange mouth, and rhythmically waves his wings (Southern 1974). Southern has shown that the activity increases the likelihood that neighboring pairs will also begin to copulate. He postulates that this contagious effect should increase the synchrony of breeding activities within gull colonies; the advantages of such intracolonial synchrony are considered in the following section.

Some birds and mammals also perform elaborate displays after copulation—usually males do most of this displaying. Its functions may often accrue from making it easier for the pair to remain together. For example, such displays may facilitate repeated mounting, in species in which this is the usual practice. In other species, the displays may assure the female that the male will now try only to associate, and will not mount again. This may help preserve the pair bond or, if there is no bond, at least let the female be with the male where he can guard her from insemination by other males (see Parker 1974a). In some mammals, postcopulatory displays of the male may calm the female so that she remains quiescent and thus perhaps more likely to retain the sperm.

In general, mammals seem to make available more, or more elaborate, sources of information about at least their precopulatory behavior than do most birds. For instance, female mammals often have special badges or pheromones, and one or both sexes performs tactile displays before and perhaps after copulating (figure 10.4). However, only some of these touching patterns may be specialized for functions related only to copulation—many are not so restricted in their use. Mammals may also employ gesturing or posturing before and after copulating, although in some species even the display postures that facilitate mounting have additional functions (see discussion earlier in this chapter).

Many functions of displays in the sexual interactions of arthropods are similar to the functions obtained by birds and mammals (see Bastock 1967). For instance, meetings must be arranged and species identification made clear, as in vertebrate animals that do not form pair bonds. However, many arthropods find little or no need to control agonistic behavior. In such species the main problem seems to be for males to sort out those females who are more or less ready to be inseminated, and then to maneuver and stimulate them until they will

Figure 10.4. The *Laufschlag* display of the Uganda kob. Before mounting a female, and in the intervals between the several mountings of a copulatory series, a male Uganda kob *Adenota kob* touches the underparts of the female with a stiff foreleg. He performs this tactile display from behind, placing his leg between the female's hindlegs (as shown), or from the side, touching her in the flank region. (According to Buechner and Schloeth 1965.)

permit it. In some cases displays may then have further functions. For example, a male cricket stridulates both during copulation, keeping the female from moving off before he has completed, and afterward, keeping her from removing the spermatophore before it has emptied, and then again to keep her with him while he is producing another spermatophore (Alexander 1967).

GROUP COHESION AND COORDINATION Under some circumstances displays function to elicit responses that form, reform, or maintain cohesiveness in a group. Often this is by providing information that permits group members to join or rejoin each other, affirm that they are together even if not visible to each other, or coordinate their activities so that they can remain together and perhaps perform some joint activity.

For instance, birds and mammals that forage socially in dense vegetation often have difficulty in keeping track of other members of their groups. They may utter periodic "contact" calls, or may utter "lost" calls when separated; numerous examples are provided in chapter 5 under the association and seeking messages. In response to these vocal displays other individuals may join or follow the communicator, or they

may display in kind and it will join them. Thus, when a green-backed sparrow utters a plaintive note display, "it is almost always joined or followed by its mate almost immediately" (Moynihan 1963a), and when a red-legged partridge begins rally calling others usually make their way toward it if it does not come to them when they answer (D. Goodwin 1953). When the members of a group of gorillas utter hoot bark displays they tend to cluster together (Fossey 1972).

Joining and following responses are also elicited by displays performed when animals are fully visible to one another. Many of these displays are done when one is leading or about to lead the others to a new location, and they encode at least the association message (chapter 5). As an alternative, joining and closer following may be elicited even by a threat in some kinds of social structures. For instance, a female hamadryas baboon who lags behind will rush back to the male of her group if he threatens her, and his threats help to keep his harem together. When changing location with respect to another male, however, a male hamadryas may approach and display in a special "notifying" procedure that is not threatening, but functions to inform of peaceful arrival or departure (Kummer 1968).

In Canada geese, family groups remain together until the young have matured and gone off to begin their own families. In winter each family is part of a very loosely organized larger flock, often numbering thousands of birds that come and go on local flights, and individual members of families can easily lose track of one another. They are reluctant to separate, and if not suddenly alarmed all family members will stay together continuously, letting the gander lead their group. When he is ready to fly to another site he does a head-tossing display a few times, the others repeat it, and they soon depart (see chapter 5, under locomotion message; the display also appears to encode an association message). When other family members initiate the display they may continue it for long periods before the gander responds; if he does not respond they will not fly away. Raveling (1969) watched one immature goose in seven different events; it averaged 181 head-tosses in periods averaging 27 minutes before the gander of its family would fly. When seen separated from its family the same immature head-tossed only once before flying. Raveling found it typical of single geese in a flock that they perform this display much less than do members of families; of course, for them it cannot have the function of maintaining family cohesion, even if it relates in some degree to maintenance of the larger flock.

There are various reasons why it is adaptive for individuals of some kinds of animals to keep together. Flocks, herds, or schools can make detecting and avoiding predators more likely, as discussed above in

considering agonistic events and earlier in this chapter. Groups can also be used to avoid attacks by intra- or interspecific competitors, effectively overwhelming the latters' attempts to hold exclusive feeding territories by invading en masse (see Gibb 1956 on wintering rock pipits, Drury and Smith 1968 on wintering gulls, and Barlow 1974 on surgeonfishes for examples).

By banding together, individuals can also increase the likelihood that they will find food, for instance if it is distributed irregularly in space and time but locally concentrated. Individuals can search in a "stretched flock" (Fisher 1954) by spreading out on a broad front that sweeps forward until some individuals find food and stop; the others then converge on them. For many species the main specializations for communicating that are relevant in this behavior seem to be badges (such as the highly visible white plumages of most gulls and many herons) that render their wearers more conspicuous; no displays seem to be peculiar to this function, although at least vocalizations (such as what are often termed "flight calls" and "contact calls") are probably involved in some species. Flocks of individuals who forage together by day may join other flocks when roosting; these communal roosts are apparently "information centres" (Ward and Zahavi 1973) that give very large numbers of individuals access to each other's information about feeding sites. Birds and flocks that have exhausted sites and failed to find new ones can join the luckier ones, following them as they leave in the morning. Again, no displays specialized to provide information about food seem to be involved, although specialized evening group performances that advertise the locations of the roosts are sometimes used (Ward and Zahavi 1973).

Flocking may facilitate not only finding but also exploiting food. For instance, some species appear to increase the efficiency with which they exploit food by increasing the patchiness of its distribution by thoroughly depleting each site they use in a single visit, rather than by reducing food evenly throughout a region, which makes finding it ever more difficult. Local depletion procedures leave untouched patches for later use. Other species live in regions in which food supplies are renewed as plants continuously come into fruit. Species in both kinds of circumstances need behavioral mechanisms that govern the ways they move through regions while foraging. Individuals must join flocks in the first kind of circumstance to keep from searching where other individuals have fed. In the second circumstance, being in a flock helps individuals to adjust the intervals between their joint visits to different locales to the renewal times of the food resources (Cody 1971, 1974). In either case, behavior within flocks must be regulated, for example, by controlling the directions and distances flown by a flock within a

bounded area as it moves between successive foraging sites. The functions of displays in these flocking mechanisms must be primarily to enable the members of a flock to keep together as they forage and move between sites.

The synchronous breeding behavior of colonial gulls was mentioned in the immediately preceeding section ("sexual encounters"). In most species this synchrony appears to be beneficial because it keeps most of the members of a flock available to each other in the colony region. For the duration of the breeding season all pairs must work out from the same site as they seek their localized and moving sources of food. The definite advantages of this behavior have been demonstrated by Hunt and Hunt (1976) in a study of a colony of western gulls *Larus occidentalis*. Most of the food these gulls brought to their chicks was obtained from schools of fish and squid, and when the flock had found a school more individuals within the colony received food than when only items requiring individual search were available. Not just gulls, but many other kinds of birds with dispersed and changing food sources appear to have found similar advantages in synchronous colonial nesting (see, for example, Ward 1965; Horn 1968; Ward and Zahavi 1973; Krebs 1974; Emlen and Demong 1975; also see figure 10.5). The synchrony may be at least loosely determined by seasonal or other changes in the environment, but in some if not most species owes its precision to the use of display behavior. Thus in weaver finches (see Hall 1970) and in the ring-billed gulls described in the previous section, responses of pairs to the pair-bonding and copulatory displaying of their neighbors appear important in synchronizing the whole group.

As an additional or alternative benefit, synchronous colonial breeding can sometimes reduce the vulnerability of very young animals to predators. Predators can be swamped by the number of prey individuals available at the peak of breeding, but they can take a great toll of those born later (see Ashmole 1963; Kruuk 1964; Patterson 1965; R. G. B. Brown 1967). A comparable antipredator advantage from synchronous breeding may be available to herd-living mammals such as the migratory wildebeest, in which the presence of numerous calves can make it very hard for hyenas and wild dogs to single out those which would most easily be caught (Estes ms.). (These predators, incidentally, also depend on group cohesion and synchrony, because they hunt in packs. Estes and Goddard (1967) have described how packs of African wild dogs rouse themselves for a cooperative hunt through a period of playing, greeting, and milling about in escalating excitement.)

Yet other advantages can be found for individuals who stay together, even in socially undifferentiated groups. For instance, flocks of migrating birds may serve as mechanisms for averaging information, thus

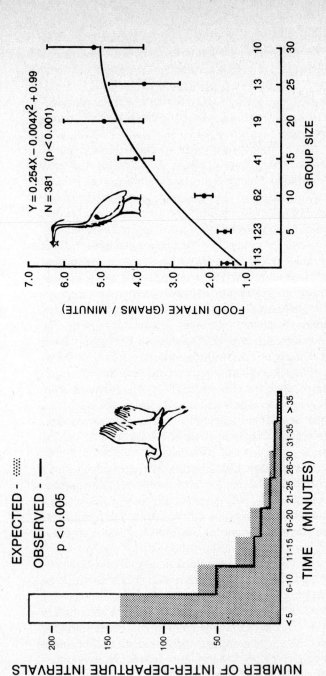

Figure 10.5. Colonial nesting and the exploitation of food resources by great blue herons. The great blue heron *Ardea herodias* is a colonial nester. The herons from a colony feed in different areas on different days, suggesting that their food supplies are locally ephemeral. The birds tend to leave the colony in groups, which would be expected if they were following each other. They are attracted to forage in areas where there are other herons and more attracted by a group than by a single bird. Birds in groups obtain more food than do solitary birds.

Left: Intervals between successive departures from a colony by adults making foraging trips (open histogram). The shaded histogram shows the expected distribution of departures predicted from a negative exponential distribution; the observed departures are "clumped" with respect to this prediction. Right: Rate of food intake of adult herons as a function of flock size. The vertical bars represent standard errors of the means of grouped data; the regression equation describing the relationship was arrived at by stepwise multiple regression. (From Krebs 1974.)

reducing orientational errors (W. J. Hamilton 1962, 1967). The intermittent vocalizations of nocturnal migrants could, in this view, enable
each individual to keep track of and accommodate to the navigational
headings of the others. This would result in a flock achieving a compromise course based on the integration of the individually preferred headings, a course in which the unsystematic errors of the individuals would
tend to cancel each other out.

Vocalizations may also be used by fin whales *Balaenoptera physalus*
to maintain contact with each other, although in this case the low-
frequency vocalizations must be audible at great distances (see chapter
12) and should permit members of the group to be widely dispersed.
R. S. Payne and Webb (1971) estimate, in fact, that fin whales may live
in "range herds," each of which extends through "the entire deep-water
range of the species in any one ocean." By maintaining audible contact
the whales should be able to rendezvous for breeding or foraging at
times and places that change unpredictably from one year to the next.

Social insects, as well as vertebrate animals, have specialized responses that lead to coordinated group behavior. Some of these responses involve display behavior, although many may not. The
phenomenon has been distinguished by E. O. Wilson (1971a) as "mass
communication," and defined as "the transfer, among groups, of information that a single individual could not pass to another."

The ants and other social insects in which Wilson recognized mass
communication are highly specialized animals in which the efforts of
individual workers must contribute efficiently to the maintenance and
growth of their whole colony. No individual in a large colony can be
aware of much of what is going on, and its responses are determined by
the limited stimuli it can encounter. Evolution has prepared social insect
workers with sets of response patterns and with thresholds adjusting the
probability of the occurrence of each response; in any given event a
statistical order in the behavior of a group emerges from the frequencies
with which particular behavior patterns are selected by individual
workers.

Two basic features of this mass communication, the adjustment of
signaling behavior to the number of responding individuals that can be
used efficiently by a colony, and the use by individual social insects of
multiple communicators as sources of information, can be seen as fire
ants deposit and respond to odor trails (see chapter 2). These trails are
laid as foragers return from a food source to the nest; each trail becomes
the cumulative result of displaying (trail-laying) by numerous individuals. The more trail pheromone laid down the more attractive the trail,
but after an initially exponential increase in the number of foragers
attracted to a new food source, trail-laying by further recruits largely

ceases. Newcomers can no longer reach the food mass because of the crowd there and, unable to collect food, lay no odor trails as they leave the site. An equilibrium is thus reached between rate of recruitment and the size of a food mass when the number of attracted foragers reaches crowd proportions: the size of a crowd is basically a linear function of the area of the food mass, although it takes fewer ants to produce a crowd if the food is relatively unattractive. As the odor trails of each successful forager evaporate within minutes and must be replaced by the trails of subsequent forages, the amount of pheromone available decreases to zero as the food is harvested (E. O. Wilson, summarized in 1971a).

Army ants, like fire ants, employ pheromones for mass recruitment of foragers. Although their procedures are basically similar, they have modified the system to maximize the *rate* of recruiting. A scouting forager (in the genus *Eciton*) who has discovered a food source does not run back to the nest, but only to the nearest raiding column. Here she dashes back and forth making contact with other ants, leaving repeatedly to return to the food. Responses are immediate, and 50 to 100 or more ants are usually recruited and adding to the trail within the first minute. The adaptive value of this rapid mass recruitment is evident when the food source, the ants' prey, is a wasp nest: the wasps can throw small numbers of ants off a nest, but abandon their brood to the raiders if overwhelmed by a massive attack (Chadab and Rettenmeyer 1975). As in the case of fire ants, trails to a depleted food source are abandoned within about 10 minutes.

Some of Wilson's other examples of mass communication reveal how statistical order is achieved in the behavior of quite large groups of insect workers. Order is evident, for instance, in the behavior of the foragers of a honeybee colony when a fundamental change is needed, such as a change from collecting nectar to collecting water. Rising temperature within the hive is controlled by caretaker workers by means of evaporative cooling, for which they must obtain water from the foraging workers. When they need water badly they meet foragers at the hive entrance and virtually mob those who have it while refusing to accept nectar. Foragers unable to give away their nectar will switch to collecting water, then reverse to collecting nectar when water is no longer so eagerly sought and they must search at length within the hive for a caretaker willing to accept it. Nest construction is another example that Wilson proposes is an ideal behavior with which to study the complexities of mass communication. Each species of social insect has an elaborate nest of very predictable form, and the job of building it may span the consecutive lifetimes of many workers. What each builds is determined by what has been built before. Display behavior need not

be involved, but information is being made available by the combined actions of many individuals.

In Wilson's definition of mass communication, the information must originate in the behavior of a number of individuals and must affect the behavior of a number of recipients in such a way that a whole group acts in a concerted fashion. These essential features are also characteristic of some of the coordinated behavior of groups of vertebrate animals. For instance, in the breeding of a colony of black-headed weaver finches, simultaneous nest building by many males is necessary to get females to join them; only when females are present does nest advertisement by the males sweep through the colony. This displaying attracts further females and increases the likelihood that they will pair with the males and breeding will begin throughout the colony (J. R. Hall 1970).

In another case of comparable mass communication, there are species of birds in which infant siblings can communicate with each other to synchronize their time of hatching! A mother quail can incubate a large clutch of eggs, but can brood and care for the chicks that hatch only if they are able to stay together. Although the incubation period for each egg in her clutch differs by at the least two or three days, the whole clutch may hatch within a relatively brief period of a few hours. During hatching the chicks make sounds by moving, breathing, tapping on the shell, and (in many species) vocalizing. Vince (1969) has shown that in such species as the bobwhite quail some young accelerate their hatching by up to 49 hours if they can hear these sounds from the other eggs, while, to a lesser extent, others retard their hatching. The mother quail may also be important in providing stimuli, particularly if she becomes increasingly active or vocal on the nest in the last days or hours of incubating (Orcutt 1974).

Although the behavioral phenomenon of mass communication is not restricted to social insects, its adaptive significance differs greatly for invertebrate and vertebrate animals. Unlike the sterile insect workers, each individual in a group of vertebrate animals has the capacity to breed—to pass its genes along to subsequent generations. Thus each individual, not the group as a whole, is the unit on which natural selection operates. Further, although a vertebrate social group may contain individuals who are relatives, and whose fates must be considered in determining the contributions of the behavior of each of its inclusive fitness, most individuals are unlikely to be as closely related as they are in a hymenopteran or termite colony. In other words, the adaptation driving evolution of vertebrate mass communication is not the emergence of efficiency at the level of the group alone, as it is in the social insects. It is instead the benefit to each individual from membership in the coordinated group.

Humans are also said to engage in mass communication. At least three aspects, however, are very different from the mass communication of other species. First, the information we provide through such "mass media" as newspapers or television need not be about the behavior of a communicating group. This is not peculiar to mass media communication: the same diversity of subject matter is available in much human communication. A second difference results from our use of technology: the signals of human mass communication can emanate from a single individual. This communicator may make available information that a whole group has collected, but still can disseminate it alone to a group of recipients. The latter receive the signals simultaneously. They then often begin a chain of events that, although far from understood (see DeFleur 1970), involves sharing information between and among groups of recipient individuals. At this level the diffusion of information resembles Wilson's model. Third, the single communicating individual is able to act as a very selective filter on the information being made available to the first group of recipients. To the extent that they lack other sources of information for comparison, they are unable to check it for completeness, generality, or distortion. Such selective filtering may be to the advantage of only some individuals, and even if negative feedback responses occur the mass media system may not be sufficiently responsive to reach a stable equilibrium that serves the needs of all its users. Thus, although nonhuman mass communication evolves because it has adaptive consequences for all or most of the group, human mass communication can be used to have adaptive conseqences for very few of its users. The effects of its potential for selfish filtering and distortion may be very large, and are as yet unmeasurable. We should thus be cautious in using the term mass communication or at least in seeking similarities between its characteristics in human and nonhuman species.

PARENTAL CARE AND OTHER PROVISION OF SERVICES   Communication between parents and their offspring based on displays that warn and defend against predators, control aggression between individuals, maintain group cohesion, and have many other functions, does not differ in most ways from communication among adult animals. However, the relationship, at least while the offspring are very young, is not reciprocal; it is a one-way flow of services. A flow of goods is also involved and, with the notable exception of the social insects, this feature is not common in interadult relationships of nonhuman animals.

Parental animals are often much more able to detect predators than are their offspring: adults are more alert, more experienced, and in most species move about more widely. Because parents are also usually more capable of fleeing or of defending themselves, it is not surprising that the

displays they perform on detecting a predator elicit from their young largely selfish responses such as hiding or scattering and hiding while the adults attack, distract, or at least monitor the movements of the predator. Young that are captured may utter vocal displays that function to bring their parents more directly to their aid, and which also cause their siblings to scatter, perhaps confusing the predator. In some species parents also display in the defense of their offspring in intraspecific situations: for example, a mother elephant seal will display at another adult that is hurting her pup and causing it to call (Bartholomew and Collias 1962).

After parents and their offspring have become separated in responding to predators or in less noxious incidents, a parent may employ displays to call the offspring back and reestablish the family group. Alternatively, lost offspring may call until found.

In some species young must be able to recognize their own parents, or mutual recognition is necessary, and displays function to provide identifying clues. In birds, such clues may begin to be learned before hatching as the young call in their eggs and their parents answer (Tschanz 1968; Norton-Griffiths 1969; Beer 1970), leading to an imprinting of the young on individual parents (Hess 1972). With increasing age the young may discriminate more finely among the parental calls they recognize; in laughing gulls (Beer 1973, 1975b) this differentiation enables two forms of the long call to diverge for adult-adult and adult-offspring functions.

The virtually helpless infants of many species of animals are fed and warmed or shaded by their parents, and have displays with which they solicit this care. In other species the young are precocial and fend for themselves, at least in finding food. The procedure of the young displaying (begging) for food is reversed in some of the latter birds, the parents drawing the attention of their young to special foods by displaying. Galliforms, for instance, perform their tidbitting display as a means of offering food (Stokes 1967; Williams, Stokes, and Wallen 1968). Although their young rapidly find and learn the visual characteristics of various food classes, they can fail to learn some of the high-protein foods if they initially flee from, for example, a wriggling worm and subsequently avoid similar things (Hogan 1965, 1966). The parents literally teach them that such objects are edible. As a further function, species such as the oystercatcher perform some food presentation and display as a means of keeping their mobile young with them, where they are more readily protected (Norton-Griffiths 1969).

Displays that adult galliforms use to "offer" food may also be performed by young birds themselves and serve to attract their siblings (Kruijt 1964). Perhaps the main functions of this are to share learning

and to keep the group together. Tidbitting continues to be used to invite other group members to share food in the adult life of these species, but it appears that sharing food, when it does occur (the display is sometimes done in the absence of food), is then a means to achieve other functions such as reaffirming bonds, appeasing, or initiating copulation.

Vertebrates rarely employ displays that have resource-sharing as a regular consequence, and even more rarely have displays that provide information specifically about food. An animal may display when it has food, which may elicit approach from other individuals, but as Marler (1967a) has argued, such approach is not a sufficient criterion for assuming that the display itself implies food or that sharing is one of its main functions. For instance, an adult herring gull with a shoreline feeding territory will attempt to exclude other gulls. If a passing immature gull sees some copious source of food such as a large dead fish, it will intrude and stand near the adult uttering the vocal display with which gull chicks beg to be fed. The adult will not feed it, but may tolerate its presence. Other gulls that hear the immature or see the event will approach, intrude, and eventually so distract the owner that each, including the initial immature, may get some part of the fish (Drury and Smith 1968). Thus in this circumstance the immature's call does have the consequence of enabling food to be shared, although sharing is a last resort for both the territory owner and the first intruder. (There may be more to this case than this, although our work has failed to find anything. A study by Frings et al. in 1955 purported to show that the adult herring gull has a "food-finding call," but the description of this call is not clear and we have found no vocalization with this apparent usage. Their experiments to demonstrate the function of this call were marred by the inclusion of other information sources, including the begging vocalization.) In some circumstances, some functions of displays in other species depend on food-seeking responses. Because the displays are also performed when food is not at issue, they do not specify food or feeding behavior. For example, Venezuelan *Cebus nigrivittatus* monkeys periodically repeat combinations of "huh" and "yip" vocalizations while feeding, which appears to have the consequence of attracting other troop members to a food source (Oppenheimer and Oppenheimer 1973), although the calls' functions probably have to do with group cohesion.

11

Origins and Differentiation of Display Behavior

The tools that are specialized for communication are displays and other formalized actions, badges, and other physical structures. Up to this point I have treated them only descriptively, examining the process of communicating and taking the existence of the tools as given. Like any adaptations, however, their forms are not arbitrary. They evolve either genetically or culturally in response to the requirements of the process they serve, within the limitations of opportunities and opposing selection pressures. Now that the behavioral involvements of both communicators and recipients in the process of communicating have been considered, and the pragmatic-level topic of adaptiveness broached, we must examine these tools more closely. Several questions arise. What are the evolutionary precursors, the origins, of these tools? How are the tools changed in the course of their specialization for communicative functions? What exigencies guide natural selection or its cultural counterpart in determining their adaptive features? Finally, is our existing concept of formalized behavior adequate to encompass what we now have described and to guide further research?

The first two questions are considered in this chapter. The evolutionary origins of displays, and the processes of formalization (that is, of "ritualization" or of "conventionalization"; see below) by which these precursors become modified into displays, will be discussed. It also deals with causes and consequences of a crucial evolutionary characteristic of the display repertoire available to each species: its severely limited size.

Our understanding of the evolution of displays is limited by the methods we must use. As behavior does not fossilize, precursor acts have been postulated on the basis of comparisons among the displays and nondisplay acts observable in related groups of living species, a

method reviewed by Tinbergen (1962) and J. M. Cullen (1972). Based on their interpretations of the nature of precursors, ethologists have also considered the processes that alter these precursors. Despite the limitations of such indirect procedures, there is now considerable agreement on the probable sources of many kinds of display behavior and on how the initial acts have been changed through evolution.

Origins of Displays

The structural and behavioral adaptations of organisms, however novel and distinctive, do not arise de novo. Modern theory holds that evolution is gradual and proceeds through successive stages of transition. Even highly innovative kinds of change must begin as slight modifications of previously established features. Innovation is a matter of modifying suitable precursors in ways other than those that drove their earlier evolution. A feature is suitable to become the precursor of a new kind of adaptation if its existing characteristics give it even a slight potential to accomplish novel tasks, or old tasks in novel ways. It is then said to have "preadaptive" capacities that can gradually be enhanced as natural selection acts through subsequent generations of the bearer's descendants.

Acts that have evolved for direct functions (that function for their performers without the mediation of other individuals) can be preadapted for formalization if some of the information that they make available is socially useful. Most classes of displays must have evolved from acts that had previously been specialized only for direct function; the informative innovations proceed without interfering with the already established functions of these acts (Mayr 1960, 1963; Bock 1959). Dual function would have persisted, except in cases in which the precursor acts were eventually rendered superfluous, or in which the two kinds of functions (direct and indirect) became separated into different acts. Once animals had evolved repertoires of displays tuned to each of their senses, however, much of the further evolution of displays could gradually develop from previously existing displays.

Ethology is a branch of evolutionary biology. Since Darwin (1872) a great deal of discussion has been devoted to the origins of displays. The topic has been reviewed by, among others, Tinbergen (1952, 1953a), Hinde and Tinbergen (1958), Moynihan (1955a, 1955b, 1955c), D. Morris (1956a), and Andrew (1963). Several basic categories of behavior that might provide suitable precursors for at least visible display patterns have now been identified.

First, directly functional movements are often incompletely performed: they are begun and then aborted either before they are fully

under way or at least before they have accomplished their functions. Because these movements represent the preparatory phase of an action, or even just some components of this phase, they are called "intention movements" (Daanje 1950; the term is meant only to be descriptive). For instance, a species in which biting is part of the behavior of attack may evolve a display that emphasizes a widely gaping mouth (figure 11.1). Through formalization, the display movement is changed from an actual attempt to bite into a signal that the animal may attack. When more likely to flee, the same animals may have displays formalized from components of movements used in turning away from their opponents.

Many actions are more complex than biting or turning and provide richer opportunities for the evolution of displays. For instance, many bird species have displays that appear to have evolved from intention movements of taking flight, and Daanje pointed out that different displays often resemble different components of the preparatory movements. To take flight a bird first flexes its legs, lowers its breast, raises its tail, and pulls in its neck. Then it extends its legs, thrusts out its neck, rapidly depresses its tail, raises its breast, and extends its wings. It may abort the process at either the crouching or the springing phase, providing for ritualization a variety of different postures (such as crouching, with the neck withdrawn) and movements (like raising and then depressing its tail). In displaying, the movements need not maintain the

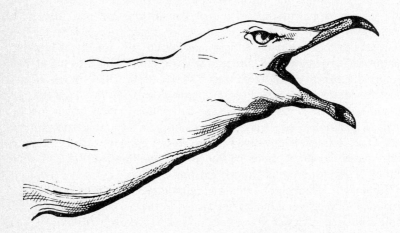

Figure 11.1 The gaping jab display of the ring-billed gull. A disputing ring-billed gull *Larus delawarensis* jabs with its wide open bill, conspicuously revealing a vermilion-colored mouth. Although the gaping jab display involves an attack movement, the communicator does not actually attempt to make contact with its opponent; it is threatening and not fully committed to attack behavior. (From Moynihan 1958a.)

same relations to each other and to the postures that they do when the bird actually takes flight, so the potential for developing different displays is large. Figure 11.2 shows the incorporation of these components into the full forward display of the green heron, and compares this display posture with various directly functional postures of the species.

Aborted intention movements of alighting have also been formalized by some kinds of birds. Red-winged and yellow-headed blackbirds, for instance, sing as they fly around their territories in postures that are gross exaggerations of the movements made in approaching a perch: they beat their wings slowly and deeply, and spread and lower their tails (Orians and Christman 1968).

Many kinds of directly functional acts may be aborted well after the preparatory phases. For instance, several species of parulid warblers have a display called "circling" (Ficken and Ficken 1965), which appears to have evolved from an aborted attack flight followed by withdrawal to a nearby perch. For instance, a chestnut-sided warbler will fly in the direction of a territorial opponent and then turn back toward its initial perch, or may instead make a flight that is an almost unoriented arc. The orientation of circling flights done by yellow warblers varies, but apparently they are nearly complete circles that bring the performer back to its original perch more often than do the chestnut-sided warbler's flights. The American redstart's display flight is almost always a full circle. Further, it is done with modified wing beats, and tends to elicit a very similar flight from an opponent. A comparison of the three species suggests that the very formalized redstart performance is probably homologous with the much more variable performance of the chestnut-sided warbler, in which we can see the close relation between the display flight and an aborted approach to attack or challenge.

The formalization of circling has incorporated a component of attack behavior (approach) in sequence before an escape component (withdrawal), and attack and escape messages appear to be encoded as alternative behavioral selections (Lein 1973). In other birds, displays that encode both these messages may be based instead on features that are common to both kinds of behavior, and hence not on oriented acts but on preparations for flying: postures of crouching, wing-raising, and so on. Andrew (1956) designated such display components "compromise behavior," although Hinde (1970:404) has pointed out that such a term would be more appropriate for acts of intermediate form than for acts that are shared by different behavioral selections.

It can be difficult to say whether the precursors from which some displays appear to have evolved were incomplete acts or not. In the extreme cases, the entire pattern of a directly functional act is apparently

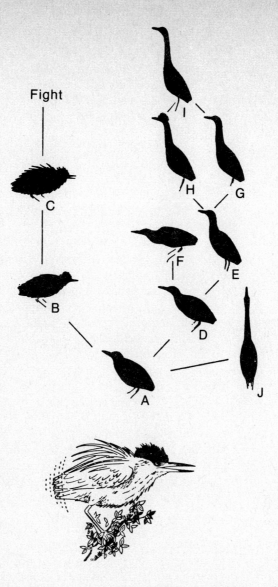

Figure 11.2. The incorporation of intention movements into a display pattern of the green heron. The green heron's full forward display (shown in silhouette in *C* and in more detail in lower figure) involves lowering the breast (a preflight movement) and flipping the tail, a repetitive action probably derived from the movements of lifting, then thrusting the tail downward in taking flight. Postures *D* through *I* are not displays, but stances that facilitate monitoring or flight. In the breast-lowering preparation for flight (position *F*) the bird extends its neck back and forth as it looks for danger. Position *J* is cryptic; it is adopted when a predator appears nearby very suddenly and capitalizes on the resemblance of the heron's plumage to the appearance of marsh grasses. (From Meyerriecks 1960.)

performed, even repeated several times, but done short of the place where it could have direct effect. For instance, eastern kingbirds have a tumble flight display in which they go through all the motions of grappling with and beating on an opponent that is not in their grasp, and often not even in sight. A kingbird will tumble and fall down as much as 3 or 4 meters while in this mock fight before it resumes flying, then will gradually rise and repeat the act (W. John Smith 1966). Some shorebirds such as black-tailed godwits have comparable performances (Lind 1961). Two confronting bull tule elk of nearly equal size and rank will stand a meter or more apart, twisting their antlers at each other as if they were physically engaged in a sparring bout (McCullough 1969). All these display performances are formal mock fights, incomplete only in that blows do not land on opponents.

Incomplete acts should make useful sources for the evolution of displays because they are especially likely to occur at times when their performer is indecisive. Its next activities can be difficult for other individuals to predict unless they have some indication of the alternative options it is considering. When it performs incomplete acts, it provides clues to some of those alternatives.

That animals actually are attentive to the intention movements of their companions was shown by Davis (1975). He observed that when a pigeon is startled and takes flight, the pigeons foraging in association with it also fly. On the other hand, a pigeon can also fly alone from the midst of a flock, causing no disturbance. Detailed examination of videotaped recordings in both kinds of cases revealed no differences in the movements of taking flight, but did show variations in the preceding behavior. If a pigeon made flight intention movements within 10 seconds of flying its companions apparently expected flight and did not usually follow it. If it performed no flight intention movements in that period they were much more likely to react to its departure by flying too. Preflight intention movements enabled them to distinguish a "planned" flight from a startled flight, and thus forestalled unnecessary alarm. The evolutionary elaboration of such intention movements into preflight displays has happened in some other kinds of birds, such as the pelecaniforms (van Tets 1965; see also chapter 10, this book).

Second, although incomplete acts often seem to have been the precursors of visible displays, completed acts can also provide material for formalization if they occur consistently in appropriate circumstances. For instance, the forms of some facial displays of mammals appear to differ little if at all from movements with direct functions. Andrew (1963) offered several examples of display movements that are also protective acts: lowering the eyebrows helps shield the eyes from being struck, and flattening back the ears gives them some protection from biting and tearing attacks. These eyebrow and ear movements are

Figure 11.3. Tongue showing displays of humans and gorillas. In *A*, a Cuna Indian child reaches to touch a more assured friend in response to the stare of an adult stranger who is photographing her. She averts her eyes, and shows her slightly balled tongue between her lips. (This is not a defiant sticking out of the tongue.) In *B,* a young captive gorilla gives a comparable performance. It has retreated to a ledge while avoiding overly active play with a human, and seeks touching reassurance from that person with a hand-offering display. Its eyes are averted and its slightly balled tongue is easily seen. (From Smith, Chase, and Lieblich 1974.)

often performed when their protecting functions are not required. Similarly, humans, gorillas, and perhaps other species of at least primates have a display of tongue showing (figure 11.3) that apparently derives from acts of ejecting food or a nipple from the mouth when an infant is being fed (Smith, Chase, and Lieblich 1974). After infancy, the visible tongue showing display is done much more often in interactions that have nothing to do with feeding than in those that do. It still indicates that interaction, or some aspect of interacting, is being shunned by the communicator, although objects against which the tongue can push need no longer be available to be rejected.

Among the clearest examples of acts that have served in their entirety as precursors for displays are autonomic responses with thermoregulatory functions. These acts have probably been the sources of many displays that involve movements of fur or feathers, for instance, in circumstances in which muscular activity or tension is increasing the heat load, or passive inactivity is leading to chilling.

Figure 11.4. The crouch display of the crimson-backed tanager, with and without feather ruffling. The typical posture without ruffling is shown at the top. A more extreme form of the posture is shown below. The bird's legs are more flexed and its body axis is tilted farther forward. It is using the gaping display, and has very conspicuously ruffled the feathers on its head, neck, upper back, breast, and belly. In this species ruffling tends to involve most of the body plumage, although often omitting the rump feathers of the lower back. In other species display ruffling can be restricted to fewer tracts of feathers. (From Moynihan 1962d.)

D. Morris (1956a) and Andrew (1956, 1972) were the first to point out that the conspicuousness of the plumage adjustments by which a bird can increase or decrease the rate at which its body surface loses heat makes these actions highly suited to be display precursors. A bird's feathers customarily enclose some air over the skin, and this entrapped air acts as insulation. The thickness of the insulating layer can be increased somewhat by fluffing the plumage, which is often done by quietly submissive birds. In fluffing, each feather is raised only to the pont at which its tip still falls back far enough to lie on more posteriad feathers. Feathers raised beyond that point are ruffled. Because they separate and no longer entrap air, the bird cools rapidly. Ruffling is often conspicuously incorporated into agonistic displays (see figure 11.4), coming and going in correlation with behavioral changes sufficiently rapidly to indicate that it is probably not serving just as a thermoregulatory device. Finally, a bird can also decrease the insulating effect of its feathers by depressing them, giving its body a sleeked appearance. Sleeking, like ruffling, also appears in the agonistic displays of many species.

Mammals show similar pelage adjustments to fluffing, ruffling, and

sleeking—for example, the electrified, hair-on-end appearance of a terror-stricken alley cat. Two examples of primate piloerection displays are shown in figure 11.5.

Third, acts that are completely performed but "redirected" onto substitute objects have served as the precursors of many visible displays (Bastock, Morris, and Moynihan 1953). The mock sparring by bulls of the tule elk could be viewed as a possible example, for instance. In addition to being performed incompletely with the antlers pushing only at air, as mentioned previously, it can also be directed onto an intervening bush. Instead of his opponent, each bull thrashes the hapless plant. Similarly, many species of gulls have a so-called grass-pulling display in which they vigorously jab and bite at nearby grass, seaweeds, or bits of flotsam or other debris, then often rip up the material with sharp twists of their heads and throw it aside. Performed in agonistic encounters, the display appears to have been derived from attack movements that were redirected from an opponent, much as a man may use his fist to strike the palm of his other hand or to pound on a desk. Gulls redirect yet other forms of attack behavior in their aerial swoop and soar display, which is often performed by an unmated male in response to a female intruding into his territory. It is not directed toward her but toward some gull flying so far in the distance that it takes no notice (Moynihan 1955a). The apparent function is not to threaten the distant bird, of course, but to provide the closer gull with the information that although the male may attack, he may instead do something incompatible with attacking— not a very warm welcome for a prospective mate, but a step that can help by letting an uncertain female know how things stand.

Fourth, another class of activities that has been proposed as a potential source of precursors for displays contains what ethologists call "displacement" behavior patterns, acts which appear to be irrelevant in some situations in which they occur, and yet may be performed quite consistently in these situations.

This category is difficult to recognize because it is up to the ethologist to evaluate relevance (see Hinde 1970). "All too often, in actual practice, the fact that a behavior pattern has been called a "displacement activity" is nothing more than an indication that the human observer who saw the pattern did not expect to see it when he did" (Moynihan 1962a:160). If overly impressed by exigencies that an animal cannot meet in some circumstance, a human observer may fail to realize that the animal may be behaving more appropriately in turning to some activity of secondary importance than if it did nothing at all. Further, the observer usually lacks unambiguous evidence about the evolutionary precursor of a display, and may suspect only an apparently irrelevant source without thinking of a relevant one that makes as good

Figure 11.5. Piloerection displays of the rufous-naped tamarin *Saguinus geof-froyi*. The upper drawing shows a typical leaping posture, with tail-ruffling. One of this species' most common displays, tail-ruffling provides the information that escape behavior is more probable than some form of friendly inter-action. The tamarin in the lower drawing shows the much more conspicuous general ruffle display, in which it has raised almost all the hair of its head, body, limbs, and tail. This individual is more likely to attack or behave inde-cisively than it is to do other alternatives. (From Moynihan 1970a.)

or better a postulate. The precursor of the agonistic grass-pulling display of gulls, for instance, was originally attributed to displacement nest-building acts because materials are often picked up and tossed to the side (Tinbergen 1953b). As suggested above, the close similarities of the display's movements to jabbing and tearing are more consistent with an origin in attack behavior, a source that is not at all irrelevant in agonistic encounters.

Displays derived from what have been thought to be displacement acts often resemble such "comfort movements" as preening or shaking plumage to rearrange feathers or to remove water droplets (for detailed examples see reviews of comfort movements of waterfowl, McKinney 1965b; of penguins, Ainley 1974b). In many species of ducks, for instance, a courting drake will point at brightly colored feathers in his wing with a mock preening display movement. In different species he may or may not actually manipulate the feathers with stilted, abortive, preening actions of his bill. Because the reasons for self-preening during social interaction are not obvious, it has been thought of as displacement behavior. However, the other displays of courting disarrange feathers (Moynihan 1955c), and because the exertion of courting creates body heat, feathers become further disarranged as they move in adjusting to the needs of thermoregulation. Thus preening may often be necessary in close temporal conjunction with courting. (People tend to self-preen by touching hair and clothes just before or after interacting; Adam Kendon has suggested to me that this is relevant at such times as a means of assuring that one's appearance or "front" is in order.) Moynihan (1955b, 1955c) noted that close temporal conjunction with more obviously social activities is all that is needed to render preening informative and therefore a good candidate for ritualization. Because comfort movements can be available to be developed into displays for quite straightforward reasons, it is probably wrong to postulate that they were originally irrelevant in social events.

Are displacement patterns ever the precursors of displays? We won't know until we know more about displacement behavior. Until then, it is best to be cautious about postulating that acts regularly occurring in natural circumstances are irrelevant.

It is apparent that diverse kinds of acts offer likely sources for the evolution of visible displays, and that most categories into which they have been grouped intergrade with one another: incomplete movements differ in degree and may be virtually complete, acts may be redirected only part of the time, and so forth. Perhaps the most important conclusion is that abundant behavioral material is readily available to contribute the evolutionary precursors of visible displays; there seems to be no need to postulate mysterious sources or processes.

Because we lack the behavioral analogue of a fossil history, the exact precursors of particular displays must remain conjectural, but uncertainty at this level does not imply that the general range of possibilities is not reasonably obvious. Further, by comparing the homologues of a display that evolved within groups of related species we can often find steplike differences that accord with postulated precursors (the notion of homology holds that the forms of acts of different species are similar because the acts have been inherited from a common ancestral population). The warblers' circling display described above, which is increasingly more formalized in each of three species, is a good illustration. Arguments based on homology are most convincing if the species compared are closely related, of course, and if the behavioral differences among the species are slight. Other conditions can make behavioral homology very difficult to assess, and if the degree of resemblance is not examined exhaustively comparisons can be led astray by parallel or convergent adaptations, or by mimicry.

Most ethological conjecture about the kinds of acts that could have been evolutionary precursors has considered visible displays. Perhaps most tactile displays are sufficiently similar to such directly functional touching acts as holding for support, restraining, guiding, pulling, or striking that their evolutionary precursors seem evident. The origins of displays that depend on chemical clues, however, are less evident. The bioluminescent flashing of fireflies may have originally been adapted not as signaling, but as a means of illuminating an area in which a suitable place to alight was being sought. As males sometimes flashed when alighting near females, flashing could have become incorporated in precopulatory behavior (Lloyd 1968). The chemical substances involved in the evolution of bioluminescence, and also in pheromonal signaling, may have been metabolic waste products. However, we must identify the behavioral precursors that became ritualized to bring some of these chemicals into play at socially appropriate times, not the chemical precursors.

Marking with urine or feces may be in part initiated by autonomic responses to anxiety as an animal finds itself in unfamiliar terrain or situations (Kleiman 1966, building on postulates of Lyall-Watson and D. Morris). An individual that leaves its scent in an unfamiliar area should recognize that the region is at least somewhat familiar when it later encounters the odor again. By scent-marking it can know more surely when it is on ground that has been safe, and other indivdiuals can also learn of its use of the area. Thus the adaptive consequences of continuing and improving the behavior of site-marking with urine or feces are apparent. Yet what about species that mark not with autonomically controlled voiding of waste excretory products, but with specially produced scents? In particular, what initiates the rubbing be-

havior with which they deposit scent marks? The only suggestion that is perhaps sufficient to explain the evolutionary origins of some such rubbing is that it could have been done in response to local irritations produced by having glandular excretions or secretions on the skin (Andrew 1972); once such scent-marking became important to a species, the stimuli controlling its occurrence might well change radically.

The precursors of the behavior employed in releasing scents that are used other than in acts of marking remain even less clear. There are many kinds of pheromones, however, for which excretion is not controlled by special behavior. For instance, phermones indicative of readiness to breed are often made available in an animal's urine as its hormonal physiology changes. A recent hypothesis offered by Kittredge and Takahashi (1972) may account for the way release of sex pheromones began in crustaceans, and may even suggest a starting point for the evolution of sex pheromonal communication in the whole invertebrate phylum Arthropoda. Experiments by Ryan (1966) showed that some substance in the urine of females that are about to moult attracts male crabs. A male who has located a premoult female stays with her and protects her while she moults, and they copulate immediately afterward. Kittredge and Takahashi established that the pheromone is crustecdysone, the steroid hormone that controls an individual's moulting physiology. They propose that sensitivity to this hormone as an *internal* regulator of physiology already existed on the membranes of the target organs before natural selection began to develop its external use as a pheromone, and that potential recipients detected leakage into the water of crustecdysone from moulting females.

Because female crabs are receptive to copulation after their moult, increased ability of males to detect the hormone through *external* chemosensory receptors would be favored. At the same time, however, crustaceans limit their loss of metabolic products because they can attract predators; their urine production involves processes of active resorbtion of such substances. As the excretion of crustecdysone became advantageous in attracting males who would both be available for copulation at the right time and offer the vulnerable female some protection, females must have evolved physiological inhibition of the active resorbtion of this hormone from the urine—although only for that moult immediately preceding their readiness to breed. The Kittredge-Takahashi hypothesis is attractive both for its potentially wide explanatory scope and for the fact that it requires only minute evolutionary modifications in existing capabilities to promote the origin of this form of pheromonal communication.

In addition to sex attractant pheromones that evolve from metabolic

products, others may evolve through the development of mechanisms for storing exogenous chemicals. For instance, Hendry et al. (1975) propose that the sex pheromones of the oak leaf roller moth *Archips semiferanus* and various other lepidopterans are present in the species' food. Numerous isomers of tetradecenyl acetate are found in both oak leaves and in the sex attractants of female oak leaf roller moths reared on oak leaves, but are not detectable in females reared on a mixture of wheat germ and alfalfa. Further, the latter females are not attractive to males, although oak leaves themselves are: males even attempt to copulate with damaged oak leaves that are leaking the tetradecenyl acetates. By being attracted to the odors of oak leaves males are brought to plants that are suitable for rearing the species' larvae and, by seeking greater concentrations of the requisite chemicals, to appropriate females. They may learn the chemical clues while they are larvae feeding on oak leaves—in fact, their mothers may have stored in their eggs minute amounts of the chemicals as feeding cues to aid the hatching larvae in selecting the appropriate host plant.

J. P. Miller et al. (1976) subsequently found that a very specific ratio of two tetradecenyl acetates characterizes the sex phermones of moths of this species, and that not only was this ratio now altered by a female's diet, males were responsive to extracts of females reared on diverse foods. Their findings, however, do not necessarily disprove the earlier results. As Hendry (1976) points out, the second response tests were made with unnaturally high concentrations of extracts, the chemical tests performed in the two studies differed, and even the moths tested may have differed in their pheromonal response preferences. Research in this area is difficult.

One implication of a system based on storing exogenous chemicals for use as pheromones, if it exists, is that the evolution of closely related plant-feeding insect species may be closely tied to the pheromonal opportunities made available during the evolution of plant species. Females of insect populations evolving adaptations to foraging on different plant host species could gain different pheromones by ingesting different isomeric combinations of chemicals characteristic of the plants. Their release of these as pheromones would then enable males to find and identify mates reared on larval diets like their own, promoting further differentiation of their breeding populations. Again, however, this is an hypothesis about the evolution of chemicals and responsiveness to them; it does not answer how the display behavior by which insects control release of their sex attractant pheromones evolves, although in principle this need not be difficult.

The evolutionary sources of vertebrate vocal displays are unknown. Much of the diversity found today must result from the evolution of new

vocalizations from pre-existing ones. The original vocalizations, however, may have been ritualizations of either "acts of displacement breathing" (Spurway and Haldane 1953), or amplifications and modifications of the loud, deep, and often rapid breathing of terrified or sexually aroused animals (apparently what Darwin was suggesting in 1872). Displacement breathing is inhaling and exhaling between the breaths necessary for direct function. For instance, aquatic amphibians gulp air to support respiration and to provide buoyancy. Evolution may have favored some gulping in excess of that needed for these functions if the noises produced by the act were useful in social attraction and repulsion.

The behavioral sources for nonvocal sounds must be as diverse as the modes of sound production. Birds that make sounds in display flights by employing specially stiffened wing or tail feathers (as in manakins and snipe, respectively) may initially have developed special ways of holding their wings and tails in some flights as visible displays; the first weak noises may have occurred only as by-products of the flight postures. Woodpeckers presumably developed the loud drumming displays (chapter 2) with which they advertise their presence from the noises they inevitably make while using their bills in foraging and in excavating nest holes.

The electrical discharges with which some kinds of fish display apparently derive from the normal production of much weaker but detectable pulses during the intense activity of white muscle tissue that produces rapid movements (Barham et al. 1969). Because fish are thin-skinned and live in a conductive medium these pulses have always been available for modification, but specialized muscle organs had to be developed to propagate them at higher strengths.

The Differentiation of Displays from Their Precursor Acts

Following Huxley's description (1923:278) of certain highly stereotyped performances of birds as being acts used "in a ritual way," Tinbergen (1952) adopted the term "ritualization" for the process by which display behavior becomes "schematized" and differentiated from its evolutionary precursors. He defined ritualization as "adaptive evolutionary change in the direction of increased efficiency as a signal" (1959b:44; see the further discussion in Blest 1961). Despite considerable interest in the learned features of some displays (especially in local variations of bird song, see chapter 12), ritualization has generally been construed by ethologists to be a process of genetic evolution. That limitation is tacit or explicit in most discussions of ritualization or its effects, including those of Tinbergen (1952), Hinde and Tinbergen

(1958), Moynihan (1955c, 1967), D. Morris (1957), Chance (1962), Baerends and van der Cingel (1962), Kruijt (1964), Lorenz (1966), Huxley (1966), Andrew (1969), and W. John Smith (1966, 1969a). In fact, evolution can be assumed to be genetic, for example, in most displays of social insects, which E. O. Wilson (1971a) describes as genetically evolved behavior.

However, the assumption that the evolution of display forms is solely genetic is too limiting for many vertebrates. The limitations are perhaps most obvious in the study of stereotyped behavior patterns employed in human communication. Some can be very ephemeral and local, surviving as a part of a group's signal repertoire for only a few weeks. Even displays that are characteristic of all humans have some aspects of their use that are under cultural control: Ekman and Friesen (for example 1969a) have found differences among cultures in the circumstances in which it is acceptable to perform such "affect displays" as smiling, frowning, or crying. According to them each culture has "display rules" controlling usage, even though the same facial expressions tend to convey fundamentally the same information.

Because the term "ritualization" has been used extensively to represent a process of organic evolution, it would be useful to retain it in this sense. Where cultural processes dependent on learned transmission have acted to specialize behavior to be informative, they can be called "conventionalization." In addition, a more general term is needed to designate all behavior that has been specialized to be informative, without implying any particular process of specialization. The term "formalized" serves this end herein. The formalization process in this book thus includes both ritualization and conventionalization. A display is viewed as an act that owes part or all of its characteristics to formalization through either ritualization or conventionalization, or both.

Like any existing term, "formalized" is not perfectly suited to describe behavior specialized to be informative. In common use, to say that behavior is formal implies only that it adheres to a fixed, generally recognized form. Yet, although formal behavior patterns may be used as learning, mnemonic, or other devices that do not necessarily serve primarily or solely in what would usually be called interindividual interactions, their basic functions may all lie in enabling an individual to make available to a recipient information that might otherwise not appear. That recipient and communicator can sometimes be the same individual, encountering results of its display behavior after having performed the displays, is considered in chapter 10. It is there argued that even such cases are reasonably encompassed within the notion of interindividual communicating, to the extent that time has changed the communicator.

A second difficulty arises because display behavior has been described as "formal," "formalized," and the product of "formalization" by earlier writers, including at least Selous (1933, and perhaps earlier), Boase (1925), Tinbergen (1939b), and Armstrong (1947). At times the concept implied was probably as broad as that suggested here, although at other times it was more nearly identified with the implication of genetic evolution. The distinction between genetic and cultural evolution was not usually drawn, and at least Armstrong appears to have used "formalised," "conventionalised," and "ritualised" as interchangeable terms. Nonetheless, only ritualized took hold in the ethological literature and acquired the meaning of genetically evolved. The other terms, although not being introduced here to the ethological literature for the first time, have been so little used by ethologists in recent years as to be all but unburdened with tenacious connotations, and are free for use in a new effort to distinguish different specializing processes.

The formalization of a display makes it more effective in providing information. This often requires that it become a distinctive act, less easily confused with its precursor and thus less ambiguous (problems of ambiguity also require that it be made distinctive from other displays, as considered below and in chapter 12). Some or all of the precursor act's component movements or positions may become exaggerated in form, or in the speed of their performance; their response thresholds may change; the coordination among them may be altered, and components omitted; a tendency to rhythmic repetition may develop; the orientation of the precursor act may be lost; and the performance may develop a "typical intensity" of form (D. Morris 1957) that does not change in parallel with intensity changes of the eliciting circumstances.

One result of these changes is that displays can become relatively stereotyped. In fact, decreased variability of form is thought to be a very common result of formalization, although the first measurements made to show the differences in stereotypy between displays and their apparent precursors were done only recently (see Hazlett 1972, who studied agonistic displays and ambulatory movements of the spider crab *Microphrys bicolor*). D. Morris contends that decreased variability may originate in part in a tendency of animals to engage in stereotyped acts when thwarted, because most displays are performed when motivational conflict renders the actor indecisive. He points out that animals thwarted by being caged in barren environments develop actions that are individually stereotyped in just the same ways displays are stereotyped for populations (1966).

Stereotypy, exaggeration, and many other kinds of changes often render a display act more conspicuous than its precursor. Obviously, conspicuousness can be a useful quality for many displays because a

recipient's attention must be attracted. The strength of the selection pressures favoring it is evidenced by the frequency with which formalization proceeds along with concomitant modifications of structures that are involved in or revealed by display acts (badges, see chapter 9). For example, numerous species of birds have displays in which the crown feathers atop the head are raised; in many of these birds the feathers are brightly colored, elongated into a pointed crest, or both. In birds that spread and flutter their wings as a display act the wings may have conspicuous stripes or patches, or have undersurfaces that contrast with their uppersurfaces. Other examples are so extremely numerous and diverse that ethologists are accustomed to finding conspicuousness as a quality of display behavior, and have come to expect it (for example, see Hinde and Tinbergen 1958).

However, there is now concern that less conspicuous and less stereotyped displays may also be abundant, and that we have in the past perhaps sought and studied primarily the more blatant examples. Faced with an enormous need to describe the wealth of conspicuously stereotyped acts in species after species, it was perhaps natural to devote relatively little attention to acts that, if formalized, were less obvious to a human observer. More than anything else, the recent concentration on primate (including human) social interaction has begun to lead to more interest in relatively subtle communicative acts.

A final kind of effect that has been postulated for the processes of formalization must be mentioned, although as it is a matter primarily of mechanisms internal to the individual rather than of interactional mechanisms, we need consider it only in passing. This is the notion that the actions in evolving displays may become separated ("emancipated," Tinbergen 1952; Blest 1964) from their original motivational control and transferred to quite different control. An act originally motivated by aggression could, in theory, come under sexual control when developed into a display. The process, however, need involve nothing more than the kinds of threshold shifts in underlying motivational states that must characterize the evolution of all display behavior (Blest 1961; J. M. Cullen 1972). Further identifying major results of emancipation requires determining the hypothetical precursor acts for displays in cases in which this may be especialy difficult. In part because of this latter demand, the concept of emancipation has remained difficult to test (see, for example, Blest 1961; Baerends and van der Cingel 1962; Bastock 1967:118–120; and Blurton Jones 1968:152–154).

That formalization leads to changes in the form, or the employment, or both of acts suggests that at least two kinds of processes are involved (W. John Smith 1969a). In fact, at least five categories of formalizing processes can be distinguished by their different kinds of effects. Any

one display may be a product of more than one kind of process, thus embodying different sorts of clues that differentiate it from its precursor.

Novel Transformation

In novel transformation formalization alters the physical form of a behavioral precursor. Most of the displays that ethologists have studied are formalized in at least this way: for example, all vocalizations and other specialized means of producing sounds, all visible displays that are exaggerations of their precursors (including postures, movements, and enhanced color changes such as blushing) or are redirected away from their initial objectives, and all electrical discharges that are modifications of the precursor monitoring discharge.

Some acts appear to be slightly modified in form in specific usages; for instance, in waterfowl the comfort movements that occur before flying or in courtship sometimes look more stilted or hurried than when done in more relaxed circumstances (see discussions by McKinney 1961:48–50, 1965b). Perhaps some of these are weakly formalized, but if the acts are sufficiently informative without being modified they are not under very strong selection pressures to become much exaggerated or otherwise novel; the problems raised by minimal formalization are discussed in chapter 13. Alternatively, some comfort movements of ducks may be unchanged in form but be performed more frequently than would otherwise be expected, as is discussed under the third category below.

Displays in which body color is changed rapidly, special organs are inflated, dewlaps extended, hidden patches of color revealed, and the like, were called "transformation displays" by D. Morris (1956b:270). They are only one kind of product of the novel transformation process, specialized to enable a cryptic animal to display conspicuously and then revert to being cryptic. Although Morris defined "transformation" as an immediate and reversible change in appearance, I use "novel transformation" to refer to the development of behavior that differs in form (such as appearance or sound) from any act previously in a species' repertoire.

Although the form of a precursor act is often changed in the evolution of a display, acts have other characteristics that may be formalized in addition or instead: their temporal and spatial employment. Several kinds of change can affect these characteristics.

Form-Restricted Iteration

The form of a precursor act can be modified without introducing novelty, if a display becomes only a selected portion of the precursor which

is then sustained or repeated. This process narrows or restricts the complexity of an act, and extends performance of the reduced form for whatever duration its new functions require. Aspects of the precursor behavior (such as orientations that facilitate the gazing patterns with which the actor collects information) are sometimes left sufficiently unhampered that they can continue their original functions.

Ducks' inciting displays, for example, provoke their drakes either to chase off other drakes who have approached them too closely, or to lead them away from those drakes (see Johnsgard 1965). In a few species such as the European shelduck and the Egyptian goose, the female simply points her bill at the offending male while calling, and the angle of her head with respect to her body axis varies as her bodily orientation changes (Lorenz 1941, 1966). As she alternately rushes to threaten an offending male and then retreats from him toward her mate, she is at times pointing with her bill over her shoulder (figure 11.6). In the mallard and some related species the female incites only by turning her head over her shoulder, even if the offending male is not in that direction. Formalization has limited her act to a single posture that is sustained whatever her orientation to the offender, and this one posture has become more important than is actual pointing.

Human conversational behavior has provided several examples of behavior that has been formalized through this process. Scheflen (1964) has noted that the postural relationships of conversing individuals are orderly, and that different postures and orientations tend to distinguish persons who are interacting from persons who are not. For example, bodily postures can sometimes distinguish relationships among three individuals when two or interacting with the third but not with each other. Kendon (1970) has found that a listener who is interacting actively and alternating roles with a speaker (designated as being in an "axial relationship") tends to copy the speaker's posture and to move in congruence with the speaker's moves. That is, of the various postures available to a listener only a limited and definable subset is selected and sustained. Other available postures are evident in the more varied and less responsive behavior of nonaxial listeners, at least until one is about to assume an axial role. He or she then usually begins to shift into congruence with the speaker.

The effects of form-restricted interaction can be detected in many displays that are conspicuous primarily because they have also been affected by the novel transformation process. For instance, in its rub-urinating display (Müller-Schwarze 1971) a black-tailed deer adopts a novel posture and performs novel rubbing movements with its hind legs while urinating, presumably spreading the pheromone-bearing urine on its fur. The behavior of micturating has also been altered, although only by being broken into short bouts and repeated; that is, the behavior is

Figure 11.6. Inciting. On the left (*A*) a female shelduck *Tadorna tadorna* incites her mate against an opponent that she sees directly before her. The female on the right incites with head turned toward an opponent seen to one side. A female mallard always incites by turning her head (*B*), and in this case she is turning it toward the male who is bothering her. In (*C*) the offending male has taken a position more nearly in front of the female mallard, and her unchanged inciting display results in her turning her head away from him. In both (*B*) and (*C*) her mate is the leftmost bird. (From Lorenz 1958. Copyright © 1958 by Scientific American, Inc. All rights reserved.)

altered through form-restricted iteration. The whole rub-urinating display thus involves several changes from the precursor behavior, and is the product of at least two different formalizing processes.

Iteration in Whole

In some cases formalization leads entire acts to be extended longer than necessary for their directly achieved functions; they are either prolonged or repeated. A nestling bird, for instance, opens its mouth widely to accept food. It does not gape in this fashion only when food is being thrust toward it, however, but whenever it is hungry and a parent bird appears nearby. In form, all such gaping resembles that done to take food into the mouth. But in its extended performance the formalized act indicates the actor's readiness to accept food even when it is not being offered.

Formalization by means of iteration in whole can also be seen in the very complex challenge ritual with which male wildebeest on neighboring territories check each other daily (Estes 1969). One intrudes onto another's territory, they perform a ceremony of several steps together, and then the intruder returns to his own territory. While intruding and withdrawing, and at times when the two males are close together, the intruder or even both individuals adopt the head-down pose used in grazing. They may, in fact, eat. Without altering the physical appearance of the grazing behavior, formalization has increased the frequency and duration with which the act occurs during this type of agonistic interaction. Estes argues that the posture, and even the actual grazing, is elicited by the agonistic circumstance and provides at least the information that neither immediate attack nor escape are likely.

In some species, displays have evolved from behavior in which one individual collects materials and offers them to another. The recipient, who is usually socially bonded to the donor, may need the materials. For example, in some species in which the male gives food to his mate ("courtship feeding"), the food may help her to obtain sufficient energy resources for egg production (for example, in titmice, see Royama 1966; Lack 1966). Yet the male may provide more food than she needs to supplement that which she gathers herself, or he may feed her even when she is not making eggs. In a few species such as the piñon jay (Balda and Bateman 1971) the female also feeds the male, who can presumably look after himself. Such human food ceremonies as the exchange of pieces of wedding cake are interesting analogues. All such acts of giving food in excess of physiological needs represent iterative formalization of behavior that sometimes has necessary direct functions.

The piñon jay is also among the species in which males provide their mates with samples of nesting material. Little or none of this material gets incorporated into the nests of most species, so that the acts of providing it are significant primarily as displays. In fact, a male piñon jay may offer sticks to his mate even outside the nest-building season. Courting males of many species of tyrannid flycatchers take no part in building nests, but will go to potential nest sites and perform as if building. All such displays are formal iterations of directly functional behavior patterns, although some of the latter have been lost in the evolution of male behavioral repertoires and entire sequences of nest-building behavior do not occur.

It is by no means always possible to distinguish displays produced by iteration in whole from products of form-restricted iteration. When the behavioral precursor of a display has been an intention movement, it is a moot point which process was involved. An intention movement is by definition an incompletely performed act, yet an act iterated in all the de-

tail typical of an intention movement would represent the latter in whole. For most analyses it matters only that acts, including intention movements, can be formalized in whole or in part through iteration. For instance, when a budgerigar parrot is preparing to move away from another budgerigar (with a probability measured as 0.55; see Brockway 1964), it adopts a lean back display posture. This awkward-looking but otherwise inconspicuous and unelaborated display is either a whole frozen intention movement of withdrawing or a sustained version of part of the act of beginning to fly away. Other examples of displays formalized from what appear to have been flight intention movements involve repeating the movements in patterns, as when a goose rhythmically pumps or tosses its head, alternately pulling it between its shoulders as in a preflight preparation and then extending it as in taking flight.

Allogrooming and allopreening, the most often described iterative formalizations, in which one individual cleans a companion's fur or feathers much more than is necessary for hygienic functions alone, present a different sort of analytic problem. Because cleaning is accomplished, both direct (hygienic) and indirect (signaling) functions are served simultaneously, although only the latter are continuous. The problem is that such events are formalized at more than one level. The component actions of cleaning are individual tasks, as are the invitational facilitating or offering postures. But cooperation of two or more individuals is continuously necessary to manage the interactions themselves; this cooperating is regular and follows a formalized pattern. (Formalized cooperation in interacting is discussed in chapter 14.)

Finally, the products of formalized iteration can often create analytic problems by being inconspicuous, especially if there has been no additional exaggeration by novel transformation processes. Unexaggerated displays such as the budgerigar's lean back display may escape human notice except in unusually detailed studies. We thus do not know the extent to which iterative processes are involved in the formalization of displays. Even once the possibility has been detected, it can be difficult to be sure iterative formalization has occurred. The comfort movements of ducks are a case in point (McKinney 1961, 1965b; Dane and van der Kloot 1964). Only weakly if at all novel in form, these movements sometimes are apparently organized into short sequences that begin with bathing and occur more often than might be expected—although to be sure they are exceptionally frequent in these events it is necessary to fully understand all their direct functions. McKinney (1961) offers the reasonable proposal that the direct function of a duck's bathing is to wet and clean its plumage. Bathing occurs so abundantly in precopulatory behavior that it must exceed this function, suggesting that it has been formalized. Yet bathing must also cause rapid cooling, and precopula-

tory bathing patterns are completed more often on sunny than on cloudy days. In other words, bathing may be needed in very active social interactions to dissipate the body heat that builds up (see D. Morris's previously mentioned suggestions about feather ruffling). By this interpretation, bathing in precopulatory interactions may not be iteratively formalized, because it may not exceed thermoregulatory needs.

Temporal Patterning

The timing of an act can be formalized so that it either coincides or combines in special sequencing relationships with other acts or events. In some cases relating one act to another by modifying the timing of its performance joins both into a formalized compound. In other cases this does not happen, although the way in which the two acts relate is formalized.

Specialization of sequential and perhaps also simultaneous compounds can be seen in movements of the green heron's tail (Figure 11.1). Certainly the continued flipping of its tail is an example of formalization through the rhythmic repetition of an act—a simple kind of sequential compounding that is found in the displays of a great many species. In addition, when tail-flipping is done by a heron in the full forward posture, as in the figure, an unusual combination of acts occurs as the lowered tail coincides with the indrawn neck and bent legs. These acts appear to be formalized from the procedures for taking flight, in which the first step is to crouch, pull in the neck, and raise the tail. The tail is then lowered as the bird extends its neck and legs and leaps into the air. But in the green heron's display the neck or leg positions do not change as the tail is lowered, with the result that a brief but distinctive combination of postures is seen. The combination may not be formalized, however. That is, it may be more appropriate to consider the tail-flipping as a display separate from the crouched posture; in fact, we know that tail-flipping is also done by green herons in "alert postures" (Meyerriecks 1960). The problems involved in determining whether such simultaneous combinations have been formalized as units are considered in chapter 13, in which both simultaneous and sequential formalizations are examined.

Human speaking and conversational behavior provides unusual insight into the ways in which the relative timing in performing different acts can be formalized. When we speak we use not only words, formally sequenced by our rich linguistic grammars, but also gestures, postures, pauses in our movements, and changes in gaze directions and body orientations. The performance of each nonverbal signaling pattern is timed to coincide with a particular word or juncture, so that these acts

give redundancy, emphasis, indications of change, and other information that provides perspective for speech. One of their most important tasks is to mark the hierarchical structure of the sequencing arrangements that underlie the flow of words.

This correlation of the performance of nonverbal behavior with aspects of the hierarchical structure of speech was recognized by Scheflen in 1964. Subsequently, Kendon made a very detailed examination of a cine film of a speaking individual and was able to recognize five levels at which his body movements matched the organization of his speech (1972a). Treating the speech as a "pattern of sound and rhythm," Kendon categorized its organization, beginning with small groupings of syllables that share an intonation pattern ("prosodic phrases"), then clusters of these groups, then "locutions" (which tend to correspond to complete sentences), groups of consecutive locutions that share some feature like rising intonation, and finally clusters of these groups. The speaker's patterns of bodily movement were also found to be very detailed.

To select examples at three levels, locution clusters tended to have distinctive arm and hand gesticulations, and shifts between these clusters were marked by trunk and leg movements. Locutions, the principal units of speech, were variously marked. Among other things, each tended to show a cycle of head positions that recycled for each locution within a locution cluster; the particular form of the cycle changed between consecutive locution clusters, but the use of cycling continued with each locution. At the smallest level a distinctive movement or a distinctive position correlated with each prosodic phrase and was changed at the end of the phrase. Even other distinctions represented organizational features extrinsic to the five levels as such. For instance, very brief locutions were marked differently from longer ones, and served different functions: they were usually parenthetical insertions, or temporizers, or the like, and were less integral parts of the entire discourse. As another example, the *amount* of marking was increased for especially important locutions. The locutions that embodied the point of a whole discourse could be marked by distinctive hand gestures including a boxing-in motion, an emphatic head cycle, and several vocal devices such as stress on the key word and pauses around it.

The movements that Kendon and Scheflen have studied can involve anything from an eyebrow or a finger to the whole body. The forms of acts performed in the United States have been cataloged by Birdwhistell (1970), who refers to them as "kines" and "kinemes" by analogy to the linguists' units of phones and phonemes. A kine is a movement such as the raising and then holding up the head that is commonly done in correlation with junctures in the flow of speech. Birdwhistell recognizes

different "allokines" of this movement when it is done at different rates and amplitudes; he groups these within a single, abstract kineme defined by form. Its form and the timing of its use to correlate with other acts are aspects of a kineme that are formalized to provide quite different kinds of information, however. The form of a kineme provides information about the content of what is being said. For example, the directional vectors, amplitudes, and velocities of those kinemes functionally classified as "kinesic markers" make available such information as direction in space or time (described under the message of direction in chapter 6), rates of activities, or numbers of persons referred to. The performance of a kine in correlation with a point in the flow of speech provides information about the temporal organization of speech or speech processing. These informative tasks are sufficiently distinct from each other that kines of different forms can have the same kind of temporal relationship to the flow of speech if the content of what is being said differs.

The timing of speech-related movements provides information about the temporal organization of speaking or of processing speech, because either a speaker or a listener, or both, may make the movements. If an attentive listener who is having no difficulty in following another individual's speech moves, his movements are usually synchronous with those of the speaker. A listener who moves but fails to mirror the speaker's actions is processing the speech out-of-step with the speaker, or is processing something else—his own thoughts or a third party's signals. Coordination of these movements between individuals appears to help them to cooperate in managing their conversation, a kind of task that is considered in chapter 13. Members of various nonhuman species also employ formalized synchrony between acts of individual participants as part of their procedures for managing cooperative interactions.

Spatial Patterning

The formalization of a directly functional act may leave its form intact, but modify the way it is performed in space. Performance of the act becomes restricted to sites that are of special importance. For instance, the act of defecating has been specialized as site-marking behavior in some species. Dominant male rabbits defecate at territorial borders in regular sites where piles of feces accumulate; wandering rabbits sniff at such piles and become wary if they find them to be fresh or freshly marked with phermone (Mykytowycz and Gambale 1969). Rhinos, domestic horses, and various other mammals also mark borders with fecal piles. The quasi-wild ponies of Britain's New Forest do not have fixed territories, but each stallion marks the areas where he and his harem are

grazing by defecating or, less commonly, urinating on the piles of feces that he encounters there (Tyler 1972). His display defecation is spatially patterned in that it is done at the sites of existing fecal piles and in grazing areas that are used repeatedly. However, some species, such as domestic cattle, involve feces in display behavior without showing formal spatial patterning of the act of defecating (Schloeth 1956:91). Still other animals regularly mark specific sites with urine (see discussion of marking behavior in chapter 9).

Humans distinguish some of their interactional behavior by spatially patterned formalizations. For instance, someone approaching to join two or more others who are conversing will stop a short distance away, not intruding on the space they are sharing as they interact (Kendon in press 1). Assuming this outer position while facing the interaction announces readiness to join the conversation and usually elicits spatial moves on the part of those already participating.

The Evolution of Repertoires of Displays

Formalization makes many displays distinctive from the forms of their precursors. It must also make all the displays within the repertoire of each species distinguishable from each other—if only because, as Moynihan (1970b) has pointed out, they could not be selected for otherwise. Further, of course, they must be sufficiently distinguishable to avoid or reduce the likelihood of recipient confusion.

As distinctions develop, the most different displays within each repertoire inevitably become more and more unalike. However, there are limits to the degree to which displays can be elaborated. For instance, they must not become so complex and prolonged that they either take up more time than a communicator can afford to devote to them, or require too much attention from the recipients who must discern their forms to distinguish among them. Overly elaborate displays might also be too conspicuous, too readily attracting the interest of individuals of the wrong kinds, such as predators or competitors (Moynihan 1970b). More important, as more and more displays are differentiated within a species' repertoire some are inevitably applicable in only narrowly defined circumstances. Such displays are performed relatively rarely. Moynihan argues that there must be a point beyond which rare displays, especially if very distinctive in form, become sufficiently unexpected that they startle their recipients and bias their responses toward fleeing. Even somewhat less rare displays, though not startling, might be sufficiently unexpected to impair the speed and precision of responses. Moynihan therefore argues that evolutionary selection must work against excessive heterogeneity and unchecked multiplication of the displays in each species' repertoire.

Rare displays might be useful if recipients could take the time to scrutinize them carefully and thus recognize them, but time is of the essence in interactions. A delayed decision may be obsolete (Marschak 1965), and Bruner (1957) has argued that "the ability to use minimal cues quickly in categorizing . . . is what gives the organism its lead time in adjusting to events."

Selection for rapid categorizing has led to the evolution of a capacity that Bruner calls "perceptual readiness": the organism becomes preset to categorize and distinguish among those things that have a high likelihood of occurring, thus perceiving and identifying them more rapidly than it does less likely events. That is, perceptual readiness affects speed of response in accordance with the relative probability of occurrence of different stimuli; it leads the most probable stimuli to be effective with the least delay. Speed is also affected by the size of the stimulus array for which an individual must be prepared: a differential effect is not apparent within small sets of alternatives that can fall within the span of attention (sets of about seven items; see G. A. Miller 1956) but it is for arrays of about twenty items and larger (Solomon and Postman 1952).

The combined effects of the inability to be perceptually ready for rare displays plus the retarded response rates for large sets of stimuli should lead to selection limiting the size of signal repertoires. In a repertoire of optimal size, those rarer displays with functions other than startling recipients should still be common enough to be readily perceived.

In fact, it has been empirically established that extremely diverse kinds of animals all have rather small display repertoires of about the same size (see chapter 7). These animals differ phylogenetically, ecologically, and in many features of their social structures. They also vary in the frequencies with which they perform displays; some display almost constantly and others only in very limited circumstances (Moynihan 1970b). This makes it all the more interesting that the sizes of their display repertoires should differ so much less than do other aspects of their behavior, and suggests that the causes of this limitation must be sufficiently general to apply to all species. Perceptual capacities would have such generality.

Within the limits of their size, display repertoires do not seem to evolve into static units. Moynihan postulates continuing turnover as new displays evolve, a process he expects to be irregular but probably quite frequent on an evolutionary time scale (1970b). Because the display repertoire of each species remains small, he suggests that new displays must replace older ones in a cyclic fashion. Perhaps most displays gradually become less and less adapted to the changing environments a species encounters through evolutionary time. Or the circumstances in which any display is performed may gradually change over evolutionary

time, becoming broader through a process Moynihan envisages as a lowering of the stimulus threshold for elicitation of the display. (Broadening could also be based on the metaphorical extension of a display's referents, a process considered in chapter 4.) If displays do tend to become performed more and more often (which happens to words in speech, and often very rapidly to slang words) they will become increasingly vulnerable to shifting selective forces and hence to evolutionary replacement. Some, having passed their peak employment and come to represent a referent class so broad as to be less useful, might then succumb to the problems of increasing rarity as newer displays with more useful referents arise. Their demise might be speeded if they were redundant with the more useful features of the new displays, particularly if the new displays were formalized primarily from activities that had commonly accompanied those previously performed. Replacement of an abundantly performed display could involve incorporating it into the characteristics of displays that are given along with it. Alternatively, at least visible displays might become "deritualized," gradually reverting to an unformalized state.

Moynihan's important hypotheses about the evolutionary replacement of the displays within repertoires were prompted by a frequent finding of studies that compare groups of closely related species. Of the displays commonly performed by most species in such a group, one or more are rare, erratically distributed among individual repertoires, or even absent in other species, and novel displays crop up in a few species. Some of Moynihan's earlier papers provided detailed examples from his studies of gulls and terns (see 1962a, particularly pp. 340–342). Examples of displays that may be nearly "extinct" were described in chapter 3 for the vocal repertoire of the eastern phoebe: the initially peaked vocalization and especially the doubled vocalization of that species are rarely uttered, and the latter was not heard at all from many individuals who were closely watched. Both these displays are uttered abundantly by the two western species of phoebes, which, however, appear to lack homologues of the eastern phoebe's abundantly given twh-t vocalization. In adapting to regions not exploited by most species and genera of the essentially Andean lineage of flycatchers to which phoebes belong, the eastern phoebe seems to have shifted the correlation patterns of many of its vocalizations as it developed the new twh-t display. These changes left few distinctive behavioral correlations for its initially peaked and doubled vocalizations (W. John Smith 1969c, 1970c).

Comparable shifts and replacements have been described in diverse other genera. For instance, Dilger (1960) studied several species of the parrot genus *Agapornis* (lovebirds) that differ in the extents to which they breed colonially, and found that displays commonly performed by

some species appear to have been lost by others. Occasionally, however, an individual parrot has within its repertoire a display that is effectively extinct as a characteristic of its species. Dilger, in fact, postulated evolutionary competition among the displays of a species' repertoire, although he did not attempt to develop a general explanatory model.

Evolutionary turnover among displays is only one consequence of restricting species to small display repertoires. Another is the need to encode differing message assortments in homologous displays, even among closely related species (see chapter 7). Differing message assortments may be required because species differ in the portions of their display repertoires that have undergone replacement, or in the displays that each is beginning to perform excessively. Of course, different message assortments are also very likely to be adaptive specializations to the differing communicative needs of each species, as discussed in chapter 7. Cause and effect are inextricable, however, as such adaptive resorting of messages could contribute to evolutionary pressures for the replacement of displays.

The most significant consequences of a low upper limit to the size of display repertoires are in the directions it imposes on the evolution of behavioral selection messages. Chapter 7 offered the hypothesis that this low limit underlies the tendency for many displays to encode messages that, because they predict the behavior a communicator may select only within broad classes, can be useful in diverse circumstances. Although such messages render recipients very dependent on contextual sources of information, there seems to be little alternative. Individuals have too few displays available to be able to afford tying them down to very narrowly defined circumstances, except for eliciting responses that must be precise and immediate.

The pervasive and often very extensive dependence of display communication on contextual sources of information creates very strong selection pressures to evolve means by which the communicator can control some of these sources. Perhaps the surest control comes when the communicator provides contextual sources itself, such as through its badges and nondisplay behavior, and through performing its displays in combinations in which they act as sources of context for each other. For instance, by sustaining or repeating the performance of one display it can make available a continuous background of information against which the performance of a rarer display can be evaluated. Or it can suddenly do two or more displays simultaneously, a procedure that presumably cannot be much elaborated without taxing the perceptual abilities of recipients. Quick sequences of displays offer more freedom for elaboration, particularly if recipients develop perceptual readiness through expectations of the patterns most likely to occur, effectively

treating these patterned sequences as units (see chapter 13). Formal units compounded by recombining displays probably represent some increase in the number of formal behavior patterns available to a species in its repertoire, if having the emergent level does not require a reduction in the number of units available at the display level. However, combining displays into larger units cannot, by itself, free a species from having a limited repertoire of formal signals. If all units at all levels handle the same classes of messages (that is, behavioral selections and supplementals, and nonbehavioral identifying and perhaps locating information) then the patterned relations into which they can enter with each other are determined solely by their sequential orders of occurrence and relative frequencies of repetition.

We know of only one repertoire in which compounded units have become differentiated to handle markedly different categories of information, for example, information about things versus information about acts (loosely speaking). In this, the speech repertoire, a set of rules has developed that yields new kinds of structured interdependence among compound units. It is reasonable to suppose that the evolutionary origins of speech may have begun with the demands for context-dependency of signals that are set by limitations on repertoire size, then proceeded via the use of displays contextually to each other to the development of interdependency among displays—in some cases within compound units, and in others through less integrated sequencing. Then somehow the compound units of speech became differentiated into several lexical categories that could have relatively numerous kinds of interdependent relationships when used in sequences. By some stage, perhaps through a progressive process, the numerically limited basic components largely ceased to be performed alone; their compounds then became the main units of discourse. Somewhere, too, the intergenerational transfer of these now basic compounds came under the control of learning, and their numbers became very large. That so many of these units can exist despite the rarity with which some occur may depend in part on their use within extremely familiar sequential patterns. Each unit comes as a representative of a particular lexical category in a "slot" for which its recipients have been prepared—that is, the members of a phrase are so interdependent that any given word can be surprising in only some of its characteristics.

12

Effects of Evolution on the Forms of Displays and Accoutrements

Displays and other results of formalization must be practical to perform. Their advantages are obtained only by expending energy, and often time, and they make communicators more vulnerable. If a signaling specialization is to evolve, its benefits must, on the average, exceed its costs. The processes of formalization discussed in chapter 11 thus proceed within limitations and opportunities that are peculiar to each species.

Sources of the selection pressures that interact to determine how the evolutionary formalization of behavior and structure is optimized, and the kinds of adaptations that result, are reviewed in this chapter. What can evolve is basically determined by what is feasible for the anatomy and physiology of each species, an interrelated set of structures and processes that must be adaptive for multiple functions. The avenues open to formal specialization in each species are further determined by its range of habitats, special characteristics of the prevailing climate in its geographical range, and the relationships between individuals of both its own and other species. The openness of different environments, for instance, affects the distances over which displays can be transmitted, and the types of obstacles present determine what sorts of displays will transmit best. Different habitats provide different sources of noise, including the displays of all the other species. Local population densities and the elaboration of social organization typical of a species greatly influence its communicatory needs—for example, whether some of its signals should accumulate or fade or change rapidly, whether many messages need to be provided simultaneously, how distinct from one another some of its displays need be, and how many recipients they should or can reach without generating unacceptable levels of social confusion. In addition, differences among species in their foraging tech-

niques, in the problems each has with predators, and in many other features all affect the evolution of their displays.

Great numbers of different selection pressures must influence the evolution of displays and other signaling specializations. Although many have been identified or postulated, there are many kinds of similarities and differences among the displays of diverse animals for which we lack explanatory hypotheses. Nonetheless, our current understanding of what may be primarily the more obvious constraints operating on the processes of formalization enables us to predict many characteristics of animals still unstudied, and appears to explain many of the relationships seen as we compare species. This chapter provides an introduction to the appropriate literature and outlines the more general findings and their implications.

Categorizing and pulling apart selective pressures acting on the evolution of signaling specializations imposes an unreal isolation on each. They do not act separately. Each species' physical characteristics are a product of its genetic potential and the demands and opportunities of its environment. Its nonliving environment has determined many of the characteristics of its plant environment, which in turn have set the stage for the animal environment, intra- and interspecific. Thus the evolution of displays can be affected by a characteristic of a species' social structure, such as a particular pattern of dispersion among individuals, that may itself be an adaptation to features of the plant environment, like a low, matted growth form, that in turn are adaptations to a physical force, perhaps high winds. Not just the social dispersion, but also the matted vegetation and the noisy, buffeting winds directly influence the evolution of display form. Because the interrelationships of selective forces are complex, and can be very different for different species, it is better to begin their description by examining the general kinds of effects that, in principle, each can have. The effects each does have, in the evolution of the signaling specializations of any particular species, must be investigated within an understanding of the overall selective milieux of that species' populations.

Physical Characteristics of Each Species

Both production and reception of displays are limited for every species by characteristics of the animals' anatomy and physiology. Although this is obvious, we can too easily overlook its implications because we are accustomed to viewing the world in terms of our own biological equipment. Other animals can do things we cannot and are limited in ways that are not always readily evident; some have windows on the world through sensory equipment that is very foreign to us. Thus, in

considering the evolution of signaling specializations on nonhuman species we must discover and account for the potential of each different species.

Production of Displays

The structural and physiological characteristics of each species set limits on the kinds of displays that are convenient, economical of energy, and even feasible for its communicators to produce. For instance, an animal that frequently uses all its appendages to move about or forage cannot afford to use them extensively for semaphoring or other visible signaling. Its display behavior must be based on some anatomical system that is not preempted. Therefore, we would expect to find that audible displays are often more convenient for brachiating monkeys and birds that fly a great deal to produce than formalizations of arm- or wing-waving movements. Also, because their vocalizations are produced with little expenditure of energy, sound communication is economical for most of the functions they need.

In contrast, very small insects cannot find sound production profitable for as wide a range of uses. Because the lengths of their bodies may be only a small fraction of the wavelengths of their audible signals, they can produce sound waves of only minute acoustic power, useful only to individuals interacting quite close to each other. A courting male fruit fly, for instance, performs his low-frequency wing-vibrating display when within about 5 mm of a female—a distance from her that is about 1/400 of the wavelength of the 166 Hz tone he is producing (Bennet-Clark 1971). She detects the faint sound by using featherlike displacement receptors called "aristae" that are mounted on her antennae and are sensitive to the velocity of particles in the air (see figure 12.1).

Having a small body also limits the distances over which complex visual displays can be seen. This appears to be why the smaller species of New World primates lack the variety of facial expressions found in larger species (Moynihan 1967). As an alternative, the small monkeys have evolved displays based on patterns of erecting long hair, something visible at greater distances than are their tiny faces. At the other extreme, large body size also limits some kinds of display behavior, by decreasing a communicator's agility, for example. Among birds, the larger herons have fewer and less elaborate aerial maneuvering displays than do the smaller ones (Meyerriecks 1960). A similar trend is apparent among species of gulls (Moynihan 1956). All such trends must be interpreted carefully, however, as their causes are often not simple. In the skuas, which are closely related to gulls, the long-tailed skua *Stercorarius longicaudus* is the smallest, most agile species and has the

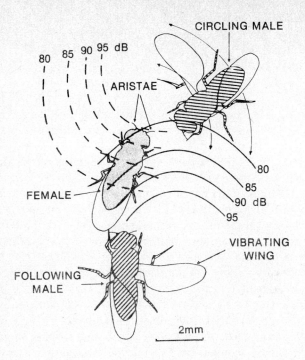

Figure 12.1. Low-frequency sound in the display behavior of a small insect. A male fruit fly *Drosophila melanogaster* courting a female. His wing, vibrating with a frequency of 166 Hz, produces pulsed sound waves which are directed at an angle of 45° to the substrate and reflected toward the female as he circles her, always well within 5 mm of her sensory receptors. These receptors, the aristae, are sensitive to the velocity of the air particles excited by the waves. The diagram shows the calculated particle velocities, expressed in dB above $0.5 \times 10^{-7} m s^{-1}$ (effectively a "loudness" measurement). Although loudness decreases rapidly with distance, it is entirely adequate at the very close proximity the male is maintaining to the female. (From Bennet-Clark 1971.)

greatest elaboration of aerial displays. However, it is also the only skua with territories so large that it must respond to intruders by displaying on the wing as it approaches them (Andersson 1971). The three species of skuas that have much less elaboration of aerial displays also appear to have less need of them.

The particular anatomical structures used by animals for direct functions and the ways these are employed influence the forms that visible displays evolve. For instance, displays providing information about the probability of attack behavior often originate in movements made in preparing to attack. Because its bill is a principal weapon with which a

bird attacks, threatening displays of many bird species incorporate a conspicuous act of pointing with the bill (J. M. Cullen 1959). Similarly, many mammals threaten by performing teeth-baring displays. High-intensity threats of fiddler crabs are ritualized from movements of gripping with their large claws (Crane 1967).

The energy expenditures required of animals at times put them on very restricted budgets, and the performance of displays figures in their economies. Rand and Rand (in press) have analyzed the interactions in which female iguana lizards *Iguana iguana* compete for nest burrows. They choose displays rather than most costly, directly functional acts, and they also select displays that can be performed with little energy rather than those involving costlier movements. It takes less energy to perform a display predictive of attack behavior than to attack. Further, it takes less energy to perform a mouth open display than a head swing; the huff, head swing with huff, lunge, and lunge with huff displays are each increasingly costly. With performance of the more costly displays the iguana is more likely to attack, and may be indicating this more precisely. But because the cheaper displays sometimes work, it can be profitable to begin an encounter modestly. The iguanas studied by the Rands had to rely entirely on stored energy reserves with which to travel to a nesting island, dig burrows and contest with each other for them, lay eggs, and fill in their burrows with soil. Further, at any time an individual could expend only a limited amount of energy before she had to cease interacting or digging and rest.

Reception of Displays

Just as the capacity to produce displays influences their evolution, so does the capacity to receive evidence of their performance. This capacity lies in the sensory equipment of each species.

The limits of this equipment also limit the forms of the information sources that can be used. To take one example, sounds with complexly changing patterns of frequency can be differentiated by the ears of birds and mammals, but the most comparable receptor of insects is based on an amplitude analyzer, the tympanal membrane. The aristae and sensory hairs mentioned above are sensitive only to low frequencies (as discussed by Bennet-Clark 1971). The tympanum can be "tuned" to particularly crucial frequencies but does not permit analysis of the distribution of spectral energies in sounds.

On the other hand, special sensory capacities that have evolved can sometimes obtain information that would otherwise have to be made available by formalizations. Such capacities thus render the evolution of displays, badges, and the like unnecessary for some functions. For

instance, many species of mammals analyze substances in the urine with an accessory olfactory system employing a vomeronasal organ (illustrated in figure 13.1): this mechanism can apparently detect chemical evidence not just of estrus, but also of proestrus. By continuously sampling a female's urine, males can thus know when she is about to become receptive before any behavioral predictors of sexual receptivity. This mechanism may in part explain why a female bovid, for instance, may visibly manifest her heat by little more than her "willingness to stand for mounting" (Estes 1972). A male of her species will by that time be associating closely with her, seeking an opportunity to mount.

Nonliving and Plant Environments

The physical environment that connects individual animals provides a medium through which the energy of visible or audible signals can often pass, the molecules of pheromones diffuse, or electrons flow from the electric organs of fish; tactile displays are the only direct contact individual animals make in signaling. In addition, the physical environment offers both sources of energy, such as through the sunlight that reflects from animals and makes most visible signaling possible, and opportunities for making the expenditure of energy more efficient. Environmental substrates, for instance, provide many insects with directional reflectors for their audible displays (see Bennet-Clark 1971). The dry leaves on which courting male lycosid spiders stand may provide sounding boards that increase the loudness of the airborne components of vibrations produced by palpal oscillations (there are also substrate-borne components that spines on the palps conduct directly into the leaves; see Rovner 1975). For all that it offers, however, inherent limitations in the environment restrict the forms that displays can take.

In some environments the functioning of some sensory receptors is seriously impaired, restricting the kinds of signaling that can be effective. Audible displays, for instance, are at a disadvantage in regions that are either excessively windy or characterized by noises created by waves washing on beaches or rivers being dashed onto rocks by rushing currents. Where such noises are continuous, animals have means of signaling other than just the production of sounds. For example, downy young of the blue duck *Hymenolaimus malacorhynchos* are raised in the roaring mountain streams of New Zealand; they have vocal displays but cannot rely on their parents hearing them as a family moves. However, each duckling also has an unusual patch of chestnut color just under its tail, and as it jumps among the rocks the patch becomes conspicuous (Pengelly and Kear 1970).

Where environmental noises are not as continuously loud, animals can make at least some behavioral adjustments to the varying amounts

of noise with which their signals must contend. For instance, mated quail utter vocalizations when separated. Potash (1972) separated pairs of Japanese quail and put each mate into a test chamber where it was subjected to varying levels of white noise. The birds altered their patterns of vocalizing by increasing both the rate of calling and the number of calls in each bout as he raised the ambient noise level. These modifications should make the birds easier to detect and find in the presence of the noise.

Both the combined use of vocalizations and visible badges and the performance of increased numbers of vocalizations are adaptations that circumvent noise by adding "redundancy." Redundant signaling repeats information. It would be unnecessary if a minimal signal would be fully received, but in practice noise always interferes with such perfect reception. Even when a signal is received, it is usually not received in its entirety, or not without distortion, which makes recipients somewhat uncertain about the signal's form (their uncertainty is known to communication engineers as "equivocation"). Redundancy is thus a fundamental necessity of the behavior of communicating, although the amount required varies with the average noisiness of different environments. Often, as in the case of the ducklings' calls and badges, redundancy is adapted for reception by more than one sensory modality, so that information not learned by hearing a signal may be obtained by seeing one, or vice versa. (The concepts of "noise," "redundancy," and "equivocation" have been quantified in the mathematical theory of communication, as explained by Weaver 1949.)

Physical limitations do not impair only audible displaying, of course. For instance, a dark environment can make visible displaying by reflected light impossible. Bioluminescence has been one solution to this problem, but most noctural animals probably do not make fine use of vision in communicating other than in twilight periods—they use other sensory organs instead. For aquatic animals, darkness need not be nocturnal but can also be caused by depth or turbidity. Even in shallow, clear water a rapid current moving over an irregular substrate can distort vision and inhibit movement. Evolved signaling specializations have compensated for this; for example, males of a small fish called the longnose dace have less mobile displaying than do males of the related blacknose dace, a species breeding in slower water (Bartnik 1970).

Some physical obstacles are universally distributed. For sound, the viscosity of the medium through which it is propagated creates distortions that increase with distance. The gaseous viscosity of air, for instance, attenuates sounds in a manner roughly proportional to the square of their frequencies so that for the same initial intensity a 200 Hz tone can travel 1,000 km and a 10,000 Hz tone only 410 m before each is reduced to $1/e$ of its initial intensity (see Konishi 1970). Production

of heat by the sound wave has a parallel effect. In addition, turbulence in the medium and reflections from various sources produce irregular amplitude fluctuations that can impair reception of higher frequencies (see Wiley in press). Thus sounds with complex frequency spectra retain their original structures only within short distances of their sources. As a result, audible displays that must function over long distances cannot usefully employ a mixture of frequencies, even though vertebrate ears can discriminate frequencies very well. Through a comparable problem, what E. O. Wilson (1968) has referred to as "medleys" of odors are limited to short-range uses. The pattern of chemical mixtures breaks up over distance because the components diffuse at different rates determined by their molecular weights.

Deep ocean water transmits low-frequency sounds over long distances much more readily than does air: loss of energy to heat is greatly reduced, and distance of propagation can be augmented by currents and by gradients in the speed of transmission that bend sound rays into paths that avoid losses due to reflections. R. S. Payne and Webb (1971) have discovered a most remarkable case in the 20 Hz sounds uttered at the extraordinary amplitude of roughly 80 dB re 1 dyne/cm^2 by the fin whale. When these loud sounds were first measured by scientists they were "thought to be everything from surf on distant shorelines to faulty electronics" (R. S. Payne and Webb 1971:113); it seemed improbable that they would be produced by animals. These sounds are also remarkably pure in tone, their energy centered at 20 Hz and confined to a band only 3 to 4 Hz wide. They are ideal for long-range signaling, and under proper conditions should be audible for a distance conservatively calculated to be between about 1,000 and 6,500 km; they are, or were, possibly audible for about 21,000 km. Although the noise of screw-driven ships must now limit their maximum range of detection to about 1,000 km in many parts of the world's oceans, fin whales evolved their 20-Hz sounds in quieter seas and have been discomfited by such ships only for 1/100,000th of their 15-million-year history. The functions of these awesome sounds remain obscure, although Payne and Webb hypothesize that they may have enabled the whales to remain in touch with their fellows over an entire ocean basin (discussed in chapter 10).

The transmission of audible displays is affected by vegetation, which acts like a complex obstacle course. Plants reflect, absorb, and diffract sound, although wavelengths that are greater than the diameters of trunks, stems, and leaves resist the attenuation better than do shorter wavelengths. This appears to have affected the evolution of some bird vocalizations. Species of birds living among the tree trunks in the shaded interiors of Panamanian forests tend to employ relatively low-frequency vocalizations, their emphasis on the average falling at about 2,200 kHz,

without rapid-frequency modulation. Those living at the forest edge or above in the treetops where leaves and branches of all sizes abound and provide much more complex obstacles oriented in all directions to the source of a sound do not specialize as a group in any particular range of frequencies (Morton 1970, 1975). Comparable results were found when birds inhabiting dense and open habitats in Africa were compared (Chappuis 1971).

The effects of atmospheric humidity on sound absorption, latitudinal and altitudinal differences in air temperature on sound velocity, and the obstacles to transmission provided by vegetation should combine in influencing the frequency characteristics evolved by audible displays. Testing their effects, however, requires comparisons of large numbers of samples from different regions, as Nottebohm (1975) did for the sparrow *Zonotrichia capensis*. Using recordings of the songs of 1,042 individual sparrows living from Costa Rica to central Argentina (sampling was most intensive in Argentina), Nottebohm found higher maximum song frequencies associated with denser, arborescent, or broad-leaved vegetation, or with such environmental factors as the absolute humidity correlated with such vegetation. He also found evidence for a nonvegetational latitudinal factor that may not have been related to air humidity, but that greatly exceeds differences that might be predicted by air temperature. Further understanding of these evolutionary pressures will require additional studies of comparable magnitude to this first survey of variation in a single species.

Although sound waves are weakened by the obstacles that plants provide, sounds (and odors) do get past vegetation that can fully block visible displays. Some animals can overcome such line of sight obstacles by taking high, conspicuous perches or by displaying in flight, either above the vegetation or in large spaces within it (see Marler 1967a; Crook 1962). For instance, the ovenbird *Seiurus aurocapillus,* a North American warbler that forages on the leaf litter of forest floors, sings its advertising songs above the ground from perches at levels where the vegetation is least dense (averaging about 9 m where the mean height of trees is about 15.5 m; Lein 1973).

If the environment is sufficiently open, visible displaying has many advantages: the location of the displaying communicator is fully evident to recipients, and the displays have a rich array of features that can be varied, such as color, brightness, form, and rate of change in form. There is the potential to encode complex or very precise information by performing the features as different, variable components in a number of simultaneous recombinations (Marler 1967a; and see chapter 13). This potential has been exploited by some species in which individuals characteristically live in groups and tend to be continuously and easily visible to each other.

The use of variable formalized components risks recipient error (see chapter 13), but is possible in open environments at least in part because there is great "continuity of context" (W. John Smith 1966:201–202). That is, when animals are nearly continuously visible to one another, each receives enough information from unformalized sources to be able to predict many of its companions' acts. Specialized signaling then becomes necessary primarily for details and surprises. For species in which social behavior is not complex and detail is not often important, open environments may correlate with the evolution of unusually simple display repertoires. The fork-tailed flycatcher *Tyrannus savana,* for instance, dwells in more open country than do closely related species, and differs from them in having vocalizations with simpler, click like forms (W. John Smith 1966).

The Animal Environment: Own Species

Formalized behavior and other specializations for making information available affect both the likelihood that individuals will interact and the courses their interactions take. Interactions with individuals of their own species probably provide most animals with the bulk of the advantages they gain from being informative, although numerous kinds of interspecific interactions are important (see final section of this chapter). Many of the selection pressures that mold the evolution of signaling specializations thus arise in the needs of intraspecific interactions.

For one thing, within repertoires of displays, badges, and the like, each specialized unit must be distinguishable from the others. Often the signals must also be capable of rendering communicators distinguishable as individuals or as members of particular social groups or subpopulations. Numerous conspecific individuals that signal while near each other may need adaptations that reduce their mutual interference; in some cases they have evolved adaptations that change the potential for interference into mutual enhancement of signaling.

Whether signals will interfere with or enhance one another, and what they must accomplish, depends of course on the social structure typical of each species. Different social structures exert a variety of influences on the evolution of signaling specializations.

Distinctiveness

In selecting an appropriate response, a recipient of a display or other specialized signal usually needs to know more than simply that a signal was made available. Recipients and communicators belonging to the same species share a whole repertoire of signals, and a recipient needs to

know which of these was performed. As a general rule, therefore, the signals within a repertoire must be sufficiently distinct that they do not usually confuse recipients into making inappropriate responses.

This general rule allows considerable leeway. Where a confusion between two signals leads only to similar and reasonably appropriate responses, errors can be tolerated and the signals may resemble each other closely. Signals may also be alike if sources of information contextual to two or more of them are too readily available to permit confusion: for example, use of the homophones "to," "two," and "too" has made this sentence awkward, but it remains comprehensible (though comparable phenomena in nonlinguistic signaling may be rare). Conversely, in ambiguous situations in which the different acts that may occur can have crucially different effects, selection will favor the development of unmistakably different signals.

ANTITHESIS Observation of strongly divergent signaling patterns led Darwin (1872) to propose a "principle of antithesis": when an event leads a communicator to be in "a directly opposite state of mind" from that engendered by a different circumstance, the animal will tend to perform formalized "movements of a directly opposite nature" in the two cases. Darwin contrasted the posture of a dog ready to attack, with its tail raised and its head slightly elevated with the neck-hair bristling (chapter 1, figure 1.1), with the slinking, hair-depressed, tail down, and wagging actions that accompany what he called "joy" in a dog's greeting of its master. (As Marler pointed out in 1959, the latter displaying is "submissive"—it incorporates a prominent escape message). Many displays have subsequently been described that illustrate this antithesis principle. For instance, the facing away displays of appeasing gulls (Tinbergen and Moynihan 1962 and figure 12.2) , in which an individual repeatedly directs its bill away from another individual, contrasts with displays encoding a greater likelihood of attack and having the bill pointed forward as a weapon. Willis (1967) formulated a "rule of angles" to describe the antithetical posturing of the bicolored antbird: when attack is relatively likely these antbirds "close" the angles of their extremities (by bending the head down, holding the wing tips in, and so forth) and "open" the angles of closer body parts (for example, extending the neck and wrists); in submissive, escape-related behavior they do the reverse. This rule is more precise than Darwin's notion of "opposite" forms (an overstatement). Darwin's notion of "opposite state[s] of mind" is also somewhat misleading (is joy or fear the opposite of aggression?), but the divergence of signaling patterns that he recognized is quite real. To restate the principle in terms more commonly used in this book, if confusing the information carried by two displays would

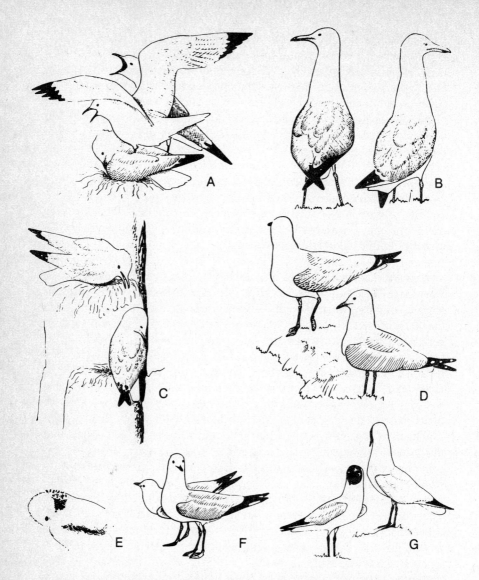

Figure 12.2. Facing away displays of gulls. Examples of displays of various species of gulls in which one individual faces away from another, or two individuals head flag, each repeatedly directing its bill away from the other. Note that kittiwakes (*A*, bird intruding on the nest of a long calling pair, *C*, lower bird, as upper bird jabs, and *E*, chick) add a downward angle to the display movement, and that the dark faces of black-headed gulls (*G*) are replaced by the white plumage of the backs of their heads. These displays differ very obviously from displays in which the bill is pointed toward an opponent who may then be attacked. (From Tinbergen 1959b.)

cause havoc or other harm (that is, where the cost of recipient error is high), the forms of the displays will diverge greatly.

IDENTIFICATION   Not only must the displays within a repertoire be distinct, but devices must evolve that can render whole repertoires or parts of repertoires of the displays of individuals distinguishable from those of other individuals. The distinctive features supply information that identifies communicators, as described in chapter 6.

The capacity of displays to identify an individual facilitates interactions between mates, members of families or somewhat larger social groups, and neighbors. Displays with characteristics that are shared among a limited number of individuals as a distinctive argot (see chapter 6) are thought to facilitate the cohesion of groups in which the members are bonded to one another and can interact on the basis of their mutual familiarity. In some species such individuals may act as resources for each other, for example, by having accumulated special knowledge of previous resource distributions in the localities they inhabit or visit. On the other hand, dialectal variants of displays may help to identify populations in which the members share evolved ecological adaptations, facilitating pair formation among like individuals (Marler 1967b; Nottebohm 1970, 1972). This is suggested by the existence of different dialectal forms in regions with different habitats, for example, within the species populations of sparrows such as *Zonotrichia leucophrys* (Marler and Tamura 1964) and *Z. capensis* (Nottebohm 1969; King 1972), and the correlation of dialectal differences with genetic differences (Baker 1974, 1975). Yet in some species differing forms of displays that have been described as "dialects" meet in apparently uniform habitats, and along extremely sharply defined boundaries—as narrow as 50 meters for songs of the splendid sunbird *Nectarinia coccinigaster* (Grimes 1974), and somewhat larger for the chaffinch's "rain call" (Sick 1939; Thielcke 1969). These cases may be of secondary origin because the populations have met after the habitats had been substantially altered by human activities. Alternatively, if such dialects are each restricted to very small regions they may be functionally comparable to argots, and keep mutually compatible individuals within their groups.

The duets of some species comprise an unusually restricted form of argot that has attracted much attention in recent years. Duetting is a practice of mutual displaying between the members of a pair, one leading and the other joining in with either the same display (polyphony) or a different one (antiphony; see Thorpe 1972). Offspring and other individuals may also join in, creating trios, quartets, or communal performances. These performances apparently have diverse functions

(reviewed by Thorpe 1972; R. B. Payne 1971, 1973a; and W. John Smith in press): for instance, in some species they help mates to keep track of one another if separated in dense vegetation, the mists of tropical mountain tops, at night, or within large foraging or roosting flocks; duetting performances in regions where seasons are either not sharply differentiated or change rapidly and unpredictably may foster synchrony between mates in the physiological states necessary for breeding; duetting may enable mates to reaffirm their bonded relationships and to coordinate such activities as incubating and caring for nestlings; and duetting appears to be of significance in the territorial relationships between neighboring pairs.

Species in which duetting individuals cannot always see one another need individually distinctive displays. In some species, such as African shrikes of the genus *Laniarius,* mates come to have pair-specific duetting patterns in which each combination of the individuals' various song forms is a stable characteristic of that pair; in most cases duetting combinations of pairs differ from those of their neighbors.

Pair-specific duets are sometimes mimicked. For instance, when neighboring pairs of the slate-colored bou-bou shrike *Laniarius funebris* countersing in territorial disputes, each pair tends to imitate the other's duets (Wickler 1972b). Superficially, imitating would appear to be a perversion of the identifying potential created by distinctive displays. In fact, it may not thwart such functions when used intraspecifically, but instead be a way either of providing information about membership in some sort of group, or of indicating, by effectively pointing to an "intended" recipient; that is, it can provide similar information as the direction of a communicator's attention described in chapter 6. Interspecific mimicking is different. It does not appear to interfere with the identification of the communicator's species in such birds as Australian lyrebirds (see F. N. Robinson 1974) and in the New World Mimidae (mockingbirds, catbirds, and thrashers), but in many other cases (such as Batesian mimicry, discussed below) it is adapted specifically to conceal that information.

LEARNING   The need for distinctive signaling repertoires creates pressures for the evolution of specialized learning processes. Dialectal variants, argots, and in some species even species-specific attributes of displays are developed in each individual by learning to mimic the characteristics of particular classes of models. The learning processes, the degree to which the characteristics of displays are left open to be filled in by experience, and the extent to which the appropriate models to be mimicked are genetically specified has been studied most intensively in experimental work on bird song.

Bird songs, even such complex examples as that of the song sparrow *Melospiza melodia* (see Mulligan 1966), can be inherited genetically. There are species, however, in which the genotype governs only some aspects of the form of songs, and provides other information that helps young birds to select appropriate song models or singing "tutors." Young males can copy these tutors' songs, and young females can learn them as a basis for later response preferences. (That females have intraspecific preferences for home dialects has been shown by Milligan and Verner 1971; although having learned these dialects does not always provide a barrier to pair formation with a male whose song is typical of another population; see Baptista 1974.) The models for young bullfinches *Pyrrhula pyrrhula* (Nicolai 1959) and various grass finches (Estrildidae; see Immelman 1969), for instance, are the songs of the male who rears them: in nature their father, in experiments even a foster father of another species. Young parasitic indigobirds *Vidua* species learn the song of the host who raises them, although their additional knowledge of the song of their own species may be genetically inherited (R. B. Payne 1973b). Young white-crowned sparrows between about 10 and 50 days old learn songs being sung by local males of their species, selecting these from the songs of a number of species that they may be able to hear (Marler 1970a).

Learning in all such species depends on the availability to young birds of songs that are appropriate models. They typically hear these and store memories of them in a critical period early in life, well before starting to sing themselves. Experiments in which young birds are permitted to hear only selected sounds or are deafened at specific developmental stages have been used to determine several parameters of their procedure: the limits of the critical period in different species, the characteristics of form that are accepted as models in this period, and the contribution that is made by a subsequent period or periods of practice and inventive elaboration. These experiments have been reviewed for the chaffinch by Thorpe (1958, 1961) and Hinde (1969), for the white-crowned sparrow by Marler (1970a), for several species of sparrows by Marler (1967b), for estrildid finches by Immelman (1969), and in general by Hooker (1968) and Konishi and Nottebohm (1969).

The process of song learning by young birds parallels that of language learning by young humans in several respects. These parallels are not unexpected, because the characteristics of signals that individuals learn to produce and be responsive to cannot be left wholly to chance, and because complex signals cannot be produced without the development of considerable skill (Marler 1970a, 1975). Both song- and language-learning involve selection from myriad available sounds of a set that is

copied, and young birds and young humans are genetically predisposed to recognize the set that is appropriate for each (reviewed by Marler in press). The learning occurs during a critical period before sexual maturity (Lenneberg 1967; Nottebohm 1970). Humans and young male birds are further predisposed to practice reproducing the sounds through a stage of "babbling" or "subsong" until their auditory feedback confirms their competence in copying the model; even though female birds of many species do not sing, they must be familiar with the sound of appropriate songs. In addition, from shortly after birth humans copy with their body movements the rhythms they hear in the speech of older persons (Condon and Sander 1974), learning elaborately organized patterns that later characterize both the rhythm of their speaking and the body movements they make during their own or other persons' speech. In birds, the rhythm and phrase structure of song can also be part of what is copied by the young (see, for example, Güttinger 1973).

Learning requires the evolution of a developmental program that is, within set limits, open. Unlike the much more widespread closed programs that limit the characteristic forms of most displays to what is specified by genotypes, open programs let in information that is obtained as a result of individual experience (see Mayr 1974b). However, open programs are costly to evolve: the capacity for storing information gained by individual experience must be increased, and mechanisms must be found to ensure the availability of external sources of guidance, commonly in the form of parental care. Although apparently too costly to be widespread among species, the ability to learn displays greatly increases the flexibility with which signaling can be adapted to local or immediate needs. Further, the increased capacity it requires for the storage of individually acquired information may well have been preadaptive in our species for the evolution of speech as a richly detailed means of communicating (Moynihan 1976).

Interference and Cooperation

Members of gregarious species are sometimes so closely packed that their displays interfere with each other. However, to the extent that intraspecific interference is confined to just some sensory modalities and some circumstances, it is probably not difficult to evolve adaptations to it. In many colonies of gulls or terns or blackbirds, for instance, the hubbub must often render vocal communication chancy if a communicator and recipient are not extremely close. Yet the highly elaborate visible displaying of such birds may compensate for much of the social interference with their calls. Further, the calls may incorporate wide ranges of frequencies and abrupt changes of amplitude, characteristics that make it extremely easy for recipients to locate individual communi-

cators, and thus to disattend selectively sounds that come from other directions (Wiley in press). Alternatively, auditory interference may be circumvented by not depending on vocalizations: the colonial brown pelican, for instance, is almost entirely silent and relies on visible displays (Schreiber in press).

Interference is not the inevitable consequence of the performance of the same repertoire of displays by two or more individuals in each other's close presence. The other side of the coin is that each animal may be able to put the other's displays to work for it. This kind of cooperative displaying has often been called "chorusing," although the calling of frogs by a pond or birds in the predawn or evening twilight— sounds that usually have contributors of diverse species, are largely disorganized, and sometimes cacophonous—have also been called chorusing. In contrast, when members of one species gather and chorus their voices augment each other, making their combined advertisement louder, more continuous, or both.

In periodical 13- and 17-year cicadas, for instance, one male will call and a second provide an answer, the two of them then alternating. Large numbers of males will gather and call in synchrony, producing the loud choruses in which most mating takes place (Alexander 1967). Even though the members of the chorusing group effectively cooperate to draw in females from a distance, however, they must compete for females who have approached. Competitive structuring of their chorus is particularly likely to be developed in species in which males remain at least somewhat dispersed, and involves patterned successions in sequences of calling by neighbors (Otte 1972; see also the discussion of the frog *Engystomops pustulosus* under the consideration of information about location in chapter 6).

Many lek-forming bird species, in which males congregate and display in each other's presence, probably obtain much the same cooperative advantages from chorusing as do periodical cicadas, and generate the same problems of competing at close quarters. In some lek birds that call loudly, the calfbird *Perissocephalus tricolor* of northeastern South America, for instance, males are careful not to overlap their calls with each other (B. K. Snow 1972). This generates a more nearly continuous sound than if they overlapped. Males of the black-and-gold cotinga *Tijuca atra* of southeastern Brazil actually achieve fully continuous calling: each utters a loud, pure whistle that is virtually sustained for nearly 3 seconds. After about 2 seconds he whistles again. If another male on the lek calls, he fills in the 2-second interval (D. W. Snow and Goodwin 1974). Because males of this species appear to chorus only when females are already present, however, their alternation is probably more competitive than cooperative.

Some weaverbirds of the African savannas produce a visible analogue

of chorusing by a "mass display" done by the males of a nesting colony whenever an unmated female comes near (Crook 1963). This mass display greatly changes the appearance of a colony and makes it much more visible from a distance. Even more spectacularly, in southeastern Asia large numbers of males of some species of fireflies gather in trees and synchronize their flashing to within 30 msec of each other (see Hanson et al. 1971, who offer a physiological model for the flash-timing mechanism).

Cooperative chorusing may also be a group device for territorial proclamation, as it appears to be for instance in the howler monkeys of New World tropical forests (Carpenter 1934; Chivers 1969). Whole groups of howlers chorus at dawn and at various other times in vocal exchanges with neighboring groups. Packs of wolves *Canis lupus* also chorus in response to the chorusing of neighboring packs. In the howling of a pack, which may be audible over an area of about 130 square kilometers, each wolf begins separately and avoids unison howling with each of the others. The complexly braided chorus that results quite possibly enables neighboring packs to assess each other's numbers (Mech 1970). On a lesser scale, the duets of mated pairs of birds (see above) are minimal choruses, with a diversity of functions that often includes territorial proclamation.

Two or more members of a species can signal cooperatively in at least one other way, apart from chorusing. They can display in a way that enables a recipient individual to sum the information obtained from each: by each offering tokens of the food it has found, for example, or by each adding its pheromonal contribution to an odor trail. E. O. Wilson calls this "mass communication," and it is described in chapter 10.

The Demands of Social Life

Animals that profit from being together, whether on special occasions or continuously, have signaling adaptations that help them to form and maintain cohesive social groups. For example, birds that flock usually have badges and vocalizations that enable the members of a flock to keep track of one another by making them conspicuous when they fly (Moynihan 1960; and see chapter 9). These adaptations are not found in solitary birds: for instance, in the South American woodpecker *Colaptes melanochloros* a whitish rump and gold breast patch are badges revealed in flight by individuals belonging to the social populations of open country, but are absent in the populations of forested regions where woodpeckers do not flock (Short 1971). Herd-living ungulates have comparable flash pattern badges, revealed or made conspicuous when they run. Vocalizations are uttered during movements of

groups of monkeys and apes, as well as chickadees, jays, weaverbirds, tanagers, mynahs, geese, and many other kinds of birds (see examples in chapter 5 under the behavioral selection messages of locomotion and association).

Of course, many species of gregarious animals are more than simply gregarious. They have evolved elaborate social behavior and live in stable social groups structured by diverse kinds of interindividual relationships based on age, sex, ability to dominate, and differences in individual preferences and experiences. Especially complex mammalian societies include those of many primates (for example, see Crook and Gartlan 1966; Crook 1970; Kummer 1968, 1971; Jolly 1972; Rowell 1972; and Eisenberg, Muckenhirn, and Rudran 1972), a ground squirrel called the black-tailed prairie dog *Cynomys ludovicianus* (King 1955; W. John Smith et al. 1973), and such carnivores as lions and wolves (Schaller 1972; Mech 1970). Among birds with complex social organizations are gulls (see, for example, Tinbergen 1953b), the Australian magpie (Carrick 1963), and jays (J. L. Brown 1974)—particularly the piñon jay (Balda and Bateman 1971). The behavior of social insects has been reviewed by E. O. Wilson (1971a).

Management of relationships within the social groups of such animals requires the evolution of displays that can do more than just indicate that the communicators are moving—information must be provided about the likelihood that individuals will interact, or interact in particular ways. Yet these kinds of information are also provided by the displays of species that are *not* persistently social. We have no good measures yet of the extent to which elaboration of social structure has influenced the evolutionary elaboration of messages about interactional behavior, but the limitations on display repertoires that are discussed in chapter 7 do appear to have restricted opportunities here. On the other hand, the *form* of displays and badges, and the *extent* to which they are performed do appear to be modified in accordance with differences among species in their degree of sociality.

CONTROL OF AGONISTIC BEHAVIOR   The more a species is specialized for social life, the less it can afford the disruptive effects of fighting and fleeing. Comparisons have therefore been made of the extent to which the displays and badges used in agonistic encounters have been elaborated among species that differ in the complexity and persistence of their social behavior.

Among passerine birds, for instance, Moynihan (1962c) has noted that the species that do not have particularly well-developed sociality are often those with the most highly elaborated and conspicuous agonistic displays—as if the evolution of greater sociality had required the

A

B

Figure 12.3. Agonistic displays of lovebirds. In *A*, a male Madagascar love-bird *Agapornis cana* approaches his mate with the long, rapid strides of an aggressive walk display, to which she responds submissively by ruffling. In this noncolonial species such an encounter can lead to a violent fight. In *B*, a pair of peach-faced lovebirds, *A. roseicollis*, bill-fence in a similar kind of event, parrying and thrusting with their bills and aiming nips at each other's toes. In this highly colonial species, nipping is restricted to the toes and the mates will not fight. (From Dilger 1962. Copyright © 1962 by Scientific American, Inc. All rights reserved.)

damping down of inflammatory signaling. Dilger (1960) found some-thing similar within the parrot genus *Agapornis,* the lovebirds, in which both the likelihood of attack and the extent to which displays are exaggerated and modified from their precursor acts decrease with in-creasing sociality (figure 12.3). Lovebirds that are relatively solitary when nesting have agonistic displays that are conspicuous ritualizations of intention movements of fighting (such as gaping, carpal-flashing, and an exaggerated aggressive walk with which an opponent is approached); the occasional fights of these birds are violent. On the other hand, the more social lovebirds have formalized the fighting performance itself as a formal bill-fencing interaction: thrusting with the bill at the oppo-nent's toes, parrying thrusts, and counterthrusting just as in fighting. This interaction does not develop into serious attempts to bite. Further, among these more social species an individual may induce its mate to move along a perch by gently nipping or making intention movements of

nipping at its toes. The nipped individual moves unhurriedly along until its mate ceases. Formalization in this case is slight: the biting is gentle and directed specifically at the toes. Yet these modifications of aggressive bites are enough to enable the biting to be treated not as an attack but as signaling, at least between lifelong mates who are well adjusted to one another.

A different kind of trend in agonistic signaling, correlated with a greater frequency of reaffirming and testing the relationships between individuals, is seen among the family Canidae, which contains wolves, dogs, and foxes. Canid species differ in the extent to which members of each are persistently social and in the kinds of activities in which they become communally involved. For instance, communal hunting, although well developed in wolves and Cape hunting dogs, is uncommon in the canid family. Communal feeding is somewhat more widespread, occurring both among the pack hunters and among scavengers. Sleeping and resting together occur in all species in the breeding season, but all year long only in the more gregarious species. Unlike lovebirds and the passerines Moynihan described, however, the relatively solitary species such as the foxes possess "uncomplicated" visible signaling patterns. They apparently rely heavily on vocalizations and pheromones (Kleiman 1967). The more persistently gregarious species such as the wolf and Cape hunting dog have more elaborately developed visible signaling based on facial expressions and, especially, tail postures and movements. This appears to be very important in the daily management of agonistic interactions in which the dominant-subordinate relationships between individuals are invoked or tested.

SEASONAL SOCIALITY   Being social only seasonally, such as for breeding, poses its own problems for the evolution of signaling specializations. Species whose members are customarily solitary need mechanisms that bring some individuals together in season, and mechanisms that help others to avoid contact that would lead to disputes. The specializations must therefore be effective when the distances between participants are great; examples already presented include the wind-borne sex attractant pheromones of many insects and the loud songs of other insects and many amphibians, birds, and some mammals.

Long-distance displays with different degrees of elaboration have been shown to correlate with different sorts of social behavior in the many species of African weaverbirds (Crook 1962, 1964). Members of species dwelling in forests do not flock, and a solitary male sings and roves about when searching for a mate. In habitats that provide ecological advantages for social life, however, males are grouped and are able to have louder, more conspicuous, and more continuous performances

by cooperating in chorusing, as described above. The way food is distributed in African savannas makes flocking advantageous for the species of weaverbirds who live there rather than in forests. They nest in dense colonies in which each male builds a nest at which he sings and does bowing displays, receiving considerable stimulation from his neighbors. Within these social units, the extent to which the displaying of each male is tied to his nesting site—and hence the forms of the displays he performs—are fixed by the ecological regime. In species such as *Ploceus niggerimus,* males are free to fly off in conspicuous display flights. Their absences from their nests slow the pair-formation process more than weavers in more arid country can afford. For instance, *P. cucullatus* has to breed in a briefer season, and its males have no displays with forms that would require them to leave their nests (Crook 1963).

Seasonal sociality not only forces the development or renewal of social relationships within a limited period of time, but arrangements of at least two sorts must often be developed simultaneously, for example, relationships between mates and between territorial neighbors.

Differences in habitat among related species can determine the relative amount of work necessary to manage each sort of relationship, and hence the emphasis of the social organization and the ways in which displays must be invested by a species. For instance, Lein (1973) found that the unpredictable availability of the chestnut-sided warbler's nesting habitat creates much greater competition among neighboring males in the defense of their territories than is characteristic in the much more stable habitat of the ovenbird. Further, the ground foraging habits of ovenbirds leave them exposed to view on the forest floor, whereas the foliage-gleaning chestnut-side is hidden among sun-speckled leaves. These habitat differences correlate with differences in the ways in which the two species have developed their social behavior and their signaling specializations. The terrestrial ovenbird must be cryptic, and the inconspicuous brown plumage of both sexes makes them look alike. This makes courtship a fractious process, worked out during a great deal of mutual association between a male and a female. A male ovenbird is left with little time available for establishing relations with neighboring territorial males, but needs little. Chestnut-sided warbler males could not afford such a time-consuming courtship because they must spend a great deal of their time establishing and maintaining boundaries with each other. Their elaborately patterned plumages are sexually dimorphic, however, making life with mates easier and freeing both time and displays to be used in developing intermale relationships. Mates spend less time together in this species, and chestnut-side males utter more song and a larger number of different types of songs in territorial proclamation than do male ovenbirds, even singing while they forage.

Their increased number of song types correlates with the need of chestnut-sides for fewer displays in interactions with their mates. Male ovenbirds have invested much more of their repertoires in sexually distinctive displays for intrapair use, and because the two species have about the same number of displays this leaves ovenbirds with few that are specialized primarily for male-male communication. The different social and signaling adaptations of the two species are summarized in table 12.1.

TERRITORY   The use of territory as a means of organizing social relationships makes various demands of signaling specializations. For instance, signals must be suited to the size of a species' territory and to the structure of the population within which territoriality operates. Where territories are large, some displays must carry far, and others may form a graded series used primarily when intruders are not readily deterred. Species whose members crowd onto smaller territories may have to display more often, and the displays with which they respond to intrusions have to escalate much more rapidly and blatantly toward attack. The two Australian ravens *Corvus coronoides* and *C. mellori* differ in just these ways, the former living year-round on large territories and the latter coming out of nomadic flocks in the breeding season to form neighborhoods of small territories (Rowley 1974). In species whose territories are packed unusually closely, territorial advertising displays may be unusually inconspicuous, as are flight displays in the locally dense nesting groups formed by western sandpipers *Calidris mauri* (Holmes 1973) and in the dense colonies of tricolored blackbirds

Table 12.1. Differences in social organization and signaling specializations of two species of warblers.

Feature	Ovenbird	Chestnut-side
Median courtship duration (days)	8	7
Percentage of time associated with mate	83.4	16.4
Chases of female by male	frequent	rare
Territorial encounters between males	infrequent	frequent
Male and female plumages	identical	different
Number of nonsong vocalizations		
Male only	6	2
Female only	3	1
Both sexes	1	3
Total	10	6
Number of different song types	1	5
Percentage decline in singing at pair formation	50	12
Singing while foraging	rare	common

(From Lein 1973.)

(which have lost the conspicuous flight displays of the closely related but less densely colonial red-winged blackbird; see Orians and Christman 1968).

Of course, territory size is only one source of selection affecting the carrying power of displays, a property that is probably severely regulated. Because "social confusion" (Moynihan 1967) can so readily result from the reception of displays by inappropriate recipients (and because predators can be attracted, see below), most vocal displays are probably adapted to carry no farther and persist no longer than is usually necessary.

The Animal Environment: Other Species

Among the different species inhabiting most regions are some with similar sensory capacities. Members of each such species are exposed both to the others' displays and to the others' responses to their own displays. Opportunities for interspecific interactions then abound, and although members of any one species have little direct commerce with members of most others, they do interact regularly with some.

Evolution of the displays, badges, and other sources of information that are specialized by each species is thus determined in part by the characteristics of other local species. The kinds of effects are as numerous and diverse as the kinds of relationships that can develop between species. Resemblances are affected: each animal must be distinctive enough to be recognizable to other conspecific individuals, and yet it may be able to profit by appearing similar to other species. Detectability and locatability are also affected: the responses of members of other species give each animal either the need to conceal itself or advantages from making its presence known.

Adaptations to Noisy Neighborhoods

Neighborhoods in which there are many signaling animals can be thought of as noisy. First, similarities among the signals of different species can confuse their recipients, each of whom has the most to gain by attending only or primarily to signals provided by members of its own species. Second, the signals themselves can obscure one another, adding to the confusion.

DIFFERENTIATION   Recipient animals, their senses bombarded by displays, scents, and other sources of information, need to choose the appropriate signals to attend. Conversely, they must avoid being misled or unduly distracted by inappropriate signals from animals with whom

they do not need to interact. If the signals of their species are insufficiently distinctive, recipients have to pay too much attention to other species. They then waste time and may even make mistakes, becoming involved in interactions with members of the wrong species. Thus an important effect of the evolutionary processes of adaptation is to reduce confusion by making the signaling repertoire of each species distinctive.

Formalized sources of information should evolve species-specific identifying characteristics whenever there can be any doubt about the identity of a communicator. Of course, identity is not usually in doubt in some situations. Nestling birds and other dependent infants that must remain at fixed sites, for instance, can be identified from knowledge of the sites—and the begging vocalizations of nestling passerines of many species are very similar. Among adult animals, only those who know each other as individuals and share a close social bond may be able to come together closely enough to permit the exchange of tactile displays, very faint vocalizations, or (in some cases) surface-adhering pheromones, and these displays and chemicals may then evolve little or no species-specificity. Further, where specific identification is necessary it need not be provided by all formalized features. The formalized movements of a visible display need not be specific in form if a species-specific badge is visible when they are performed: birds that raise specialized crest feathers do not need to differ in the movements with which they raise them if each species has a crest of a different color or shape. Finally, in some circumstances (discussed below) animals of different species converge in the evolution of some displays or badges that enable them to obtain advantages from living in mixed-species social groups, from dealing with members of competing species, or from influencing predator-prey interactions. Despite exceptions of these kinds, the greater part of each species' display repertoire, most or all of its badges, and most of its pheromones usually evolve characteristics that identify the species.

There are different extents to which the forms of displays and formalized accoutrements are restricted by the needs of species to be distinctive. Some differences correlate with the apparent functions of different displays, and the most distinctive signaling specializations in each species' repertoire often are those employed in pair formation or the initiation of interactions in which eggs are fertilized. For instance, although more than half the approximately 1,000 species of orthopteran and homopteran insects in North America and Europe have been examined, no sex-attracting acoustic displays of species active in the same regions at the same times have been found to be identical or confusingly similar (Alexander 1967). Similarly, the call by which males attract mates in each of the fourteen species of Puerto Rican frogs of the genus

Eleutherodactylus is unique in the length and number of its component notes, and in the lengths of intervals between the notes (Rand and Drewry ms.). Many, many such information sources have this function of facilitating species discrimination in mate selection. They are called "prezygotic isolating mechanisms" (see Mayr 1963), devices that prevent interbreeding between species before fertilization is attempted, thereby preserving time and energy, and even the chance to breed successfully without having wasted gametes in liaisons with an inappropriate partner. These formalized mechanisms seem to function well. Members of different species scarcely ever try to interbreed in nature, even if they can be shown to be capable of hybridizing under special conditions in captivity.

The more species from which the members of a population must distinguish each other, the more narrowly specialized for differentiation should be the formalizations they evolve. In geographical regions in which many species coexist, the adaptations evolved for specific distinctiveness may exhaust the potential of some aspects of form to permit discrimination. A species coming at a later time, expanding its range into such a region, presumably has to modify its displays within the limits left open. For instance, the northernmost populations of the titi monkey *Callicebus moloch* were apparently among the last monkeys to inhabit Colombia and eastern Ecuador (Moynihan 1966a). This species lacks the loud, continuous roaring vocalizations with which troupes of howling monkeys *Alouatta* species call and answer each other. Moynihan postulates that they may have been unable to develop or retain such roars because the howlers had preempted that "acoustical niche." Titi monkeys apparently had to evolve other display patterns to communicate among their troupes, and developed unusually complex and elaborate patterns of song phrases.

Because small islands have fewer species than do adjacent mainlands, Marler (1960) proposed that there should be a relaxation of the selection pressures that force the songs of each bird species into narrowly defined patterns. The sound environments of island birds should permit more variability of form than on the mainland. The hypothesis is testable, although not yet adequately tested: the size of recorded samples of songs necessary to compare variability among populations is much larger than has generally been assumed (Thielcke 1969), and other evolutionary effects that island environments may have in addition to permitting increased song variability complicate the comparison.

SELECTIVE TUNING   The task of a recipient individual in determining which communicators are appropriate for it to attend can be greatly simplified if it has sensory and perceptual capacities that filter external

stimuli (Marler 1961b), that is, which are selectively tuned to narrowly defined ranges of stimulus characteristics. Virtually anything passed will then be worth attention. Further, if they are largely filtered out, other potential stimuli will be much less likely to obscure the appropriate ones. Not surprisingly, then, evolutionary opportunities for selective tuning have been exploited by many kinds of animals.

Insectan sensory hairs, feathery aristae, or antennae, all of which are excited by the displacement of particles in the medium caused by sound waves of low frequency, can be tuned to vibrate at particular frequencies. For example, the wingbeat frequency of a female mosquito sets the antennae of a male of the same species vibrating (Bennet-Clark 1971). Some other kinds of insects, including grasshoppers, have a tympanal membrane capable of detecting higher frequency sounds that range from a few hundred Hertz to 60 kHz or more. This tympanum responds to changes in the amplitude of sounds, for example, in their pulse rates and durations, but does not analyze a sound's frequency spectrum (see Alexander 1967; Konishi 1970; Bennet-Clark 1971). Species of grasshoppers do differ in the frequencies of their songs, however, and each species' tympanum is probably tuned to respond to the necessary frequencies through innervation by neurons with narrowly peaked frequency sensitivities.

An analogous membrane tuning has been discovered in frogs. Bullfrogs have two different auditory receptors each tuned to different frequencies characteristic of adult male bullfrog vocalizations (Capranica 1965, 1968). The different frequency characteristics of the calls of immature male bullfrogs actually inhibit tympanal responses, so that the tuning of the receptors excludes irrelevant calls whether they are intra- or interspecific (Capranica 1968; Frishkopf, Capranica, and Goldstein 1968). Tuned auditory sensitivity even varies geographically in the cricket frog, accommodating dialectal shifts in the frequency of the species' mating calls (Capranica, Frishkopf, and Nevo 1973). Among the fourteen species of *Eleutherodactylus* frogs of Puerto Rico mentioned above, those species usually calling in the same localities at the same time typically employ separate and isolated frequency bands with little or no overlap (figure 12.4). We hear these largely as differences in pitch, but because frogs apparently hear them simply as frequency-dependent amplitude differences, only the calling of its own species will be loud to each individual (Rand and Drewry ms.). In fact, it has been shown that in at least one of these species (*E. coqui*) each of the two notes of its call is important to recipients of only a single sex (Narin and Capranica 1976). Males answer the calls of intruding males and answer similarly playback of a synthesized copy of only the "co" portion of the call; they will approach and attack a close source of such calling.

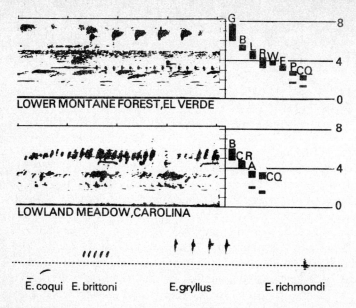

Figure 12.4. Differential use of frequency bands by *Eleutherodactylus* frogs. Recordings made in a forest and a meadow in Puerto Rico reveal that the numerous species of frogs calling at each site are separated by their use of different frequencies. The range of frequencies spanned by the call of each species is diagrammed to the right of each sonagram: CQ = *E. coqui*, E = *E. eneidae*, G = *E. gryllus*, L = *E. locustus*, P = *E. portoricensis*, R = *E. richmondi*, and W = *E. wrightmanae*. The bottom portion of the figure depicts the forms of the calls of four species so that their contributions to the complex sonagrams of the field recordings may be more easily seen. (From Rand and Drewry ms.)

Females also approach the sources of calling, and approach playback of synthesized versions of only the "qui" portion of the call but are unattracted by the "co" portion alone.

People have been found to have a different sort of auditory tuning in which certain changes in sounds that are physically continuous are heard as discontinuous. That is, discrete categories are perceived that do not exist as such physically; our ability to detect some continuous differences is limited. This phenomenon, called "categorical perception," determines how we hear sounds that are involved in discriminating speech (such as the phonetic differences between *b, d,* and *g*), but does not affect our ability to hear similar physical continua when they occur in nonspeech sounds (Mattingly et al. 1971). Categorical discriminations, however, are also made of some musical sounds: changes in the

rise time of such a sound are categorized by a listener as if coming from a plucked or bowed stringed instrument (Cutting and Rosner 1974). Our ability to both categorically and continuously process essentially similar sounds may be fundamental to the evolution of speech, and may depend on cerebral dominance—the two types of processing occurring in the dominant and subordinant hemispheres, respectively (see Marler 1975).

Research on this tuning phenomenon is of recent origin, and it is not known yet to what extent species other than humans also impose the perception of categories upon physically continuous phenomena. Other species apparently can perceive human speech in the categories we use, which has been shown for rhesus monkeys by Waters and Wilson (1976), although their categorical boundaries may be less resistant to shifting than are the boundaries used by people in perceiving speech. That nonhuman species may categorize continuous sounds used in their own vocal repertoires has been shown for pygmy marmosets by Snowdon and Pola (ms.), and the phenomenon is being sought in other primates and in birds (Marler, Dooling, and Zoloth, research in progress). It may be a general feature of the perception of continua for the recipient to obtain a reference, perhaps learning the midpoint of a series or being presented with a point anchored by external considerations, and then to divide the continuum into two categories on either side of the reference point (M. Wilson 1972). This shows up in an ability to discriminate stimuli better between than within the two categories that are created. The need to perceive categories may also be a result of using limited kinds of neural detectors for the features in which stimuli can vary, for example, feature detectors responding to upward frequency shifts, downward shifts, and both kinds of shifts in sounds (Stevens 1973).

Tuning biased to special sources that are made available in social communication also occurs in both vision and olfaction (see review by Marler and Hamilton 1966). Visual tuning, for instance, can be based on differential sensitivity to different wavelengths, on the fact that numerous combinations of wavelengths can excite the same set of color receptors and provide perception of a single hue, on form detectors in the retina and brain, and so forth. In some moths, olfactory tuning to sex attractants is sufficiently narrow that specific proportions of the geometric isomers of their pheromonal molecules are required for optimal behavioral responsiveness (Minks et al. 1973; Klun et al. 1973). And just as bullfrog auditory receptors are inhibited by certain frequencies, the olfactory receptors of some moths are inhibited by the sex attractants of related species and narrowly tuned to their own attractants (see Tumlinson et al. 1974).

AVOIDANCE OF BEING OBSCURED   Selective tuning of recipient systems by no means eliminates the interference from noises that obscure animals' signals. Tuning cannot completely shut out sources of white noise or, usually, stimuli closely similar to the appropriate ones. And inflexible tuning has drawbacks: narrowly tuned auditory systems, for instance, must miss or deemphasize sounds that could be very relevant, such as noises made by approaching individuals of one sort or another, including predators. Fortunately, animals have additional options. For instance, communicators can reduce the degree to which reception of their displays or other signals is interfered with by the activities of other species by signaling when the others are not.

A dawn chorus of singing birds can be almost deafeningly loud at the season when males of diverse species are most actively establishing territories and seeking mates. The amount of song lessens during the morning, but there can be considerable acoustic interferenec for most species through much of the day. The worst interference is to some extent avoided: birds of different species join in and drop out of the dissonance of the chorus at different times, some beginning to sing at the first feeble predawn light, and others not until the sun is well above the horizon and the earlier birds have left off continuous singing for other pursuits. A staggered relationship may be maintained during the daylight hours, for instance, if members of two species each vary cyclically in the frequency with which they repeat songs, and cycle asynchronously with each other. The two most common species in the dense, shrubby California chaparral, the Bewick wren *Thryomanes bewickii* and the wrentit *Chamaea fuscata,* cycle this way. The former is the first to sing in the morning and later has the more regular cycles of singing, with less minute-to-minute variation in numbers of songs than does the wrentit. Wrentits apparently begin to sing only after the overall swell of Bewick wren singing has begun to subside, and then sing less as the wrens begin a new outburst. The result is that "each species sings most frequently when singing in the other is at low ebb" (Cody and Brown 1969, who recorded the number of songs by each species per minute and used a computer time-series analysis to test the cycling).

The advantages of asynchronous cycling, while mutual, are not equal in this case. Cody and Brown suggest that the Bewick wren is able to be relatively independent of the wrentit because its greater diversity of song forms provides a relatively unpredictable, and hence confusing, source of variation for the song environment of the wrentit. A similar interspecific arrangement may be indicated by field experiments of Littlejohn and Martin (1969), in which recordings of the calls of one species of Australian frog, *Crinia victoriana,* were played back to individuals of a second species, *Pseudophryne semimarmorata.* The latter species was

inhibited from singing when the former's calls were sufficiently loud, but sang very rapidly afterward—suggesting that in dense populations males of *P. semimarmorata* may sandwich bouts of their calls between those of *C. victoriana*.

The advantages to compensating for each other's singing are also mutual and unequal in the case of red-eyed vireos *Vireo olivaceous* and least flycatchers *Empidonax minimus* in Minnesota forests. When a male of each species has a male of the other as a neighbor, their singing appears to be based on precedures comparable to the time-sharing that controls access from two or more terminals to a computer: each sings more or less continuously, but their active moments interleave. That is, they avoid overlap of individual song units, even though each sings about as often as the other. In five sets of neighbors, a statistical analysis of the overlap of songs by a male of one species with those of the other revealed that each flycatcher tended to insert his short songs into the pauses between the longer songs of his vireo neighbor (Ficken, Ficken, and Hailman 1974, and see table 12.2). Three of the five vireos also made some attempt to begin songs when the flycatchers were silent, even though at least part of each of their songs would always be free of interference.

In the wren-wrentit and vireo-flycatcher cases, the songs of the two species are markedly different in form. Where species with similar songs live together they may also tend to avoid simultaneous singing. Moynihan (1963b) discovered a complex case involving four species of nectar-

Table 12.2. Timing of least flycatcher songs relative to singing and silent periods of a nearby red-eyed vireo.

Flycatcher (5 individuals)	Total songs	Number of songs begun during vireo song		Number of songs begun during vireo silence		χ^2	Probability of chance difference
		Predicted	Actual	Predicted	Actual		
1	275	66.7	23	208.3	252	37.796	0.001
2	519	130.1	23	388.9	496	117.729	.001
3	226	49.0	18	177.0	208	25.035	.001
4	43	11.3	1	31.7	42	12.690	.001
5	146	35.5	11	110.5	135	22.336	.001

Predictions about the amount of overlap to be expected between songs of these neighbors were made on the basis of the proportion of time the vireo was vocal during the total time recordings were made, and the χ^2 test was used to compare predicted and actual occurrence of overlap. The test showed that overlap was less than expected on a random basis, suggesting that each flycatcher avoided beginning a song while a vireo was singing. (From Ficken, Ficken, and Hailman 1974. Copyright 1974 by the American Association for the Advancement of Science.)

and insect-eating honeycreepers (*Diglossa* species and *Conirostrum cinereum*) in the shrubby paramo of the higher Andean regions of Ecuador. The similar foraging requirements of these species frequently bring individuals into mutual proximity, although they avoid approaching each other closely. As they are not interspecifically aggressive, their territories often overlap extensively. Thus if two individuals of different species uttered their similar songs at the same time, the various possible recipient individuals might become considerably confused. Moynihan found that these honeycreepers do tend to avoid such simultaneity. They have no comparable response to the singing of other local species, all of which have forms of song that are relatively distinct from those of the honeycreepers.

In summary, adaptations to noisy neighborhoods are of two sorts. First, there is differentiation: the development of specifically distinctive repertoires of signals that reduces the chance of confusing them with the signals of other local species. Second, there is the evolution of means of reducing the level of interference engendered by other species' signals. The development of receptor systems or perceptual apparatus tuned to respond only to the appropriate signal forms has this function, as does the behavior of signaling when interference is momentarily low.

Providing Information without Becoming Prey

Animals inevitably make information available by everything they are and do, which is, of course, a disadvantage when it exposes them to predators and parasites. Therefore selection limits the provision of information that increases vulnerability and maximizes the availability of any information that will deter enemies. In addition, there are advantages in being misleading and in distorting information.

APOSEMATISM AND MIMICRY   Much has been written about "aposematic" badges and their mimicry. Animals that are distasteful or dangerous to predators usually evolve conspicuous badges that call attention to their identities, for example, patterns of boldly contrasting yellow and black markings. This is aposematism. Members of two or more aposematic species sometimes evolve resemblances to each other. This is Müllerian mimicry, advantageous because it enlarges the pool of individuals from which visually hunting predators take samples while learning the kinds of animals to be preyed upon or avoided. On the other hand, animals of one species will often evolve resemblances to those of another species when only the latter are distasteful or dangerous. These Batesian mimics take advantage of their models, and their

duplicity misleads potential predators. A general review of kinds of mimicry and their evolution is provided by Wickler (1968).

Batesian resemblances are evolutionarily unstable because both predator and model species may continuously evolve counteradaptations. Those models survive best who look least like their mimics, and predators profit from recognizing the differences. However, models must also be conservative: to the extent that a model's badges are effective in deterring predators it is under pressure to retain them. The evolution of characteristics differentiating it from its mimics is slowed (Brower and Brower 1972), while the pressures of predation keep mimics continuously evolving to resemble their models.

Nonetheless, interspecific mimicry is not purely advantageous for all mimics. Because it confounds the task of identifying the mimic, it can lead to difficulties with intraspecific communication. This may explain why mimicry is sex-limited in some species of butterflies: only females mimic the color patterns of species that predators reject; the males' color patterns are badges that clearly identify their species and sex. Brower (1963) postulated that one sex must be visibly identifiable during the courtship of these species, and therefore must forego the advantages of Batesian mimicry.

CRYPSIS Whereas aposematic badges and behavior and mimicked aposematism enhance conspicuousness, badges and behavior used by other species in mimicking plants or inedible objects render their wearers cryptic. Crypsis, like Batesian aposematic mimicry, is thus the evolution of sources of misleading information. In it, too, the adaptations of one species can influence the evolution of those of another. If all cryptic species in a region mimic the same environmental objects in the same ways, it becomes too easy for predators to find the basic clues and form a single searching image that is useful in finding all the available prey. Thus the cryptic insects hiding in grass will not all evolve identical longitudinal patterns of yellow and green stripes, but different species or subpopulations will employ a variety of patterns. Their "aspect diversity" (Rand 1967; Ricklefs and O'Rourke 1975) makes it difficult for a visually hunting predator to form a consistent image of the camouflage of its prey.

On the other hand, related species may be forced to have very similar cryptic adaptations. If many different kinds of predators hunt the same species, they make it difficult for the prey to evolve cryptic badge patterns that are suited to all predators. Its cryptic adaptations thus become very conservative, unable to be changed in any feature unless the change is adaptive with respect to diverse kinds of recipients (Moynihan 1975). For instance, Moynihan has found that related species of

squid differ little from each other in the displays with which they change their color patterns when alarmed by predators; squid are preyed on by a wide assortment of fishes, birds, and mammals.

MINIMIZING LOCATABILITY   Most display behavior cannot be cryptic because its job is to be informative, and it must be noticeable if it is to facilitate social interactions by providing information that would otherwise not be available about an animal. Some displays, such as bird songs and flight performances, are specialized to make communicators easily locatable by conspecific individuals that may be at great distances, and these displays expose their users to predators. Locatability is also a crucial inherent feature of odor trails, and those of ants and termites, for intance, make them vulnerable to insectivorous snakes (Gehlback, Watkins, and Kroll 1971). However, the extent to which displaying provides information about a communicator's location can sometimes be modified so that exposure to predators is minimized. This can even be true for song-type advertising. As mentioned in chapter 6, lone males of the Panamanian frog *Engystomops pustulosus* appear to rely on patient search by females and utter their least locatable form of song; the same males give more locatable forms when competing with other nearby males in choruses (Rand ms.).

The ease with which the sources of audible displays can be located usually depends on physical characteristics of sounds that can be compared as they are received by each of a recipient's ears. A very detailed review of evidence for spatial localizing abilities in all sorts of animals has been prepared by Erulkar (1972); basically, if there is a detectable difference in intensity, or in the timing of the arrival of phase points or frequencies between the ears, a binaural recipient orients toward a communicator by turning its head until the difference disappears. Different species probably use different clues, and at least some birds may respond primarily to intensity differences (Erulkar 1972). For instance, in experiments in which barn owls *Tyto alba* were trained to strike targets that were audible but not visible, Konishi (1973) showed that the simultaneous presence of a wide frequency range is very important because it provides information about intensity differences at many frequencies. Konishi has questioned the use of binaural comparison of phase differences by birds, although Hailman (1975) has argued the need for further experiments to distinguish between perception of direction and of distance. Whatever clues are most important, though, most communicators emitting audible displays are readily located.

Most, but not all. For avian predators with heads of any particular width some sounds with simple waveforms are both too high in frequency for binaural comparisons of phase and too low for differences in

intensity to be appreciable; if these sounds lack abrupt changes in frequency or amplitude and are not repeated their sources are very difficult to locate. Marler (1955a) pointed out that small birds tend to have two classes of calls with just these characteristics. First, there is an "alarm call" which a member of a flock will utter on first detecting a dangerous predator. In flocks that customarily include more than one species, the different species may have converged on (or failed to diverge from) a single form for this call, a case in which the usually advantageous interspecific distinctiveness would become a disadvantage. Second, there are the calls of nestlings, individuals who are at a site known to their parents and who are unable to flee should a predator locate them.

Ultrasounds are useful for communicating at close quarters without attracting predators, because they are absorbed very rapidly in air (see Griffin 1971) and are scattered by even such small objects as grass blades (Sales and Pye 1974). Like the calls of nestling birds, the ultrasonic vocalizations of many rodents, for example, of neonates (Zippelius and Schleidt 1956; Sewell 1968, 1970; Noirot and Pye 1969; Colvin 1973; Brooks and Banks 1973) or copulating adults (Sewell 1968; Barfield and Geyer 1972; Brooks and Banks 1973), are both faint and hard to locate, and presumably represent antipredator adaptations. Infant rodents produce ultrasounds when removed from the nest (Zippelius and Schleidt 1956) and in other circumstances, as when a sleeping mother rolls onto a pup. Maternal rats have been shown to be aroused from sleep especially easily by such cries (Zoloth 1975), and respond by retrieving young to the nest or changing their position or activities within the nest in ways that alleviate the discomfort of the young (Sales and Pye 1974).

Readily locatable vocal displays are uttered by some species of small birds when individuals are alarmed. The calls do not betray their communicators' locations because several individuals cooperate in the pattern of their displaying: the flock generates a "confusion chorus" that makes it difficult for a predator to resolve individual targets. This mechanism is an audible analogue of the confusing visual effect that results when attacking predators cause animals to scatter widely and erratically.

Confusion choruses were first described by Grinnell (1903) as the response of a flock of tiny bushtits to a nearby aerial predator. Each bushtit remains absolutely stationary and repeats a "shrill quavering piping." This calling would be easily located except that it comes at irregular intervals from different members of the flock, its source now here, now there. The resultant chorus becomes an "indefinably confusing, all-pervading sound" that Grinnell recognized as a "composite

protection device." Field evidence that chorusing is successful was later offered by A. H. Miller (1922) who "half a dozen" times saw a dangerous sharp-shinned hawk pass just over or through a chorusing bushtit flock without making a strike.

The shrill vocalization used by chorusing bushtits is not strictly an alarm call. In fact they utter it, prefixed by simple notes, in many circumstances that do not involve predators. Their antipredator specialization is not the display, but their group performance of it—a simple kind of formalized interaction (see chapter 13).

Although effective, confusion choruses do not appear to have been developed by many species, even those living in the densest cover who could use them best. Another case has been reported by R. B. Payne (1971) on the basis of an observation of African arrow-marked babblers *Turdoides jardineii:* a group of about ten birds chuckled and babbled asynchronously in chorus when approached by humans, and no one individual could be located. Perhaps many cases remain to be described. In fact, the repetition by numerous individuals of calls at odd intervals may have evolved often in flocking birds, although not as a response specifically to the presence of a predator. Instead, such calling may be a compromise between the need on the one hand for individual members to be aware of and provide each other with information about the location of their flock, and on the other hand for each such individual to have its precise position concealed from predators. Erratic calling by all the flock can accomplish both. For example, foraging goldcrests *Regulus regulus* and titmice (*Parus* species) utter high-pitched calls at irregular intervals while moving through the trees in dispersed flocks. Because the unpatterned succession of calls from different individuals seems to come "from all directions" (Voipio 1962) only the general whereabouts of the flock is apparent. This, however, is sufficient to keep members from straying away from the flock. Further, the effect is for each such foraging flock to generate continuously a confusion chorus; in this chorus the rate of repetition of vocal displays is slower than when a hawk attacks a flock of bushtits, but the choruses are otherwise comparable.

A quite different sort of chorusing may be less confusing than it is overwhelming. Male periodical cicadas *Magicicada* species gather and sing in earpiercing choruses that get louder as more males sing. The noise appears to repel birds that prey upon less densely massed cicadas, as very few approach and stay when a chorus is loudly blasting (Simmons, Weaver, and Pylka 1971).

MINIMIZING DETECTABILITY   The location of a communicator performing visible displays is apparent if the displays are seen. The likeli-

hood of detection by predators can be reduced, however, by limiting the amount of time during which displaying renders a communicator conspicuous.

Many species use what D. Morris (1956a; discussed in relation to my term "novel transformation" in chapter 11) has called "transformation displays." These show off badges that are concealed or inconspicuous when not in use, for example, the collapsible colored dewlaps of *Anolis* lizards, or the bright flash patterns revealed on wings and rumps as some birds take flight (chapter 9).

Some species are made especially vulnerable to visually hunting predators by the openness and lack of hiding places in the habitats within which they forage. Some, like the plain-brown woodcreeper *Dendrocincla fuliginosa,* are inconspicuously colored and evasive, and seem to have very few displays—interacting instead by avoidance, chasing, and fighting (Willis 1972). The reduced display behavior of the Steller's eider *Polysticta stelleri* can be compared directly with the displaying of the related common eider *Somateria mollissima* to show the effects of differences in the ease with which prey species are visible to their predators (McKinney 1973). These two species of sea ducks obtain their food by diving to the ocean floor, the latter in deep water and the former in shallow. Deep water is concealing, but shallow water leaves Steller's eiders exposed to aerial predators which can watch, then take them as they come to the surface. The defense of individual eiders is based on what amounts to hiding warily among the many other individuals in a flock, a procedure known not only from these and other kinds of birds but also from fish schools and herds of ungulates (Williams 1964; Hamilton 1971; Vine 1971). A Steller's eider rarely leaves its flock. It also rarely spends much time preoccupied with displaying even when within the flock. Its courtship displays are abbreviated and silent compared with those of the larger common eider, and its precopulatory displays are very reduced and inconspicuous. A pair of Steller's eiders must separate from the flock to copulate without interference, but they leave for the briefest possible period and spend little of that time displaying. Differences between the two species in feeding, displaying, and responses to approaching birds of prey are summarized in table 12.3.

Communicators producing audible displays can also reduce the ease with which they are detected by displaying less, especially if they live in dense environments or are nocturnal. Even when trying to obtain mates, individuals may not display at all for long periods if they have another option. For instance, male field crickets *Gryllus integer* attract females in the night by singing, but their sounds also attract parasitic tachinid flies of the species *Euphasiopteryx ochracea* (Cade 1975). Some males

Table 12.3. Major behavioral differences between Steller's eider and common eider.

Characteristic	Steller's eider	Common eider
Body size	Small	Large
Feeding habitat	Shallow water	Deep water
Food	Small invertebrates	Large invertebrates
Diving flocks	Large, densely packed; synchronize dives	Small, not densely packed; dives not synchronized
Response to birds of prey	Flocks fly up readily	Fly less readily; may dive to escape
Precopulatory behavior	Brief; single shake; final display invariable	Up to several minutes; conspicuous wing-flap; many shakes; some calls by male; final display variable
Postcopulatory behavior	Pair flies back to flock; male displays silent	Male calls
Courtship behavior	Preflight movements; short flights; aerial pursuits; displays rapid; males silent	Underwater pursuits; displays slower; some compound displays; males give cooing calls

(From McKinney 1975.)

go to an area where other males are singing and walk about silently, attempting to intercept and copulate with females attracted by the latters' sounds. Relative measures of mating success are not available, but the noncalling males are parasitized much less than are the singers: Cade found fly larvae on only 1 of 17 examined, but on 9 of 11 singers.

SECONDARY EFFECTS OF NEST-SITE ADAPTATIONS　Unusual adaptations to avoid predation on their nests and young have had marked effects on the perfomance of displays by some species. Among the best illustrations of this phenomenon are E. Cullen's studies (1957) of the kittiwake gull. Unlike most other gulls, it avoids many terrestrial predators by nesting on cliffs. Of approximately 25 peculiarities of the kittiwake that appear to adapt it to cliff-nesting, 7 relate to displaying—for example, the performance by kittiwake chicks of the facing-away display as an appeasing gesture. In other gulls only the adults do this display, but appeasing is more important to kittiwake chicks who must avoid being knocked from their narrow nesting ledges. On the other hand, kittiwakes do not perform an upright posture in their aggressive encounters; disputes tend to be at very close quarters with little room for maneuvering

and often involve birds whose nest sites are at different levels on the cliff face, which requires different extensions of the participants' necks just to fence with each other, let alone display.

Two species of arctic gulls (N. G. Smith 1966b) and one species in the Galapagos (Snow and Snow 1968) also nest on cliffs, and although they share some of the kittiwake's adaptive peculiarities they have not converged in their performance of facing-away and upright. Nonetheless, facing-away by chicks is among the behavioral convergences of yet another cliff-nesting seabird belonging to a different taxonomic order, the gannet (J. B. Nelson 1967).

RESPONDING TO DISCOVERY Displays and badges are also important in behavioral responses to predators that have already discovered their potential prey (see review by M. H. Robinson 1969). These responses may startle or confuse a predator (Humphries and Driver 1967; Driver and Humphries 1969; Moynihan 1975), or frighten it by mimicking a more dangerous animal (M. H. Robinson 1969; and see figure 12.5). Some distract predators, drawing them away from nests or young toward the displaying communicator (also shown in figure 12.5). Others deter further approach by letting a predator know that its potential victims are aware of it, and will therefore be very difficult to catch, or prompt the predator to begin its pursuit from too great a distance (Smythe 1970). Yet others attract other animals that join in mobbing the would-be predator, and drive it away.

Disguises of Predators and Parasites

Being inconspicuous, hard to identify, or misleading can be just as useful to a predator or a parasite as it is to its prey or hosts. There is no monopoly on formalized deception.

Many predators are cryptically colored and patterned, camouflaged to fit unseen into their habitats. Others rely on the technique adopted by a wolf in sheep's clothing—they mimic a harmless species. The plumage of the zone-tailed hawk *Buteo albonotatus,* for instance, resembles nonpredatory vultures; it will soar with them, dropping from a group of vultures onto its unsuspecting prey (Willis 1963). The various forms this "aggressive mimicry" takes have been reviewed by Wickler (1968). For example, some insects of the family Mantidae, the praying mantises, are colored and shaped like flowers; they visually attract insects seeking nectar as prey. Other predators even mimic their prey's prey: angler fish (Lophiiformes) and alligator snapping turtles *Macroclemys temmincki* can wriggle fleshy outgrowths of their fins or tongues and attract small predatory fish close to their mouths. A small fish that lives by coral reefs

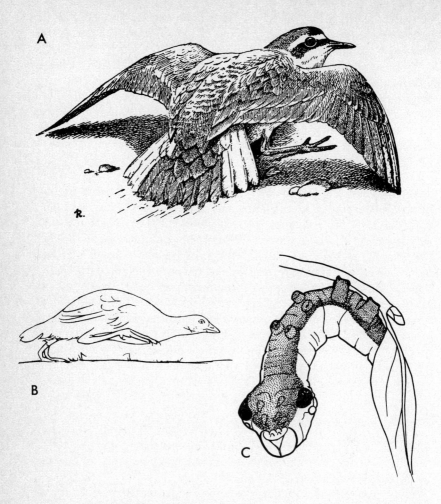

Figure 12.5. Misleading displays. *A*. "Injury feigning" by a Kentish plover as it runs erratically from a predator that has approached the plover's eggs or chicks. The displaying bird appears injured, often flopping about as if it had a broken wing, and some predators become distracted and try to catch it. (From Simmons 1951.) *B*. "Distraction run" of the Tasmanian native hen, done in a conspicuous zigzag course in the open when diverting a predator from the nest or chicks. The running bird gives the "fleeting appearance of a small mammal, such as a bandicoot, small macropod, or rabbit." (From Ridpath 1972.) *C*. "Snake display" by a caterpillar of the sphingid moth *Leucorampha ornatus*. The dilated thorax and large, false, shiny eyes mimic the head of a snake; the display incorporates swaying, serpentine movements that are thought to be frightening to potential predators. (From M. H. Robinson 1969.)

uses aggressive mimicry in preying upon larger fish. Its model is another small fish species that finds food by cleaning parasites and damaged tissue from the many larger fishes that dwell on reefs. This cleaning wrasse *Labroides dimidiatus* is identified by bright markings, and the larger species that need its services let it approach unmolested. The shape, brilliant blue and black stripes, and special swimming behavior of the wrasse are imitated by the sharp-toothed *Aspidontus taeniatus,* which employs its guise in approaching larger fish closely from behind, then bites chunks out of them (Eibl-Eibesfeldt 1959; Wickler 1968).

Among the many bizarre forms of aggressive mimicry are some ways of using bioluminescence in luring prey (reviewed by Wickler 1968; Lloyd 1971). Luminescent glow worm larvae in New Zealand caves attract midges by exploiting the positive phototactic responses that normally guide hatching midges to the caves' mouths. Females of the firefly genus *Photuris* mimic the flashes by which females of other species attract males for mating—and then eat the males they capture. A female of a Florida population of *P. versicolor* can adjust the duration and temporal pattern of her flashes in response to the flashing of passing males, and can mimic the patterns of two or three different prey species (Lloyd 1975).

Parasites, like predators, may evolve to be inconspicuous or difficult to recognize. Most species of brood parasitic birds, who lay their eggs in the nests of birds of other species, have dull brown or black adult plumages. Several species of cuckoos, less dull than parasitic cowbirds, honeyguides, and ducks, are either polymorphic or have much individual variation in color and markings (R. B. Payne 1967)—analogues of the "aspect diversity" adaptations of prey. The eggs laid by cuckoos and cowbirds are usually quite similar to those of the host species, and their young, if reared with the hosts' nestlings, may resemble them very closely. The resemblances are extremely detailed in the viduine finches reared by estrildine finch hosts. Nestling estrildines (such as the melba finch *Pytilia melba*) begging for food show a species-specific mouth pattern of black spots on the tongue, palate, and lower mandible and white or blue papillae on their mandibles; their begging vocalizations are also species-specific. Nestlings of their viduine parasites (in this case the long-tailed paradise widow bird *Steganura paradisaea*) have exactly the same gape patterns and vocalizations (see Friedmann 1960; R. B. Payne 1967; Wickler 1968). The vocal mimicry of the parasitic greater honeyguide *Indicator indicator* is extreme in another way. A single honeyguide nestling, which is often reared by African bee eaters (Meropidae), can sound like a whole brood of bee eater nestlings—and elicit feeding not only from its foster parents, but even from neighboring pairs of bee eaters (Fry 1974)!

Unusual signaling specializations are found in some parasites of social insects, as might be expected from the diversity of insectan adaptations. For example, certain species of rove beetles (Staphylinidae) are parasitic on ants. Adults of the beetle species *Atemeles pubicollis* gain entrance to *Formica* ant nests in the spring, and mate and lay their eggs there. The ants not only feed and groom the parasitic beetles, they even rear their young. Hölldobler (1971) has described how a beetle is "adopted" by ants. It approaches an ant, taps it with its antennae, and raises the tip of its abdomen. The ant licks a pheromone from there, then another pheromone from glands along the sides of the beetle's abdomen. Then the ant picks up the beetle and carries it into its nest. Later the beetle's larvae are adopted when they secrete a pheromone that attracts ants to groom them. The larvae also perform formalized signals to elicit feeding, tapping the ants' mouthparts with their own. Other species of staphylinid beetles have fewer specializations for communicating with the ants they parasitize, although some can appease attacking ants by releasing the abdominal tip pheromone.

Some ants parasitize other ants by invading their nests and stealing their worker pupae, which then become workers in the parasitic slavers' nests. Slave-maker ants of the *Formica sanguinea* group use great quantities of their alarm pheromone against defenders of the nests they raid (Regnier and Wilson 1971). The pheromone acts as a "propaganda substance," dispersing the defenders without hindering the slavers, although the latter do respond to it as an alarm pheromone when it is used at their home nests.

Competing Interspecifically

Sympatric species—those which live in the same region—with similar requirements for food or other resources are ecological competitors. Most evolve ways of partitioning environmental gradients of their resources and coexist, or compete until their closest competitors become extinct. In geographic regions where populations of different species meet and overlap, they may sometimes have evolved greater dissimilarities than elsewhere in their ranges. This is "character displacement" (Brown and Wilson 1956; the concept is reviewed and the evidence criticized by P. R. Grant 1972). As an alternative means of coexisting they may evolve "character convergence" (Cody 1969, 1970, 1973) and become more alike, especially in many of their signaling specializations. (Actually, convergence is not necessary if the species are initially alike and then fail to diverge, or evolve in parallel; see P. R. Grant 1972.) Competitors that have converged or remained alike become interspecifically territorial, spacing themselves out in their common

habitat just as members of each species space themselves. They have divided the contested resource by evolving a system based on the way each partitions spatial access to it territorially.

Based on behavioral observations and territorial measurements, Cody (1973) offered examples of species that may have converged, and suggested various other pairs of strikingly similar species that may also represent cases of character convergence. Some of these species may not have converged, but instead *retained* similarities they already had, as Grant suggests; this possibility does not alter the picture of adaptation for interspecific competition. On the other hand, for some of the species, convergence or parallelism may have been based on cooperation rather than on competition (see next section). Nonetheless, among Cody's examples are birds known to be interspecifically territorial in some regions in which their geographic ranges overlap. For instance, some of the songs of the green towhee *Chlorura chlorura* and the fox sparrow *Passerella iliaca* converge where these species meet on a ridge of the San Gabriel Mountains in California (Cody 1973), and some songs of two *Pipilo* towhee species may converge on Cerro San Felipe in Oaxaca, Mexico (Cody and Brown 1970). Other songs of these species remain distinct even in regions of overlap, and presumably facilitate identification for intraspecific mating.

For that matter, species identification in territorial encounters can be based on other specializations than song. For instance, eastern and western meadowlarks *Sturnella magna* and *S. neglecta* treat each other as territorial rivals where they overlap along the long thin fingers of riverside vegetation that extend into the grasslands of the American Great Plains, although their songs do not converge. However, visible characteristics that are important in postural displays used in their territorial defense do converge: body size, the yellow of the breast, and a black V-shaped breast badge (Rohwer 1972, 1973).

The development of cryptic signaling specializations may also help in avoiding competitors where territorial behavior would be impractical. For instance, Galápagos gulls *Larus fuliginosa* are the color of the lava shorelines on which the species scavenges. Their dark plumages may hide the gulls from such competing scavengers as graceful petrels *Oceanites gracilis* and frigate birds *Fregata species* (Hailman 1963).

Fostering Social Groups of Mixed Species

Many of the advantages that come from banding together intraspecifically, such as enhanced ability to find food or a greater likelihood of detecting and avoiding predators (see chapter 10), can also be gained from appropriate interspecific sociality—and without the disadvantages

of increasing the number of competing individuals, if species are sufficiently different. Many birds, mammals, and fish form interspecific groups. These groups are often large, but are more organized than are the aggregations that occur around localized resources such as a waterhole or carrion. Their organization persists in the absence of concentrated resources.

The basic signaling adaptations fostering the cohesion of such groups are the badges and displays employed in intraspecific groups, for example, flash patterns (chapter 9) and vocalizations carrying messages of associating and locomotory behavior (examples in chapter 5). These devices are adapted for interspecific use by what Moynihan (1960, 1968b) calls "social mimicry"; they evolve convergently. Signaling convergence in social mimicry thus promotes interspecific association, whereas character convergence (see above) promotes the spatial segregation of species. In part, the two functionally defined forms of convergence affect different badges and displays. In part, however, they do not. Social mimicry appears to have led to very extensive similarities in the appearances of different species that flock together (examples in Moynihan 1960), and Cody (1973) points to cases of apparent character convergence that are at least as extensive. Some convergent signaling specializations (or specializations that have been kept from diverging evolutionarily), like the plumages of *Muscisaxicola* flycatchers of the Andean mountain chain, may even serve alternately as social mimicry in facilitating mixed flocks in the early spring, and as character convergence for interspecific territoriality later (W. John Smith 1971a; Cody 1973).

Members of mixed-species groups appear to have developed (and converged in) more signaling adaptations than just those that make them attractive to each other. Many of them have also become adapted in ways that must reduce interspecific hostility. Moynihan (1960, 1962c, 1968b) has discovered that species of birds typically found in mixed flocks display in intraspecific agonistic situations less often and with less elaborate displays and also fight less than do closely related species that are found less regularly in mixed flocks.

Some of the intraspecifically gregarious birds that are most attractive to the largest numbers of other species in mixed flocks have unusually dull plumages (apart from their flash patterns). They are typically restless and vocal, easily seen and followed when in motion, yet lack badges that might agonistically arouse other species. They are neutral in coloration, less distinctly different from diverse other species than if they had more striking plumages. Moynihan suggests that their drabness is a comprise where social mimicry of divergent associates would be impossible.

The active little plain-colored tanager *Tangara inornata* is one of

these neutrally colored species. It is dull gray and black with blue flash patches on its wings; it frequently utters brief vocalizations and rattling sounds. In central lowland Panama it is an important "nuclear" species in mixed bird flocks.

Moynihan (1962c) has shown that the species regularly or occasionally found in its flocks are adapted for interspecific sociality to different extents and in different ways. Among other nuclear species, the palm tanager *Thraupis palmarum* is drably colored and the blue tanager *T. episcopus* is not, but is dull blue with bright blue wings and tail. Moynihan describes these two as "active" in interspecific flocking because they join and follow other species; conversely, the plain-colored tanager is "passive" because it joins and follows only intraspecifically, despite its attractiveness to other species. Many additional "active nuclear" species are elaborately patterned: they join and follow plain-colored and other tanagers, but although they help make the flock conspicuous, they are much less often joined and followed by other species than is the plain-colored tanager. Still other species contribute less to the flocks: they have few signaling specializations for mixed-species flocking, and do not attract other species to join and follow them. Moynihan calls them "attendant" species.

Interspecific social groups can contain very few, often just two, individuals. Many represent remarkable adaptations for mutual advantage, each individual providing special services for the other. Their association may be brief, as between the small honeyguide bird (genus *Indicator*) and a mammal (usually a ratel *Mellivora capensis,* baboon *Papio cyanocephalus,* or a human being). Repeating a churring vocalization and fanning its tail (which has white outer feathers) as it flies and pauses, a greater honeyguide (*I. indicator*) will lead a mammal to an active bees' nest and then wait while the nest is ripped apart (Friedmann 1955). The bird is the finder, the mammal the preparer of their feast. And they share—the mammal eats the honey, and the bird beeswax. Whether any of the honeyguide's signals are specialized for this interspecific function is not known, but their performance in leading is clearly a specialization.

A pistol shrimp *Alpheus djiboutensis* and a gobiid fish *Cryptocentrus cryptocentrus* have a much more durable relationship. Two shrimp and a fish live together in a burrow dug by the shrimp. The shrimp are almost blind, and depend on their fish to warn them of predators. A shrimp will emerge from its burrow only when its fish is nearby outside. They exchange tactile signals, the shrimp within the burrow touching the caudal region of the fish with one antenna, and the fish responding by undulating its tail with a rhythm of about 1 cycle per second (figure 12.6). The shrimp then emerges, but maintains constant antennal contact with the fish (Karplus, Szlep, and Tsurnamal 1972). They may

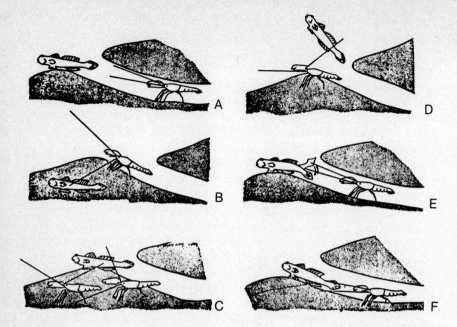

Figure 12.6. Positions of the fish *Cryptocentrus cryptocentrus* and the shrimp *Alpheus djiboutensis* as they emerge from their retreat to their burrow. The fish, near the entrance (*A*), is approached by the shrimp. The shrimp will not leave its burrow unless it can maintain contact with the fish. The fish emerges farther from the burrow (*B*), followed by the shrimp which keeps one antenna constantly in touch. When both resident shrimp are out of the burrow (*C*), each maintains contact with the fish. This contact is maintained whatever the spatial relationships of fish and shrimp, and both species keep their tails oriented toward the burrow (*D*). The fish rapidly flicks its caudal fin on being disturbed (*E*), for example when it sees a predatory fish. The shrimp and the fish then retreat into the burrow (*F*), the shrimp maintaining antennal contact. (After Karplus, Szlep, and Tsurnamal 1972.)

remain outside all day, but if the fish is frightened it retreats to the burrow and the shrimp retreat with it. They remain inside as long as the fish does, although they may push it to venture out again. Apart from having a burrow provided for it by the shrimp, the fish has its body surface cleaned by them. It attracts them to it chemically, although the full extent of pheromonal communication in their relationship is not yet known.

13

The Concept of Formalized Behavior

Although any detectable act by one individual can be informative to another individual, some behavior is specialized to be informative. The latter is called "formalized" behavior throughout this book, whether it evolves through genetic evolution (which ethologists call "ritualization") or through cultural or individual changes ("conventionalization").

Formalized acts achieve their ends indirectly. That is, the actor does not act physically to alter things to suit his needs, pushing or dragging other individuals about, beating them into submission, or the like. Instead, the actor's behavior provides other individuals with information, and the actions that *they* take on the basis of this information lead to any functions that are obtained.

Displays

The main kinds of behavior that ethologists have traditionally recognized as products of formalizing processes are displays: behavior specially adapted in physical form or frequency to subserve social signal functions. This definition and the ways ethologists have viewed the concept are discussed in chapter 1; the diversity of display behavior is surveyed in chapter 2. Displays have been the acts most studied by ethologists interested in communication.

The display concept and the ways it has come to be used have advantages and disadvantages. The principal advantage is that it defines an important class of activities in a general way, without narrowly and arbitrarily restricting the properties to be expected of this class. The resultant disadvantage is that it does not specify the range of criteria by which a unit of display behavior can be recognized. In practice there are

several kinds of behavior for which it is difficult to determine whether a particular act is a display, and there are ranges of variable display behavior that are hard to tease apart into appropriate units and their relationships; further, some formalized behavior patterns fit the definition but are not displays in the usual sense of the term. At least six problems can be defined:

1. Formalized and unformalized behavior patterns differ in the extent to which they are distinct from one another. Some acts are only slightly formalized; on the other hand, behavior that is specialized for ends other than communicating is sometimes very striking and thus resembles displays.

2. It is not clear how many individuals within a population should have a particular formalization in their repertoires or be prepared to respond to its use by other individuals before it can be called a display.

3. Many displays vary noticeably in some or all characteristics of their form, some of them over considerable ranges. Some displays not only vary but also intergrade with each other.

4. Some acts that appear to be displays are performed in simultaneous combinations with each other, and no consensus exists among ethologists as to whether the combinations or their recombinable components (at some level) are the fundamental units of communicating.

5. Similarly, some acts usually or only occur as components of sequential combinations.

6. Some formalized patterns of behavior cannot be performed by single individuals. Two or more individuals cooperate in these patterned interactions, each assuming a well-defined part or succession of parts to play.

Most of these difficulties can be resolved without greatly altering the display concept, although it is necessary to modify it through the addition of explicit criteria defining display units. We must also recognize that some compounding of displays can be formal, and that formalization can yield not only acts that can be defined as units, but also rules that generate relatively complex behavioral products which may be very flexible in the details of their form. The sixth problem, however, requires us to recognize a kind of formalized behavior in which the units are not displays or other acts of individuals, but specialized modes of interacting: "formalized interactions."

Redefining the display concept is the main goal of this chapter; chapter 14 is a preliminary examination of the further concept of formalized interactions. The study of communicating has now reached a stage at which we must be more concerned with the nature and relationships of the fundamental behavioral units with which we are working, and with the limits of the concepts these units represent.

Difficulties with the Display Concept

Distinctiveness of Displays

The first difficulty with current formulations of the display concept is that they specify what displays are, but do not specify how displays may always be distinguished in practice from nondisplay acts.

The processes through which displays evolve lead them to become distinct from other kinds of acts and stereotyped in at least some characteristics, as discussed in chapter 11. Their distinctiveness is not always apparent to investigators, however. Their perceptual abilities are not identical with those of most other species, a problem considered further under the topic of variation and intergradation of displays in this chapter. In addition, they may encounter at least two other kinds of problems in recognizing whether or not an act is a display.

First, because evolution may begin with a nondisplay precursor act there are stages in the process of formalization at which little change has occurred. This stage may be transitory, or may not: in some cases natural selection or cultural pressures may not be strong enough to drive formalization far. This state must be possible when the precursor act is often sufficiently informative even without change: running at an opponent in an agonistic encounter will usually be interpreted as predictive of attack behavior, and perhaps be effective only slightly more often if flourishes are added. The selection pressures to develop and fix such flourishes as displays are thus weak. Alternatively, even where formalization could markedly increase an act's potential for being informative, the process may be stunted if the circumstances in which the act occurs scarcely warrant the cost of developing a display; that is, if they do not provide sufficient adaptive advantages.

Behavior that has begun to be formalized might be thought of as "nascent display," but whatever it is called it is only slightly and perhaps inconsistently distinguishable from its nondisplay precursors. Acts of this sort may occur in the display repertoire of the eastern phoebe, described in chapter 3; likely candidates in that species are the wing shuffle display and perhaps some sleeking and fluffing of the plumage.

The evolutionary converse, displays that are becoming less formalized, provides the same problem. Senescent displays in the process of becoming "deritualized" (see chapter 11; Moynihan 1970b) can be as difficult to distinguish from nondisplay acts as are nascent displays.

The gray area that results from these evolutionary limbos is very difficult to assess. Is it a brief, evolutionarily unstable twilight zone, hiding few displays from us? Or is it populated by numerous variant behavior patterns, some evanescent and some stable: unproductive

genetic accidents, marginal cases, and budding or senescent displays, each exposed to the tests of practicality and not, or not yet, winning the game? We don't know.

In principle, it should not matter very much with respect to the effective communicating behavior of a species at any given time, because signals that are hard to differentiate from nondisplay acts cannot be much more effective than the latter in making information available. Further, if they are not striking actions, they may often be either missed or misidentified by potential recipients. Yet, although displays should be less readily missed or misidentified than are the unformalized precursor acts from which they are derived, it will rarely be practical to measure such differences as indicators of formalization.

The immediate task is to seek and recognize a class of acts on the fringe of any species' display repertoire, acts that are not necessarily striking and may or may not be formalized. Our traditional and somewhat haphazard criteria for recognizing displays have led us to concentrate on the conspicuous, missing most slight formalizations. As a result, our current estimates of the numbers of displays in the repertoires of most species are undoubtedly conservative (some implications of this are discussed in chapter 7). Seeking a class of marginal displays and comparing its properties among species will at least give us a broader perspective.

A second problem an investigator encounters is that not all stereotyped acts, not even all conspicuous ones, are displays. Indeed, displays are merely one class of the acts that ethologists call "fixed action patterns." Fixed action patterns have diverse functons and include, for instance, the form and pattern of the "fanning" behavior by which a stickleback fish aerates eggs (Tinbergen 1953a), and the movements and their sequence with which a female canary constructs her nest (Hinde 1958) or with which a goose retrieves an egg that has rolled a short distance away from her clutch (Lorenz and Tinbergen 1938). Relative fixity of behavior must be economical in the service of many kinds of functions in addition to communicating.

Though many fixed action patterns obviously achieve their functions by direct effects (effects not mediated by individuals other than the performers of the actions), and thus are not displays, others can appear to lack direct effects and thus be much harder to understand. A classic example is the *Flehmen* response of many mammals, which is very conspicuous in some bovids and equids. In a typical performance (figure 13.1) that may be prolonged for up to a minute, "the animal stands open-mouthed with head extended and elevated, while the upper lip is retracted, wrinkling the nose and baring the gum" (Estes 1972). This very conspicuous act is done by males virtually whenever a female

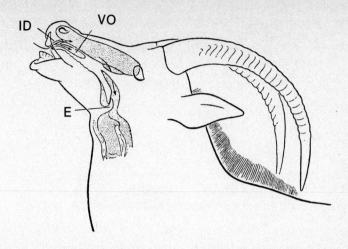

Figure 13.1. A male sable showing the *Flehmen* response. The sable's con-spicuously up-flung head and retracted lips resemble a display posture. These movements are not specialized as a display, however. They permit air to be sucked through the incisive ducts (*ID*) into the vomeronasal organ (*VO*) while the lips block the nostrils and the epiglottis (*E*) is closed. (From Estes 1972.)

urinates, but especially when she is in heat; on some occasions it is a response to a male urinating. As Estes makes clear, there is *no* evidence that *Flehmen* is a display, and its performance does not readily permit prediction of the actor's subsequent behavior. Estes presents strong anatomical and physiological evidence that *Flehmen* is "purely and simply a motor pattern associated with the olfaction of certain sub-stances," especially sexual hormones dissolved in the urine. The con-spicuous lip-retraction and nose-wrinkling components probably block the external nares and help to fill, via the incisive ducts, another set of sensory receptors in the vomeronasal olfactory organ (figure 13.1); present evidence indicates that the vomeronasal system is crucially involved in the detection of pheromones important to at least sexual behavior (Powers and Winans 1975). Thus although this was not realized clearly before Estes' investigations, the bizarre-looking behavior of *Flehmen does* act directly; it brings pheromones in contact with the appropriate sense organs. The form of the *Flehmen* act is apparently not a result of specialization for communicative functions.

Another example of a behavior pattern that might appear to be formalized can be seen in the conspicuously stern-high swimming pos-ture of small grebes. The wings are tilted up, exposing the erected white feathers of the lower back (figure 13.2); these feathers have black

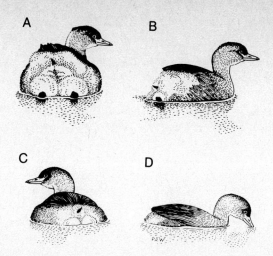

Figure 13.2. Sunning and pre-diving postures of the least grebe. A sunning least grebe *Podiceps dominicus* swims conspicuously stern-high, tilts its wings up, and exposes and erects the white feathers of its lower back (*A* and *B*). Although this posture is strikingly different from other swimming postures (such as those in *C* and *D,* as the grebe prepares to dive), it appears to be not a display but a means of increasing the amount of heat taken in from the sun. (From Storer, Siegfried, and Kinahan 1976.)

bases, and the skin surrounding them is also black. The grebes adopt this posture while orienting away from the sun, and the fluffy portion of the erected feathers can presumably be penetrated by solar radiation, which is absorbed on the black skin and bases of the feathers (Storer, Siegfried, and Kinahan 1976). Despite its resemblance to display, the behavior is almost certainly a form of sunning that permits the grebes to add to their energy budgets by taking on heat. As in the case of *Flehmen,* then, a strikingly conspicuous posture is linked to anatomical specializations and serves a physiological function.

Undoubtedly many further stereotyped acts of dubious function resemble displays. Yawning is one such pattern that people share with many other species. It appears to have a physiological function in emptying accumulated carbon dioxide from the lungs of an individual that is inactive and breathing only shallowly. Yet it has a social significance of some sort, because it is so often "contagious" among the members of a group and because it does not occur at random. However, most yawning individuals turn their heads away or otherwise partly conceal the act, and at least partially lid their eyes. Perhaps both these acts are displays, and both suggest that an effort is necessary to prevent yawning from being mistaken for the open-mouth threat displays it resembles. This in turn suggests why it is contagious: once one individ-

ual has yawned without being misunderstood, the "climate" of a group can be seen to be appropriate if you must get the need to yawn off your chest. Yawning is at least momentarily established as an expected and acceptable kind of behavior. Yawning thus appears not to be a display, but a respiratory act that has interactional implications because it resembles a threat display. Still, it may trade on this resemblance at times and be used as a display, for example to indicate disaffection or boredom; the status of yawning is equivocal.

Conspicuous stereotypy can arise under aberrant conditions, and it appears as if this behavior can also be functional. For instance, D. Morris (1966) has pointed out that animals confined in cages usually have too little sensory input and very often develop highly stereotyped performances which they repeat frequently. The functions of these acts may be to obtain an increased and predictable sensory input. At the other end of the scale, Rimland (1964) has proposed that children suffering from early infantile autism find too much stimulation in normal environments and adopt stereotyped, repetitive patterns to bring some of their confusing sensory input under predictable control. (The Tinbergens advanced a somewhat similar hypothesis in 1972, proposing that fearful children will lock out social stimuli from certain sensory modalities; however, they and Rimland disagree on many issues concerned with childhood autism; see letters to the journal *Science* by Rimland 1975; Tinbergen 1975.) Whether to increase or decrease sensory input, however, stereotyped actions arising under aberrant conditions resemble displays and can appear to have no direct function because they do little to alter the external environment. Yet they are not adapted to be information sources, however much they may reveal about the individuals who develop them.

Some minor problems also have arisen in defining certain stereotyped acts with forms specialized for mimicry, acts that function to mislead predators, parasites, prey, or hosts (see chapter 12). The form of such an act, and of structures or color patterns revealed by the act, may make a communicator resemble part of its inanimate background: an alarmed young squid *Sepioteuthis sepioidea,* for instance, may adopt an "upward V-curls" posture and a blotchy, barred color pattern that render it cryptic among pieces of drifting *Sargassum* weed (Moynihan 1975). Conversely, if suddenly jabbed by a bird's bill a hawk moth may break its immobile, camouflaged resting posture and abruptly spread its wings, exposing on the hind wings spots that resemble an owl's eyes (Blest 1957). The bird is often startled, and the moth may escape. The squid, moth, and a great many other prey species employ such stereotyped acts that have a misleading influence on the responses of their predators, but it is the mimicking form that misleads.

These acts actually do permit reliable prediction of a performer's behavior. For instance, a plover that drags about on the ground seeming to feign an injury (as in figure 12.5) will show locomotion (both intermittent running and eventually flying), remaining with site behavior, and the incompatible selection of escape. These behavioral selections correlate with the display quite regularly; it is the specialized flopping, dragging *form* of the performance that can mislead a predator into attempting to catch the plover rather than to search for its eggs or chicks. With experience, a recipient can learn that such birds are quite able to escape and perform distraction displays when their offspring are threatened with discovery. The displays presumably evolve primarily because their mimicking forms mislead recipients of predatory species to the reproductive advantage of the communicators, although in species in which both parents care for the young the messages of distraction displays by one parent may bring its mate to help cope with a predator.

These misleading behavior patterns are formalized and fit Moynihan's original definition of displays (1956; reviewed in chapter 1), if one accepts social interactions as including those between predator and prey, parasite and host, and the like. Nonetheless, because the acts function to mislead, it can be useful in some kinds of analyses to recognize them as a special class of behavior. Moynihan (1975) has proposed the term "antidisplays" to include at least those used by prey species. However, because the concepts of display and antidisplay are defined by functions, both terms can sometimes be applied appropriately to describe a single formalized act. For example, a squid may use a single kind of act as a cryptic antidisplay in response to a predator and, in other circumstances when it is seen against different backgrounds, as a conspicuous display in response to another squid. It must be kept in mind that only one fundamental class of behavior is involved: acts specialized to be informative, whether their forms mislead or not. (Comparable considerations apply to badges, which constitute a single class of nonbehavioral specializations, whether misleading or not. Dual use of a single badge configuration is possible. For example, Barlow, 1972, argues that stripes and bands can simultaneously conceal fish from their predators and make social behavior more predictable to conspecific individuals.)

Number of Individuals Sharing the Code

Second, the display concept does not specify how widely an act should be used within a population to qualify as a display.

Displays permit individuals to share information to the extent that each individual assigns the same messages to the same formalized acts— that is, to the extent that codes are shared (a code is a set of rules

relating messages to displays). The codes usually dealt with by ethologists are shared within large groups: whole species or at least major subpopulations. Variants of display form may be peculiar to single individuals, pairs, families, and local flocks, but all such variants of a display continue to carry the same behavioral messages. Their peculiarities only identify particular segments of the populations, and even this information is presumably intelligible to all individuals as indicating either in-group or out-group membership.

Nonetheless, acts are sometimes formalized only within the repertoires of particular dyads of individuals. They make information available according to a code, but only two individuals employ that code. Private information can be said to be shared by these acts, but it is not widely shared. Most members of the larger populations are excluded.

Dyadic codes are perhaps most likely to be found when two individuals interact frequently in a limited kind of circumstance. They are based on the special expectations each participant develops of the other in their repeated mutual experiences. The circumstances may involve persistent and stabilized competition, such as that between neighbors who compete for space and establish mutual conventions over a common territorial boundary possibly known only to them. Perhaps more commonly, however, they are formed between bonded individuals who share a closer and partly cooperative relationship. For instance, the precopulatory T-sequences described by Golani and Mendelssohn (1971) for golden jackals may involve dyadic formalizations: the relative positions adopted by a male and a female in the T alignment as one faced the other's side, whether or not uplifted tails were waved during the T alignment, the actions with which mutual chases were performed, and other patterns were relatively stable for each pair studied and quite different among pairs. However, though highly developed sets of mutual expectations were involved in each pair's performances, it is not known to what extent different procedures became formalized within particular dyads. To demonstrate dyad-limited formalization would have required showing some form of change (like exaggeration or simplification) of acts within the history of a pair, coupled with an increased effectiveness of the changed acts in eliciting appropriate responses. Although a general sequence of seasonal changes in the frequency with which different acts were performed was observed, it was similar in all pairs and probably preprogrammed.

Formalized dyadic conventions are known to be built up in the interactional sequences that are much repeated between a human infant and its mother. Bruner (1968) labeled them "interaction codes." Various features of nondisplay behavior that correlate time and again with the needs of a child and the response preparations of its mother can

assume signal function through mutual experience, and may subsequently become simplified and made regular (the formalizing process) until the two share a convention together. (These interaction codes exist in addition to the two kinds of widely distributed "ritual behavior" Ambrose described for mother-infant relationships in 1966: those which are based on culture-specific patterns of child care and on the performance of such displays as smiling during greeting or withdrawal.) Although it remains to be shown, humans probably continue to develop dyadic systems of formalization throughout their lives.

Behavior that resembles these mother-infant codes has been elicited experimentally from rhesus monkeys by Mason and Hollis (1962) and R. E. Miller (for example 1967). Each project involved dyads of monkeys in situations in which only one could see a stimulus, and only the other had access to an apparatus for responding. Responders (the "operator" monkeys) could see the "informant" monkeys either directly (Mason and Hollis) or via closed circuit television (Miller). Operator monkeys learned to judge the location of the stimulus, in the first project, from the informant's spatial positioning. No formalized clues may have been involved, except for one informant who developed a special back-flip idiosyncracy that it performed as its operator partner initiated a correct response; the operator learned to use this odd performance as feedback, and would watch for it while tentatively initiating a response. In the second project, operators learned to distinguish between stimuli correlated with either rewards or punishments by watching their informants who characteristically made many brief glances toward and away from the aversive stimuli but stared with wide-eyed fixation upon a positive stimulus with very few head or eye movements. Again, whether or not formalization occurred is not clear, but the circumstances could readily favor it. Training as either an operator or an informant did not facilitate an individual's performance if given the other task (Mason and Hollis), so that whatever codes existed were not fully shared. Interestingly, only monkeys that had been reared socially could be used, as those reared in isolation were deficient both as senders and receivers of clues (Mitter et al. 1967).

Although much remains to be learned about dyadic formalizations, there is no reason to think that they differ in any fundamental respects from display behavior. They are usually short-lived, because few probably outlast the initial dyads without becoming more widely used, but various argot and individual-specific characteristics of display form are probably equally evanescent. That the code of dyadic formalizations is shared only by a subpopulation of two individuals does not alter the fact that these acts do facilitate sharing information. In short, the display concept should be phrased to admit any group of two or more individuals as a sufficient population for which to define display behavior.

Variation and Intergradation

Third, the display concept provides no indication of the extent to which formalization does or does not impose rigid forms on display behavior.

EXTENT OF VARIATION   Although many displays are thought to differ from their precursor acts in part through being more stereotyped, they differ considerably among themselves in the extent to which each is rigidly fixed in form. The twh-t vocalization of the eastern phoebe was chosen to be the first display described in chapter 3 in part because its form varies little, and it rarely intergrades with any other display (sometimes with a single chatter unit). On the other hand, the kit-ter display of the eastern kingbird, although having the same kinds of messages as the twh-t (see chapter 3), varies greatly in several aspects of form.

Form variation is now known to be the rule, but it was the apparent fixity of displays that led them initially to be classified as "fixed action patterns," and ethologists subsequently emphasized fixity as a basic characteristic of displays. Barlow and Green (1969) argue that this view culminated in the concept of "typical intensity" (D. Morris 1957), which holds that the form of many displays is essentially invariant even though communicators perform the same display when in a wide variety of motivational states. Yet it is not really known just how invariant most displays are. Both Barlow (1968) and Hinde (1970) have cautioned that not only displays but most fixed action patterns appear "fixed" primarily in relatively constant environments. Barlow argues convincingly that "the fixed action pattern is . . . a graded response," and is better termed a "modal action pattern." He points out that even in the displays of goldeneye ducks, the durations of which have been measured (Dane, Walcott, and Drury 1959) and are commonly accepted examples of extreme fixity, most durations have a standard deviation of 10 or 20 percent of the mean value. Although some components of the very complex strut display of male sage grouse *Centrocercus urophasianus* have subsequently been shown to be less variable among individuals, their precision seems to be unusual and perhaps extreme (Wiley 1973).

Probably no display is invariant. Some vary simply by being abbreviated or prolonged, others are consistent in some aspects of form and variable in others, and some intergrade with other displays. In principle, ability to vary displays should present the opportunity to make available a wider range of information, and ability to intergrade displays to pinpoint precise points along continua linking different message contents. In practice, however, liabilities must accrue as recipients experience difficulty in assessing particular nuances of form. Minute differences are hard to perceive unless immediate comparisons with a readily available standard of comparison can be made. Often they can be made,

however: a display can be varied while its performance is sustained or repeated, providing relative standards for comparison within a bout of displaying. Under at least those circumstances, then, much variation in form may be functional.

Some displays may be defined by a single basic procedure such as raising a crest or lifting a tail, and the procedure may be variable (see figure 13.3). A simple case is the sleek display which Ainley (1974a) describes for Adelie penguins. A penguin depresses its plumage when wary and ready to escape, when stealing nest stones, for example, from the territory of an absent neighbor. As the thief walks away from the pilfered nest it becomes progressively less sleeked and less wary. Only the degree (amplitude) and duration of sleeking are varied. Many other displays have more elaborate forms and are more elaborately variable (that is, they can vary in several characteristics). Vocalizations, for instance, may be variable not only in amplitude and duration, but also in frequency and harshness (however, at least some aspects of the form of vocal displays are tied up in stable configurations that encode nonbehavioral messages).

The most detailed quantitative study of the differential performance of variant forms of a display or set of closely related displays has been done on the vocabulary, particularly the coo sounds, of Japanese macaques (Green 1975). Each coo is a simple, tonal utterance, a continuous sound with no abrupt shifts in pitch and no breaks or noisy sections. Coos vary in pitch and duration, and their fundamental frequencies can shift, being highest early or late in the call and sometimes having a slight dip. These differences are not very pronounced, and to sort coos reliably Green used a Sona-Graph to analyze the distributions of frequencies in time. He obtained seven categories that could be arranged from low to high pitch, and in accordance with other aspects of form.

The free-ranging monkeys he recorded used coo sounds when tending to seek, maintain, or encourage what Green termed "affinitive contact," although sometimes they behaved very hesitantly. From Green's descriptions, it appears that as a class coo sounds may provide messages about interactional behavior with a depressed probability of attack and indecisive behavior. However, the different forms of the sounds refine the information that is being made available about behavior. A communicator uttering low-pitched coos is usually inactive, not in contact with other monkeys, and appears relatively unresponsive, whereas communicators uttering the higher forms actively solicit contact from dominant monkeys and are agitated and very responsive (that is, aspects of the intensity of their behavior have changed markedly; see chapter 6). The intermediate forms are uttered by dominant individuals when calmly seeking or maintaining dyadic interactions with subordinates, and by

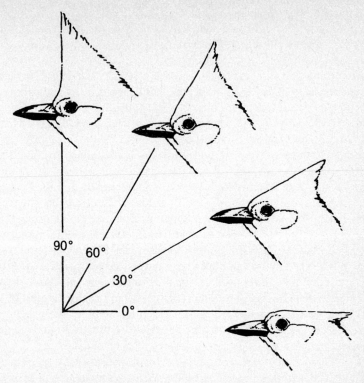

Figure 13.3. Continuous variation in a crest-raising display of the Steller's jay. Different degrees of elevation of the crest correlate with differences in a communicator's readiness to flee or engage in alternative acts. (From J. L. Brown 1964.)

very calm individuals who are interacting with others only vocally. Green's statistical analysis indicates that a single rule may relate changes in form of the coo sounds to changes in the behavior of the vocalizing animals. If so, it should be possible to formulate reasonably concise descriptions of the relationships between minor variations in the form of coos and shifts in their behavioral messages.

INTERGRADATION   Displays that not only vary but in varying intergrade with variations of one or more other displays along common parameters compound the difficulty of defining display behavior. Intermediate performances may be relatively uncommon or the intergrading may be so frequent and intricate that it is difficult for an ethologist to describe part of the display repertoire. A few species of animals even have repertoires made up predominately of intergrading displays. Such repertoires have made ethologists very wary of the definitional problems posed by inter-

gradation, although the condition has been described thus far for only a few persistently social animals such as yellow-rumped tanagers (Moynihan 1966) and certain primates (Marler 1975).

Extensively intergrading vocal displays have been a serious problem for primatologists. For instance, Rowell (1962) has suggested that captive rhesus monkeys utter only one basic agonistic sound which is "crude and structureless, but which can be modified in any combination of at least five ways" that vary independently of each other: "shrillness," loudness, duration, "scream quality," and "syllabic quality." She named nine "particular noises" and illustrated with sonagrams how they intergrade. In describing the full vocabulary of sounds heard from their captive rhesus, Rowell and Hinde (1962) again emphasized the "almost infinite series of intermediates." Itani's study of free-ranging Japanese macaques (1963) also revealed a great deal of vocal intergradation, much of which has now been mapped by S. Green (1975, extensive sample reported by Marler 1975). Marler (1969) encountered very extensive intergrading in a preliminary analysis of chimpanzee vocalizations, and found an intergrading vocal repertoire in the red colobus monkey (1970b); he has postulated (Marler 1975) that evolution of an intergrading vocal repertoire was an essential step in the origin of human speech, although it had to be followed by the additional step of categorical perception—a process discussed in chapter 12.

Nonetheless, although intergrading of vocal displays may be extreme in some primates with large and complex social organizations, it is by no means characteristic of all primates. Calls even of colobus species other than the red are relatively discrete (Marler 1970b), and although Moynihan (1967) reports a great deal of variability in the vocal displays of New World primates, it does not lead to extreme intergrading. The *amount* of vocal intergradation that occurs in natural situations is less than that occurring when dealing with rhesus in captivity or with Japanese macaques or chimpanzees around sites where they depend on experimenters to provide food for them, and it is intriguing that the vocalizations of dependent infants of many mammal species intergrade in ways similar to that described for at least the rhesus and chimpanzees (Ristau 1974; and personal observations). However, even if extensive intergrading of displays is unusual, it can make it impossible for us to define display units solely on the basis of studies of the form of display behavior.

CATEGORICAL PERCEPTION   Other definitional problems are suggested by the phenomenon of categorical perception, discussed in chapter 12 under the topic of selective tuning. Humans can hear discrete categories of sounds in speech even though the changes in the physical properties

of the sounds are continuous—categories are imposed, and not inherent in the stimuli. Other species are being found to be capable of similar categorizing but may not fix the boundaries between categories as rigidly as we do, at least when they are categorizing our speech sounds. This raises the possibility that different species cannot readily estimate the numbers of sounds in each other's display repertoires. As we use our ears to distinguish among the vocal displays of nonhuman species we may hear continuous gradation where their perceptual systems are attuned to making categorical distinctions (thus we underestimate their repertoires), and may even hear false categorical distinctions in continua they do not split (thus overestimating their repertoires).

CRITERIA FOR RECOGNIZING DISPLAY UNITS   Variation in form of display behavior raises several difficult problems for ethologists who wish both to describe display behavior and to determine the number of display units in a species' repertoire. New criteria are required.

Criteria that can permit classification of highly variable displays can be abstracted from what we know of repertoires in which variation is moderate and intergrading is absent or nearly so. Within such a repertoire, the displays available to an individual all differ either in their behavioral selection messages, or in the values encoded by their behavioral supplemental messages, or in both ways. Although the form of any given display may differ slightly among individuals, permitting the identification of different infraspecific classes, this is usually without concomitant differences in information about behavior. The exception is where only one infraspecific class (such as one age group or one sex) performs a particular display; then the behavioral messages of the display usually do differ from those in other displays in the species' repertoire.

Differences in the information encoded about behavior are thus characteristic of different displays, and are probably the main source of adaptive value in having different displays within a repertoire. This suggests that analyses of behavioral messages can be used to augment analyses of form in attempts to recognize unit displays. Different unit displays can be recognized (1) if variation is discontinuous, and a break in display form coincides with a difference either in the kinds of behavioral selection messages encoded, or in the characteristics specified by behavioral supplemental information (such as differences in the relative probabilities or intensities of the behavioral selections that are specified); or (2) if there is a continuous range of form within which the kinds of behavioral selection messages change.

The first criterion is an elaboration of the old familiar one based on discontinuities of form. When applicable, this criterion is still the most

convenient basis for making initial distinctions among displays. By adding reference to behavioral selection messages this new statement of the criterion circumvents many errors that could arise due to species-specific differences in categorical perception, although the criterion remains vulnerable to such errors where it relies only on differences specified by behavioral supplemental messages. Such differences do usually occur within the range of variation of single, continuously variable displays.

The second criterion is novel, suggesting that more than one display can be recognized within a range of continuous variation if behavioral selection messages change. For example, many displays encode the behavioral selection messages of attack and escape, and the relative (or absolute) probabilities of these selections differ as the displays vary in form. If at some point one variable pattern of displaying ceases to encode attack as an alternative to escape and instead begins to carry the more general information of the interaction message—that some unspecified kind of interacting (such as attack, associating, or perhaps copulating) is the alternative—we can recognize a disjunction. In this example, we should recognize *two* variable and intergrading displays, one carrying information about the relative probabilities of escape and attack selections, the other of escape and interaction selections. Should several displays intergrade along different parameters there might be intergrading regions in which one kind of variation had yet to run its course before another appeared. That is, messages from different sets might overlap locally. Nonetheless, the more extreme form variants would segregate by message sets and could be recognized as displays; the remaining forms would have to be recognized as intermediates.

Determining the number of displays in a species' repertoire is only one of the objectives an ethologist may have in studying display behavior, of course. Another must be to understand the potential of form variation for making information available, even when the variation adds no new messages. For instance, if an increase in the shrillness of the sound of a vocal display correlates with an increase in the probability of escape behavior, this rule needs to be described as an adjunct to the display repertoire. A full description, then, requires both a listing of display units and an account of the rules for varying display form, as well as an analysis of the messages of both the displays and the variation rules.

T-SEQUENCES   The final issue to be considered in this section is newly developing research on the geometrical structure of behavior by Golani and his colleagues (Golani and Mendelssohn 1971; Golani 1973; Golani in press) that deals in detail with the form of positions and

movements. This work has demonstrated great variability in certain minutely analyzed precopulatory interactions of the golden jackal when these interactions are viewed as mosaics of discrete acts (Golani 1973). However, if the movements of the interacting animals are instead viewed as governed by the need of each individual to respond continuously with simultaneous reference to several aspects of its situation, rules are found that constrain the geometry of particular acts (Golani in press). The latter finding implies that much of what an observer scores as "variation," at least in an animal's visible movements, may be superficial; though no two interactions may look alike in all details, they may represent the same behavior to their participants.

Although Golani has reported his analyses in terms of the display concept, however, it remains unclear to what extent the techniques he employs will reveal variation to be characteristic of display units.

What Golani has taken as his subject matter is "a precopulatory display, termed . . . a 'T-sequence' " (Golani 1973). The T-sequence he describes is a complex set of acts performed cooperatively by a *pair* of jackals. It is *not* a "display" as the term is used in this book (that is, it is not a formalized act performed by an individual), but is at least partly a formalized interaction (see chapter 14). As such, T-sequences or portions of them are formalized through a series of rules that guide the mutually adjusted behavior of their participants. Most of the acts performed in any given T-sequence are not themselves formalized. Instead, they are labile procedures that the two participants employ in adapting their behavior to each other as they adhere to a characteristic, formalized, rule structure. The forms of most of the acts are varied in accordance with immediate circumstances.

The display acts which contribute to T-sequences have yet to be analyzed by Golani's techniques, as he indicates in saying that "facial expressions, ear movements, tail movements, bristling, and vocalizations" have still to be examined (Golani in press). Thus the contrast that Golani (1973:110) draws between the enormous variation in behavior that is revealed by his analyses and the recognition of more or less fixed display units in my analyses results primarily from our different uses of the term display. Therefore, Golani's provocative findings about the behavior used during T-sequences are more relevant in considering formalized interactions (see chapter 14).

Simultaneous Combinations

A fourth difficulty with the display concept is that most animals can combine and recombine some of their formalized acts in simultaneous performances. Which is appropriately termed a display unit: a combina-

tion or a component? Or are two levels of formalization involved here?

Examples of differing combinations are abundant. A phoebe can utter at least two different vocalizations during a flight display, and an eastern kingbird more than two (see chapter 3). In a forward oblique posture a herring gull can be silent, mew, or utter a long call note (Tinbergen 1959b). Agonistic green herons (Meyerriecks 1960), Steller's jays (Brown 1964), Adelie penguins (Ainley 1974a), and a great many other birds can raise crest feathers with or without other visible and vocal displays. The threat display of Siamese fighting fish includes various combinations of orientations, gill cover erection, pectoral fin-beating, tail-beating, and tail-flashing, each of variable duration or frequency (Simpson 1968). Even invertebrates have this capacity to combine formalized acts: a fire ant can simultaneously release alarm and trail pheromones, thereby both exciting and attracting its nestmates (E. O. Wilson 1971a); a squid can combine and recombine postures, movements, and color changes (Moynihan 1975).

Recombining formalized acts in various ways is common in at least agonistic interactions as opponents adjust their patterns of displaying in response to each other's moves. Stokes (1962) made a particularly thorough investigation of the displays of European blue tits in agonistic events at birdfeeders: of 47 possible combinations of 5 main elements (presence or absence of erect crest, erect nape feathers, raised wings, body horizontal, and facing opponent) 20 combinations were observed, 8 of them commonly (see chapter 10). Some of these elements are less striking than most acts that are called displays, and in many studies comparable acts have not been accorded the status of display units.

Similarly minor-looking acts figure as components in the actions that are traditionally labeled as the displays of gulls. Each such component appears to have its own consistent range of employment. Each therefore has its own messages, although existing descriptions do not always make it fully clear what these are. One recombinable component for which a message interpretation probably can be made now is the slight to moderate lifting of the carpals by a gull who may engage in attack behavior. This inconspicuous act figures in the descriptions of up to six displaying patterns per species of gull (Tinbergen 1959b); Manley (1960b) found it at times combined with *all* the more obvious postural "displays" of the black-headed gull. Although it has not been termed a display, carpal lifting is formalized: the carpals are not lifted just in the act of striking, but are extended during prolonged periods in which blows are rarely struck. Carpal lifting thus appears to make a consistent contribution of information about attack behavior, and does this contextually to the performance of other formalized acts. Comparably minor acts, also with

consistent patterns of use, that are usually viewed as components of gull displays rather than as displays themselves are the angles at which the bill if held during both upright (Tinbergen 1959b) and hunched (Delius 1973) postures.

The names given to these postures derive primarily from the neck position (extended or retracted) of the communicator, and bill angles and carpal positions are subsumed as variable features of the postures. Only the postures are traditionally named as displays, but it is not obvious that their formalized neck positions should be assigned more weight than the bill or carpal components. The nomenclature, in short, does not accurately reflect the distribution of display units in a gull's repertoire.

The complexity of the units that are termed displays in analyses of compound behavioral patterns determines much about the ways in which variability of performances is described and categorized. Two issues are thus involved, one concerned with the description of form variation and the other with the nature of the display concept itself.

Wiley (in press) has concentrated on the problem of variation in form, and proposed that we distinguish between displays that vary "unidimensionally" and those with "multidimensional" variation. All variation in what he terms a unidimensional display is monotonic along a single dimension. For instance, all parameters may increase or decrease simultaneously, which happens when several feather tracts are simultaneously depressed in a display that gives a bird a sleeked appearance. As an alternative form of unidimensional variation, different parameters may be ranked so that variation in some components begins before variation in others. The display's form gets more complex through the series. The wing-quivering display of common grackles is an example of such nested components in which elevation of the wings precedes elevation of the tail, which in turn precedes elevation of the head (Wiley, in press).

In multidimensional variation, some formalized components of display behavior vary more or less independently of others, and each such component makes information available independently of the other components. For instance, a carib grackle *Quiscalus lugubris* performs at least eight different acts in its "song-spread display." Variation in the extent of two of these acts was measured, and their dependence was analyzed through a multivariate technique (Wiley 1975). A male grackle elevates his beak more when performing a song-spread near another male than when alone or near a female. Although further correlates of the performances were not studied, making a message analysis impractical at this stage, Wiley's work indicates that components of what is termed the song-spread display are themselves being

used as display units. (Further, at least the beak-elevating component may be a recombinable unit; it may occur with other formalized acts in another compound called the "bill-up display.")

A dichotomy between unidimensional and multidimensional variation assumes that the fundamental unit, the unit in which variation must be described, can be a complex of formalized components. The analytic procedure works well with the formalized behavior of grackles in which there are few such complexes and each occurs quite regularly. The approach works less well, however, when complex combinations are numerous and fluid (as they are, for example, in the display behavior of many primates; see below). The number of "displays" can then become enormous, obscuring the significance of patterns of variation in their components. An alternative is to recognize independent components, and perhaps even additive components, as the basic formalized units. This is done below in redefining the display concept. By this procedure, the unit that varies is a relatively simple act, and the kinds of variation that can appear in its performances are limited.

VOCAL QUALITY   Some qualities appear to occur as recombinable components of vocalizations. For example, harshness can be either an integral or a nonessential quality of a display. It is an integral quality of the "zeer" of the eastern kingbird; zeer is uttered in correlation with a range of probability of attack behavior, the probability increasing as the harshness of the zeer displays increases (W. John Smith 1966). Among other vocalizations in the kingbird's repertoire, two that do *not* ordinarily sound harsh and that lack the rapid and sustained frequency modulation characteristic of harshness can have the quality of harshness superimposed on them. When these displays become extremely harsh, most of the differences of form by which they can otherwise be distinguished are obliterated.

In each of the kingbird vocalizations, the addition of more and more harshness correlates with an increasing probability of attacking; its message contribution is consistent. In black-tailed prairie dogs (W. John Smith et al. in press 2) harshness has a similar distribution among vocal displays, being always a characteristic of the rasp display but a quality only sometimes added to a bark or a jump-yip. As in the eastern kingbird and many other vertebrate species, it is a quality correlating with the probability of attack behavior.

Although harshness is not a vocalization as such, but a quality that may sometimes be characteristic of vocalizations, it is formalized and carries its own message about attack behavior. It is thus comparable to a gull adopting a formalized downward angle of its bill or extending its carpals when in an upright or hunched posture.

EXTENT TO WHICH RECOMBINABLE ACTS OCCUR   Most species probably perform some recombinable vocal qualities and some relatively inconspicuous, recombinable movements or postures in their displaying, particularly in agonistic situations. The extent of the contributions of the recombinable units remains hard to assess from most studies, however, because the acts are often treated only briefly as components of more dramatic acts. Yet species are known in which the compounding of recombinable units is characteristic of the larger part of their displaying behavior. In general these are the same species that perform extensively variable and intergraded displays, for example the yellow-rumped tanager (Moynihan 1966) and various Old World primates. Considering both the variations and their striking dependence on combinatorial manipulation, the displaying of these species is very specialized.

Homo sapiens is an Old World primate species, and the human face employs very complex combinations of movements and positions of the forehead, eyebrows, eyelids, gaze pattern, lips, mouth, mouth angles, tongue, and jaw. However, although much is now being learned about human facial expressions by psychologists, anthropologists, sociolinguists and others, only one study has been completed of the messages of any recombinable act. Tongue showing provides the information that the communicator will show a depressed probability of interacting, seeking to interact, or being readily receptive to attempts to interact (see chapter 5; Smith, Chase, and Lieblich 1974). Work in progress suggests that many acts which, like tongue showing, contribute to compound "facial expressions" correlate with well-defined ranges of behavior and have discrete message contents.

Complex examples of the simultaneous combination of formalized components into facial expressions are available from several nonhuman species of Old World primates. For instance, in his field studies of the black mangabey *Cercocebus albigena* Chalmers (1968) described simultaneous compounds of displays under the names "stare," "yawn," and "lipsmacking." In stare, the communicator stares at the recipient with its eyes wide, raises its eyebrows and crest, opens its mouth but with its teeth covered, lays its ears back, and sometimes jerks its head in a vertical motion. In yawn, it gazes either toward or away from the recipient, does not raise its eyebrows and crest, opens its mouth very widely and slowly, exposing its teeth, and throws its head back but does not jerk it up and down. In lipsmacking it sometimes raises its eyebrows, opens and closes its mouth very rapidly while keeping its teeth covered, sometimes slightly protrudes its tongue, pouts its lips, and may shake its head laterally. Correlated with its stare, a communicator is likely to remain with or attack the recipient, whereas with yawn it becomes much more likely to flee, and with lipsmacking to remain with but not attack

Table 13.1. Communicator behavior correlated with different compounds and component displays of the black mangabey.

Communicator's subsequent behavior with respect to the recipient	Stare	Yawn	Lipsmacking
Attack	6	9	0
Remains with	12	16	21
Flees from	0	19	9

	With gaze toward	With gaze averted	With lateral head-shaking	Without lateral head-shaking
Flees from	4	15	0	9
Does not flee	14	11	10	11

(Based on Chalmers 1968.)

(see table 13.1). More detail about its behavior can be predicted from the components performed in these combinations, as Chalmers showed for two different versions of both yawn and lip-smacking: the communicator is more likely to flee if its yawn combination includes averted gaze than direct gaze, and if its lip-smacking performance includes lateral head-shaking than if it doesn't (also shown in table 13.1).

Van Hooff has described numerous "expression elements" that are found widely in the facial expressions of Old World primates. These elements are based on differential use of the eyes, eyelids, eyebrows and upper head skin, ears, mouth-corners, and lips (1962). For instance, the eyes can (1) stare fixedly at a recipient; (2) glance evasively, or (3) be more fully turned away; the eyebrows can (1) both be lifted, (2) be lowered and contracted to the median of the forehead in a frown, or (3) be wrinkled horizontally and (medially) vertically. Van Hooff derives thirteen "compound expressions" from typical combinations of his various elements (1967). The compounds are distinguished primarily on the basis of the mouth posture because visible displays are often performed in combination with vocalizations. Van Hooff feels that with more attention to the eyes a larger number of compound expressions would be recognized.

DEFINING UNITS OF DISPLAY BEHAVIOR   The variable combinations of elements in the displaying of captive rhesus monkeys so impressed Hinde and Rowell (1962) that they summarized modes of combining and recombining these in "general types [of] expressive movements" without attempting to name discrete displays in the traditional ethologi-

cal manner. The problem of how to delimit displays was also discussed by Marler (1965) in a review of the primate literature. He remarked that it is just as difficult to define and describe the numerous and very variable "clusters" or "constellations" of independent elements that make up the visible displaying of some primates as it is to classify their complexly intergrading vocal displays.

The question of which deserves our attention first, the compounded constellations or their recombinable components, is important. It would be convenient to study the components first, both because they are fewer and because they may represent the key to understanding the complexly variable combinations. Current indications are that each component can be shown to have a particular message content, and thus to be a formalized unit. Two questions then remain. First, are the constellations freely variable compounds, limited only by the need to avoid contradictions in the messages of their components (for example, components indicating depressed and enhanced probabilities for the same behavioral selection would presumably not be evoked simultaneously), or are there some constellations of fixed form?

Simultaneous combinations are typically described as variable. Components can come and go, whether in a gull's upright posture or a primate's facial expression. If this is always true, then form alone will not provide sufficient criteria for recognizing a level of formalization above that of the component acts. The second question then becomes crucial. Are the messages made available by constellations made up entirely by the contributions of their various components, or is further information added, information peculiar to each compound as a compound? In the latter case, formalization would exist at two levels of complexity (components and constellations), but in the former case the constellations are simply concatenations.

If constellations are simply opportune concatenations, it is their components that should be called displays because the components are the only formalized units. If in some species, formalization does exist at the level of both components and constellations, terminology becomes a matter of convenience. A consistent system, applicable to the various problems in defining a display unit, would be to define as displays at least the smallest formalized acts that carry consistent behavioral selection messages (there are probably lesser components that are informative about matters of display structure, but that have no consistent correlation with particular nondisplay behavior patterns). Whether these formalized constellations should be called displays, compound displays, or something else can better be determined when we know more about them, but it may be practical meanwhile to call them compound displays.

If component acts are usually the only units formalized in simultaneous combinations, then the problem created for the display concept by species that reshuffle components is largely illusory. The large range of combinations that vary greatly to fit the circumstances need not be categorized because their recombinable parts are the fundamental units. This can make analyses easier not only at the interactional level, but also at the level of motivational research (see Hinde 1970:373).

The issue of simultaneous combinations concerns only recombinable components each of which shows distinctive patterns of correlation with nondisplay behavior. Inseparable components, such as formalized movements of the tail and neck that always occur together, are parts of a single formalized unit. As so often happens, however, this distinction is imperfect: some components *nearly* always occur together, such as the wing flutter and kit-ter of an eastern kingbird (W. John Smith 1966), and the jump and yip of a black-tailed prairie dog (W. John Smith et al. in press 1 and 2). That these visible and vocal patterns are separable, even if rarely, implies that they correlate with slightly different referents and hence that each provides messages that are very slightly different from those of the other pattern. The differences might lie in incompletely overlapping values of behavioral supplementals, for instance. As the degree to which two formalized acts are separable increases, the importance of their differences must increase for the species that perform the acts. At some point they should be recognized as different display acts, but it is not yet apparent how we are to distinguish them.

More needs to be learned about the extent to which different animals make use of recombinable, formalized acts, each of which carries its own set of messages. The advantages to having such acts should lie in the control the communicator obtains over two or more simultaneous sources of information—that is the opportunity to perform the acts contextually to one another. Control over context, even to a limited extent, should be very valuable, enabling the communicator to provide more precise information.

Sequential Combinations

A communicator can gain control over sources of information that occur contextually to an act by producing sequences of acts. This procedure has been highly developed in human speech. To what extent is sequential compounding formalized, and what are its implications for the display concept?

REPEATING ONE KIND OF UNIT   Display acts that succeed one another naturally as an event proceeds become sources of information that are

contextual to each other, whether or not any aspect of their sequencing is formalized. In the simplest cases, a communicator may repeat only one kind of display. This may be repeated often and yet irregularly, as are calls that function to facilitate the cohesion of groups moving through dense vegetation. Or it may have a rate of repetition that is remarkably regular. Bird songs, for example, are often repeated at regular intervals. The black-tailed prairie dog has a bark vocalization that may be uttered singly or at irregular intervals, or in a continuous series with regular intervals (W. John Smith et al. in press 2). Where intervals are regular they give a pattern to the repetition of barks; the pattern adds information not present in a single bark display, information about the stability of the probabilities applying to the behavioral selection messages of a bark (see chapter 6).

Formally patterned repetition of display acts is a way of behaving that is specialized to be informative. That is, two formalizations are present: the display units, and a unit that is built of successive displays and characterized by the regular pattern with which it relates them. The second kind of formal unit is not so much a particular compound as it is the result of a particular kind of *compounding* defined by the use of specialized intervals between successive units. Stated otherwise, this kind of temporal formalization is not a fixed structure, but a fluid structure produced by application of a rule or set of rules that governs the process of combining displays sequentially into ordered sequences.

SEQUENCES OF DIVERSE COMPONENTS   Numerous examples are known in which a communicator does not just repeat a display, but performs a set of different displays in a patterned sequence.

Gannets, large cliff-nesting seabirds of the North Atlantic, respond to real or imminent territorial infringement with a bowing performance (J. B. Nelson 1965), in which several displays are done in succession. First the bird does sideways head shaking from one to five times, then it extends it wings outward at the carpals, commonly leaving its wing tips crossed behind it, and dips the head and thorax forward and downward smoothly in the bowing movement three or four times. Each dip display is separated by additional head shakes, and the gannet calls "urrah" loudly 10 to 30 times in this phase. After the final dip it folds its wings and presses its bill tip tightly against its upper breast in the pelican position display, holds this from 2 to 4 seconds, then relaxes. Most of the components of the bowing performance (head shakes, dips, extending the wings, pelican position) are widely employed by gannets, either singly or in other kinds of compound performances. For example, in the advertising display which males perform toward prospecting females, head shakes are exaggerated, dips reduced to a slight movement of the male's head toward a recipient female, the wings are kept closed, and

there is no pelican posture. The pelican posture itself occurs, without the rest of the bowing performance, in most ambivalent agonistic encounters, whether between neighbors, mates, adults, and chicks, or even between gannets and other species such as gulls.

These activities of the gannet illustrate formalized units at two levels. There are displays, such as the pelican posture and head shaking. And there are the compound performances, such as bowing and advertising, in which sequential order is fixed, although with provision at each phase for a variable amount of iteration. Whether or not these formalized compounds provide information other than that which is provided by their component displays is not clear, but at least their components, as the smallest formalized units that make available consistent information about behavior, can be termed displays.

The gannet's performances are given only at the time of courting, nesting, and territorial defense. Courtship provides examples of patterned sequences of displays in many species of animals. For instance, in the fruit fly *Drosophila melanogaster* a male proceeds in order through four steps of displaying to a standing female (Bastock and Manning 1955; Bastock 1967): he "taps" her with his forelegs, tasting for information about her identity; next he faces her, often while also circling around her; then, maintaining his facing orientation, he holds a wing out at right angles and vibrates it rapidly up and down; finally, while simultaneously oriented and vibrating, he moves behind her, licks her genitalia, and then tries to mount. His initial mounting attempts are usually repulsed, and he goes through the sequence again, beginning at the second or third step. Although sequences can be foreshortened, the order of his sequence is fixed, and so is formalized.

The female does not actively display as a male *D. melanogaster* goes through his courtship sequence. However, in many other species both individuals display, and their sequences are constructed cooperatively. Because such sequences cannot be performed by single individuals they are formalized units of a different order of complexity from displays or individually performed sequences of displays. They are treated in chapter 14.

Though the effects of interaction can lead to the development of cooperative sequences, they can also interfere with the production of any formal sequencing. This is seen in the prolonged and yet highly variable series of displays performed in the courting parties of many species of ducks. The several males of a party, each with the same goal of pairing with the female which is the party's focus, continuously get in each other's way and try to drive each other off. Their interactions, and hence their displays, change on a moment-to-moment basis (see Dane and van der Kloot 1964); the males display largely to each other rather

than to the female (shown for courting mallards by Weidmann and Darley 1971). Although pair bonds eventually do form even in this unusually chaotic courtship, privacy remains elusive. Even when a pair later attempts to copulate, competing males interrupt them: 59 percent of 180 precopulatory interactions of goldeneye ducks filmed by Dane and van der Kloot (1964) were broken off. As mentioned in chapter 9, the overwhelming disorder of ducks' courting and precopulatory events has led McKinney (1961) to propose that much of the information important in getting a female duck to become receptive lies in the persistence with which her mate displays. A female goldeneye may continue to solicit copulation by lying prone on the water for prolonged periods before her distracted, displaying mate can do the one fixed sequence of the three displays that he performs immediately before mounting.

PURELY VOCAL SEQUENCES OF DIVERSE COMPONENTS   Formalized units of several levels of complexity can be found in analyses of vocalizations. The lowest level comprises brief segments of vocalizations, building blocks such as particular forms of frequency rises or frequency-modulated sections and the like (figures 13.4 and 13.5) that are strung together to make single utterances. Each unit at this level may become part of various displays, and appears to have no particular correlations with nondisplay behavior—and therefore no behavioral messages of the sorts described in this book.

Displays are at the next level of formalization, and they enter into a more complex level as components either of fixed-sequence compounds or fluid but rule-bound sequential patterns.

The fixed-sequence compounds can be seen in the vocal repertoires of many birds in which some of the longer displays are special amalgams of some of the shorter. For example, the compound vocalization of the eastern kingbird alternates a variable chatter vocalization with a harsh zeer vocalization (figure 13.4). This species' essentially songlike regularly repeated vocalization is a fixed sequence combining nearly all the rest of its vocal repertoire (W. John Smith 1966). Songs of a number of other species of birds seem to be constructed by combining much of the nonsong vocal repertoire in special sequences: for example, in the two tanager genera Ramphocelus and Chlorospingus (Moynihan 1962d, 1962b), in tree creepers (Thielcke 1966), to some extent in skylarks and chaffinches (Thorpe 1961), in crows, cardueline finches, and perhaps many other passerines (Andrew 1961b).

The "singing" of at least one primate, the South American titi monkey Callicebus moloch, has also been described as a distinctive assortment of most of the species' other calls, in a number of different

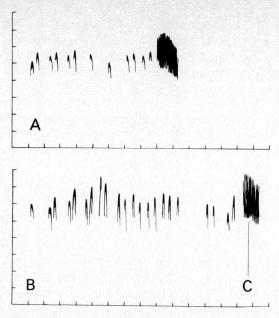

Figure 13.4. A fixed compound of vocal displays and its two component displays. *A*. The compound vocalization of an agitated male eastern kingbird, guarding his nest. *B*. A chatter vocalization, couplet variant, by a male eastern kingbird. *C*. A standard variant of the repeated vocalization, also by a male. The chatter vocalization and the repeated vocalization were uttered in separate events, each without any relation to the other display. Note that the chatter vocalization is itself a compound of lesser, repeated units, although in this case they are not uttered singly as displays. (From W. John Smith 1966.)

ordered sequences (Moynihan 1966). Compared to the bird songs mentioned above, however, this involves numerous and apparently variable combinations. Moynihan's evidence indicates that the positions different vocal displays can take within the singing sequences are rulebound.

The sequential order of components in the variable singing behavior of many species of birds is known to be rule-bound, although because in most cases no studies have been done of behavioral correlations, display units cannot usually be defined. The mistle thrush *Turdus viscivorus* does not repeat the majority of its song components, but always repeats others (Isaac and Marler 1963). Each successive component in the singing of an olive-backed thrush *Hylocichla ustulata* tends to be a variant or elaboration of the preceding one (K. Nelson 1973). The singing of cardinals *Richmondena cardinalis* shows a first order Markovian relationship between successive song forms (that is, the selection

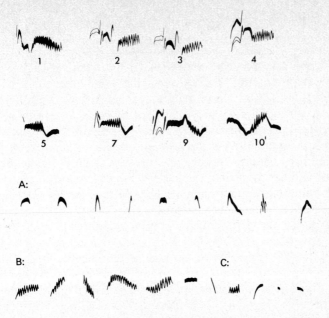

Figure 13.5. Songs of the yellow-throated vireo. These eight different forms of song vocalization were recorded from singing bouts of one male yellow-throated vireo. On 13 days sampled at intervals during one breeding season he was observed and recorded for periods of from 60 to 192 minutes (a total of 24 hours of observation), during which he uttered 1,223 song units. Each form of song unit is assigned a number in the upper two rows of the figure (gaps in the numbering reflect song forms found in the vocabularies of other individuals, but not recorded from this male). The lower two rows of the figure depict slightly stylized interpretations of approximately 20 arbitrarily recognized components which appear to be the basic building blocks of all the song forms of this and the other seven males of this species that were studied. One or more A units are the first components of a song, prefixed to type B components; a type A can also occur between type B components. Type B components are more prolonged, often sound harsh (to human ears), and one or more occur within each song form. Type C components are brief and terminal, occur in about three-quarters of the known song forms. (Based on W. John Smith, Pawleukewicz, and Smith in press.)

of each is largely determined by the one preceding it; Lemon and Chatfield 1971). Ohio song sparrows *Melospiza melodia* go through their entire song repertoires before repeating a run of any one song form (Nice 1943), and members of central California populations of this species usually employ about ten different song forms before repeating a run of any one (their song repertoires are larger than those of the Ohio birds; Mulligan 1966).

The relationship between organized singing patterns and other behavior has been studied in a few species, and their display units can be recognized. For instance, the singing of eastern phoebes, described in chapter 3 and in more detail in W. John Smith (1969c, 1970a), comprises sequences of two display units that both have some nonsong employment but are uttered primarily in singing. More important, however, the sequences are variably patterned in accordance with a simple grammar that governs the length of runs or strings of the more common unit, the less common occurring primarily singly. A comparable grammar organizes relations between the two vocal units in the daytime singing of another flycatcher, the eastern wood pewee *Contopus virens* (W. John Smith 1968 and unpublished). In this species the length of the strings of the commoner component correlates with the probability of flying during the strings: flight is very likely if the string is short, unlikely if the string is of medium length, and of intermediate probability during longer strings. As in the case of the phoebe, singing performances are distinguishable from nonsinging utterances of the display units: a singing pewee usually pauses for an interval of close to 8 seconds between song units. When pauses are much longer, the probability of flight cannot be predicted.

The singing of these two species of flycatchers involves four kinds of formalization: the display units, the subunits of which they are built, the durations of intervals between display units that provide the pattern which defines singing bouts as units, and the rules determining the sequential relationships between the two different display units of each species. The rules correlate different relationships between the units with different probabilities of particular kinds of behavior. The sequentially compounded units they produce differ among themselves in the numerical relationships of two displays, the number of times one is repeated before the other is uttered.

The formalization of sequential compounds may need to be elaborate in such singing because the performances can be prolonged, with the result that a bird can continue to sing while its circumstances change. Males in some species sing almost continuously through the day during parts of the breeding season. Some of these employ large numbers of different display units, and may need relatively complex grammars with which to organize the sequential relations between them. Individual male yellow-throated vireos have up to eight or more song units, very few of which are ever uttered in nonsinging behavior (figure 13.5). Each of these song units appears to correlate with a unique range of communicator activities, although it has not yet been possible to obtain sufficiently detailed observations to test this fully. Nonetheless, existing observations are adequate to show that the proportions of the different

song units change as the singer changes from one grossly defined behavioral category to another (table 13.2). Further, the song units are combined and recombined in specific couplets (W. J. Smith, Pawleukewicz, and Smith in press; statistical tests by Ewens in press). That is, singing is organized in accordance with definite sequencing rules. For instance, when the male whose songs are shown in figure 13.5 was at his nest or in its general vicinity, he used the following couplets more often than would be expected by chance: 2-1, 1-2, 5-1, 9-5, 1-9, and 5-5. Within this general category of behavior, if he became very inactive and sang only intermittently he used only song form 2; as he moved about more and sang more he added song form 1, forming couplets that often strung together as 2-1-2-1-2 . . . , as an alternation of these two song forms. Song forms 2 and 1, in fact, made up more than 90 percent of his singing during nest-centered behavior (see table 13.2). When he shifted to active moving about in regions farther from the nest couplets 5-1, 1-9, and 9-5 came to predominate, and he often repeated the sequence 5-1-9 as a triplet. When on the borders of his territory he added various couplets (for example 7-4, 4-3, 4-7, 7-7, and 5-2), and when singing very rapidly there in apparent attempts to confront a neighbor he would repeat the triplet 7-4-3. Seven other males of this species used similar patterns of organization in their singing, although no two sang completely alike (and no two contended with identical social circumstances). Research now in progress with vireos of species with larger vocabularies of song forms suggests that these may have even more complexly differentiated patterns of singing.

Sequencing formalizations sort song unit displays into particular combinations. At an infradisplay level of complexity, other sequencing formalizations may sort building blocks into display units. Structuring is then hierarchical. Its levels resemble those in human speech, in which words are built from stems and affixes, and phrase structures from words.

SPEECH   Speech is a uniquely human mode of vocal communication, divided into diverse languages among different populations of our species. All languages make distinctions between lexical categories, such as the classes of nouns and verbs in English (see E. Sapir 1921) and the classes that link these ("conjunctionalizations"), modify them ("adnominal" and "adverbial" classes; Martin 1964), and so forth. These distinctions permit diverse kinds of relationships among words to be organized by grammars, which can be quite complex. The distinctions are apparently purely linguistic, however. Each display, even the component displays of bird songs, appears to provide messages of the kinds discussed in chapters 4, 5, and 6. That is, displays differ from each other

Table 13.2. Songs uttered in different activities by a male yellow-throated vireo.

Behavioral category	Song form								Total
	2	1	9	5	4	7	10'	3	
Nest-centered	763(.62)	360(.29)	27(.02)	57(.05)	5(.01)	7(.01)		4(<.01)	1,223
Nonboundary excursions	96(.48)	45(.23)	22(.11)	36(.18)					199
Boundary patrolling and attempts to confront	73(.16)	62(.13)	56(.12)	101(.22)	52(.11)	72(.15)	2(<.01)	52(.11)	470
Total	932(.49)	467(.25)	105(.06)	194(.10)	57(.03)	79(.04)	2(<.01)	56(.03)	1,892
Subdivision of nest-centered category									
On nest	114(.59)	46(.23)	8(.04)	19(.10)	1(.01)	2(.01)		2(.01)	192
Approach	54(.59)	36(.40)		1(.01)	1(.01)				91
In region	503(.65)	228(.29)	12(.02)	34(.04)	1(<.01)	1(<.01)			779
Not classified	92(.58)	50(.31)	7(.04)	3(.02)	3(.02)	4(.03)		2(.01)	161

The proportions of the eight song forms sung by this male differed by activity. His behavior was classified as nest-centered if he was incubating on the nest, directly approaching the nest, or remaining within visual monitoring distance of it (whether inactively perched or actively moving about). During nonboundary excursions he left the nest region and sang while foraging more widely, but did not go to regions where he had territorial borders with a neighbor. In boundary patrolling he did go to such border regions, and either moved along them or perched and sang rapidly while facing in the direction from which a neighbor could be heard (the latter was scored as an attempt to confront). Figures indicate the numbers of songs recorded, and the decimalized fractions in parentheses indicate the proportion each song form contributed to singing during each behavioral category. (Based on Smith, Pawlukiewicz, and Smith in press.)

in which messages they make available, but not in the kinds of messages. Because of this, the rules for formalized sequencing of animal songs cannot be as complex as those of speech. They appear to serve primarily to relate displays in terms of the proportional contribution made by each display to a sequence.

Grammatical sequencing can develop without the lexical category distinctions of speech, even if the rules are simple, which suggests that some formalization of sequences may have been a relatively early step in linguistic evolution. That sequencing formalizations should be hierarchically structured both in speech and display communication probably reflects a need to compound numerous small units into more inclusive units that can be accomodated within the limits of short-term memory processes (see G. A. Miller 1956). Thus, from a recipient's perspective, a formalized sequence probably serves to make the temporal and quantitative relationships among displays maximally apparent.

Speech must have considerable evolutionary foundations. Chomsky (1965) has suggested that the rules for its fundamental grammatical processes may be inherited, Lenneberg (1967) has shown that the basic steps of learning to speak are inherited, and at least chimpanzees share with man in considerable degree the *capacity* to use langauge as a tool (Gardner and Gardner 1969; Premack 1971). Yet natural speech is specific to man, unique in the richness of distinctions among its component words and the grammatical devices that structure their relationships, unique in the diversity of information it can encode, unique in the extent of its dependency on learning for transmission (although in some species of birds many features of songs and other vocalizations have to be learned; see chapter 12). Its capacities to refer to other times and places, or to generate statements that have never been said or heard before and yet can be understood (called "displacement" and "productivity" respectively, by Hockett 1960) are, if not unique, highly exceptional. And what I have referred to as the capacity of speakers to employ metaphor, making useful approximations to intended referents when no available signal exactly fits their needs, is also exceptional in extent—Labov (1973) describes it as "our extraordinary ability . . . to apply words to the world in a creative way."

Its richness makes speech an exceptionally powerful tool. It can, for instance, permit very finely discriminating resolution of social interactions in circumstances in which bodily gestures alone are so ambiguous as to lead to "compromising situations" (Goffman 1971). Yet as people interact speech always depends heavily on nonlanguage communicating to provide contextual sources of information.

However advanced it is among formal means of communicating, speech did evolve in a primate that already possessed formalized signal

behavior comparable to that of other mammals. Speech did not replace these other signals so much as it supplemented them, and it must to some degree have been shaped by their influence. Linguistic utterances sometimes even revert to simple equivalents of displays, losing the power of their special structures. Thus "how are you" is often not a question about health but a greeting display, appropriately answered in kind. Speech is more than display behavior and more than the formal sequencing of displays, but in the communicating behavior of our species its history and its use are interwoven with these simpler formalizations.

DEFINING UNITS OF DISPLAY BEHAVIOR   The display units basic to sequential combinations are the smallest (briefest) acts that provide consistent information about nondisplay behavior. As in the case of simultaneous compounds, the recombinable components of which these acts are sometimes built should be viewed as an infradisplay level of complexity.

The basic display units, however, may not represent the only level of complexity that should be called displays. Many species have one or a few simple, fixed, sequential compounds and perform these like part of their display repertoires (such as the compound vocalization of the eastern kingbird). These fixed combinations usually have been treated as displays, and it seems convenient to continue this practice, viewing them as perhaps analogous in part to linguistic compounds such as "toothpaste" and "airfield."

Other sequential compounds, such as the strings and various sequences that appear in the singing of phoebes and vireos, are shifting patterns. The relation of these fluid compounds to displays may be less as compound units than, by rough analogy to language, as phrases.

The rules used in generating these fluid compounds should be recognized as formalizations, although behavior (acts) should be distinguished from the process of behaving. That is, the rules are not themselves behavioral units (and thus not displays), but units of procedure that govern the ways their users behave. They are grammatical formalizations. Behaving in accordance with the rules generates units that are segments of behavior characterized by a patterned structure; the pattern is based on at least regular intervals between successive displays and regularity in the order of displays. The rules may permit generation of a very large number of patterns in some species.

Where formalized sequencing rules exist in a species' repertoire they should be listed, just as grammatical rules are listed for languages. In studying languages, however, we do not try to make lists of all the phrases that can be generated. Listing the rules and the basic units

(words) is sufficient. Display behavior should probably be treated the same way, as repertoires of display units and of grammatical rules.

A practical problem remains. Different display acts may be sustained for different durations. A person may begin to blush, then smile and then laugh while still blushing. A gull may adopt an upright posture and maintain this as it goes through, one after another, all possible formalized bill angles—raising its carpals only in correlation with the downward angled bill. Probably sequences are not formalized in either of these cases, but in principle they could be. The problem is whether to treat the displays in terms of their simultaneous or their sequential relationships. The solution is to be ready to do both, if the sequential relationships are important to the kind of analysis in process. The problem appears to arise because some displays are useful in providing a sustained source of background information against which to interpret information about probabilities that shift more rapidly.

A Modified Display Concept

The current ethological definition of display behavior describes a class of activities that is distinguished by function: "all vocalizations, and all movements and postures which have become specially adapted in physical form or frequency to subserve social signal functions" (Moynihan 1956, 1960; see also chapter 1). Despite its enormous value for studies of communicating, the definition has several practical limitations: it does not offer criteria for recognizing units of display behavior; it does not deal with the levels of complexity that formalization can produce; and it does not deal with the kind of formalization represented by grammatical rules for combining acts.

Some of the difficulties now encountered in applying the display concept will be ameliorated only as we improve our ability to assess the functions of displays. With such improvements we would be in a better position to stipulate the minimal amount of formalization that qualifies an act as a display, and the degree to which two formalized acts must be separable in use before they should be recognized as separate displays. We could also distinguish more readily between displays and all other kinds of conspicuously specialized acts.

Other difficulties are largely matters of finding practical criteria for appropriate categorization and definition, for example of the number of individuals who must share a code before their formalized signalling acts are recognized as displays, or of units when formalized behavior varies in form or is involved in various combinations with other formalized acts. These issues are examined in the preceding sections of this chapter,

and recommendations are presented that entail expanding the description of the display concept, as follows:

A display is an act that can be performed by an individual and is specialized in form or pattern of employment to make a consistent set of kinds of messages available to at least one other individual.

A display may vary in form continuously, but either discontinuities in form or changes in behavioral selection messages while form is changing continuously imply the existence of more than one display unit.

Display units may be combined, either as opportune concatenations or as fixed compounds; the latter may be recognized as compound displays, particularly if they have distinctive message contents.

Fluid combinations of display units, on the other hand, are not recognized as compound displays, but may be limited in various ways by sets of formalized rules which are grammatical formalizations.

There are formalized units of behavior that lack consistent patterns of correlation with referents other than display acts themselves. These units represent an infradisplay level of specialization, and may be combined and recombined with each other to form display units. In effect, they are building blocks—units used in constructing displays.

Opportune compounds, compound displays, and grammatical formalizations are important to the behavior of communicating. They give a communicator formal procedures for structuring the perspective with which a recipient can consider a display. The communicator performs formalized sources of information contextually to each other, and recipients are very dependent on contextual information sources as they respond to displays.

Yet other kinds of formalized behavior patterns incorporate displays, formalized compounds, and often informal acts into rule-bound structures. Unlike everything discussed thus far, these patterns cannot be performed by single individuals. They are cooperative formalizations which can be performed only by two or more participants.

Consider, for example, an event in which a baboon or macaque directs a threat (for example, a staring open-mouth face; van Hooff 1967) toward a second monkey that responds with a frowning bared-teeth scream face and recoils. The formalized acts are displays in simultaneous compounds. Next, consider an encounter of two males that could develop into a fight. One individual may formally present to the other who then mounts, perfunctorily, as if to copulate. Both the presentation and the mounting are display acts, but the individuals cooperate to perform them in a dialogic ceremony, each individual playing a specialized part. Sometimes called "dominance mounting," this is a

simple, orderly sequence that is formalized as an interaction. The same individuals might later meet, perhaps again perform the mounting ceremony, and then engage in allogrooming: a formalized interaction in which the parts played by the individuals are less rigidly ordered, more modifiable, and interchangeable within a bout.

Like the displays they incorporate, formalized interactions are informative units. The messages of dominance mounting appear to deal with the readiness of both participants to reduce the probability that agonistic behavior will interfere with nonagonistic interaction, the messages of allogrooming with readiness of both participants to interact with a depressed probability of attacking. Formalized interactions can thus be considered as informative units, even though they are not, *qua* units, behavior of an individual actor. They are the subject of chapter 14.

14

Formalized Interactions

An adage of midcentury popular music conveyed the fundamental point to which we must now address ourselves: "it takes two to tango." Many dances employ formalized rules guiding the joint behavior of the dancers, as do a large variety of cooperative interactions that serve, in small or large part, as signals: greetings, embracing, making and accepting amends, and many other exchanges of both humans and nonhumans. The formalized unit in each case is a pattern of behavior that cannot be done by a single individual, as a display or a formalized compound of displays can, but only cooperatively by two or more participants. Ethologists have described formalized behavior at both individual and cooperative levels, but have not usually emphasized the crucial distinction between them. I propose that we employ the term "formalized interactions" to refer to joint performances that, *qua* interactions, are specialized to be informative.

A formalized interaction, such as a greeting or a conversation, is predictable in general pattern. What is formalized is a format, a procedural framework or "program" (Scheflen 1967), that allots each participant a predetermined set of parts or roles from which to select; each part determines a range of moves and responses, usually both formalized and informal, that must or may be performed. In being performed, the parts jointly affect each other in orderly ways that fit within the interactional format.

The circumstances in which a given formalized interaction is performed can vary, and many kinds of formalized interactions permit the participants a great deal of freedom of detailed behavior in dealing with these differences. However, the rules constrain the participants to accommodate to each other to begin, sustain, and (in some cases) to complete the interaction. The rules are specific to the organization of the

kind of interactional activity: they "frame" it, fitting the actions of its participants to its organization (Goffman 1974).

Ethologists have tended to call such joint interactions "ceremonies," and have recognized courtship ceremonies, greeting ceremonies, and so forth. There seems to be no clear definition of ceremonies as a behavioral category, however, and the term probably is too widely current in varying connotations to be worth canonizing. It has been used freely in ethology, by some even as a synonym for "display." Among pioneer ethologists, Craig (1908) defined ceremonies as what we would typically call elaborate displays or compounds, and Huxley tended to use "display," "ritual," and "ceremony" as equivalent terms (see 1966: 249). Although the ceremonies he described in 1914 for the great crested grebe involve mutual or reciprocal displaying by two birds, he adopted the name "mutual ceremony" for such interactions in the redthroated diver in 1923. Simmons (1955b) and Storer (1969) followed this lead and distinguished between display acts of individuals and mutual or reciprocal ceremonies of two birds, which is very close to the distinction I have in mind.

Other ethologists have used the term "ritual," and some have restricted it to formalized joint encounters. Thus Estes (1972) has described a very simple ritual used by many African antelopes in which a male nudges a female, who then urinates, and the male samples the urine with his *Flehmen* behavior. In a much more complex joint performance, two male wildebeest *Connochaetes taurinus* engage in a daily reaffirmation of their territorial relationships called the "challenge ritual" (Estes 1969). Although highly variable in some respects (for example, the presence or absence of certain steps or the number of times a step is repeated) the challenge ritual initiates, waxes, wanes, and formally terminates in a probabilistically predictable fashion, and employs a very fixed repertoire of component acts (briefer rituals, displays, and informal acts such as maintenance activities). Ritual, however, is another term of uncertain definition, particularly in anthropology, the discipline from which it is drawn.

Ethologists share the study of formalized interactions with those anthropologists who study rituals. Their discipline, too, is embarrassed by the lack of a fully accepted definition of the term ritual (see Leach 1966), and by having terms that overlap with it, including ceremony. There is, however, an orthodox convention, of which V. W. Turner's definition of ritual is typical: "prescribed formal behavior for occasions not given over to technological routine, having reference to beliefs in mystical beings or powers" (1967:19).

To most anthropologists, a ritual is a complex, usually prolonged group performance concerned with major events in life (for example,

puberty, marriage, or death) or in the environment (arrival of spring, completion of harvest). Such rituals resemble the formalized interactions of nonhuman species in very fundamental ways. For instance, each ritual conveys a consistent set of messages and is used in a variety of circumstances that make interpretation of these messages very dependent on contextual sources of information (V. W. Turner 1967). Each ritual incorporates smaller formalized acts, objects, or sites—the "symbols" of V. W. Turner (1966, 1967:19)—and some of these are displays, badges, constructions, and tokens in our terminology. Rituals are produced when their symbol units are variously combined according to sets of rules.

Each ritual symbol has a consistent information content and functions differently when performed in different contexts (V. W. Turner 1967; Leach 1966). For example, any ritual dealing with spirits appears to be a cluster of one or more ritual acts, each act a consistent, highly stereotyped set of gestures serving as a "meaningful unit" and useful in combination with other such units in diverse rituals (J. D. Sapir 1970). The units of Sapir's spirit-directed rituals are like displays in the following, very important way: their performance correlates with behavior (or with moves such as making a sacrifice in interactions analogous to behavioral encounters, interactions with a spirit) and not with external referents such as the kinds of spirits sought in any event.

Finally, there are limited numbers (for any one human population) of both rituals and symbols. Leach (1966) argues that because all the knowledge of nonliterate peoples must be incorporated in the stories and rituals familiar to the living generation, these must be kept few enough to be learned and remembered. Leach sees a necessary economy in "condensing" the component acts and objects of rituals into highly specialized symbols, and in then employing the same symbols in different, patterned, contextual relationships that permit each to have many "alternative meanings." His argument for context dependency in human ritual behavior is closely analogous to the argument I made for the context dependency of display behavior in chapter 7.

Particularly relevant to our interest in formalized interactions are those anthropologists and other social scientists who accept ritual as including what Goffman (1963) has called the rules and structure of ordinary social intercourse. To Goffman, a ritual occasion can be as brief as an exchange of glances used as a minimal greeting (1963), can be as loosely defined and variable as the behavior engaged in by a boy and girl while holding hands (1971), or can be as grand as a week-long conference (1967). He sees ritual behavior as "perfunctory, conventionalized," and performed to flesh out and make obvious the game plan of an encounter; he finds it in interactions, and sometimes labels it "inter-

personal ritual" (1971). One of Goffman's basic units of ritual is the "interchange," comprising "two or more moves, and two or more participants," such as "excuse me"—"certainly" (1967:20). This specialized unit of human ritual is effectively identical with the formalized interactions of other species.

Like anthropologists, Goffman tends to assume that ritual is involved with concepts of the supernatural, arguing that interpersonal rituals attest to the "sacredness" of recipient individuals (1971). This assumption may have arisen because the rituals initially studied by anthropologists were collective social occasions that somehow affirmed the place of the actors and other elements of the social structure in a wider sphere. That ritual is not rational behavior may also bear on the assumption of supernatural beliefs, particularly because ritual often imposes rigid and seemingly arbitrary formats on behavior. Indeed, the tense rigidity that can characterize the performances of participants in nonhuman animals has stirred even ethologists to imagine at least limited transcendental states on the parts of the performers. Huxley (1923:259), describing the actions by which a female red-throated diver (loon) leads her mate, found a "tenseness about the bird's attitude, a rigidity" that impressed him much as had "certain sexual dances of savage tribes—the whole thing fraught with the significance of sexual emotion, and mysterious in the sense of being emotionally charged far beyond the level of ordinary life, but completely natural and without restraint." Yet Huxley and other ethologists have seen no need to impute supernatural beliefs to nonhuman animals, or even to imagine that such species have the necessary cognitive capacities. Humans certainly have such beliefs, and we have involved them in at least some classes of our ritual behavior. But the evolution of our rituals must have begun without being dependent on supernatural concepts.

More important issues are the behavioral, and probably functional, similarities of human interpersonal rituals and many nonhuman interactions. Rituals and formalized interactions tend to isolate encounters from other events and to establish their frameworks, presumably providing the recipients with what Douglas (1966:63) has called a "specialized kind of expectancy." Goffman, like Turner, sees these joint performances as sustaining normative behavior, as helping individuals remain within the bounds of familiar relationships and activities.

Formalized interaction emerges as a behavioral category that, under one label or another, has been abundantly described by ethologists and anthropologists. Analysis of the messages of formalized interactions, their dependence on contextual sources of information, their meanings and functions, can proceed very much as it does for displays, keeping in mind that the behavioral unit is a joint performance. The following

examples, chosen first from ethology and then from anthropology and other disciplines studying man, are intended to provide a preliminary and very selective description of prominent characteristics of this class of complex behavior patterns.

Examples from Nonhuman Ethology

Initiating interaction can be hazardous when individuals are establishing, reaffirming, or testing relationships with one another. Each individual must be prepared for the possibility that the others will behave agonistically in such circumstances. Thus, as individuals join one another for the first time, rejoin after separations, or become uncertain of how the individuals with whom they are associating are predisposed to act, they commonly perform some special formalized interaction before interacting more freely. Numerous examples can be selected from behavior that ethologists have traditionally referred to as "courtship."

Courtship Interactions

Ethologists have studied the joint signaling of "courting" animals of many species, partly because the displays animals employ on coming together to form pair bonds or copulate are often among the most conspicuous and bizarre of all displays, and partly because the problems of courting evoke conflicts of ethology's traditional trio of drives: aggression, fear, and sex.

What ethologists have usually called courtship has two main functions. First, it facilitates the mutual awareness and recognition of two individuals of opposite sex, sometimes after a preliminary searching or advertising phase that brings them together. Second, it enables the paired individuals to cooperate to fertilize the female's eggs. If the timing of the second step must be adjusted to a capricious environmental variable, say the early progress of a season, then the pair may develop a prolonged bonded relationship that tides them over until the second step is appropriate. The second step is not always the last. The pair bond and actions necessary to maintain it continue past copulation if both members cooperate in raising a family. As one result of observing interactions between long-term mates that are like the interactions seen during pair formation, ethologists have described "courtship" signaling throughout the whole cycle of breeding behavior in many species.

In the formalized courting interactions of some species of arthropods, sequences of moves and responses fall into orderly chains. The courtship sequence of the silver-washed fritillary butterfly (Magnus, summarized by Bastock 1967) is described as a "reaction chain" of seven male acts

and the appropriate female responses. A male finds and begins circling closely about a flying female; then they perform a spectacular joint flight, the female holding a special straight course while the male repetitively glides below and then darts upward before her. Next the female alights, the male follows, and they mutually posture and exchange pheromones. The sequence may be foreshortened, the aerial steps omitted if a male encounters a perched and receptive female. In either case, the sequence is not an arbitrary collection of acts, but is ordered to exploit the sequence of opportunities arising in the course of searching, finding, identifying, and convincing.

Fixed reaction chains are not typical of formalized interactions of vertebrate animals, although simplified descriptions of the courting of some species have left the impression that nearly invariant sequences exist. A case in point is Tinbergen's account (1953a) of the courtship of the three-spined stickleback fish, in which a male approaches a female who intrudes into his territory, then leads her to his nest in a zigzag dance performance. Acts are frequently interpolated between the early steps of this interaction (Wilz 1970; Wooton 1972); because the male stickleback has more to do than just attend and woo the female, he periodically suspends courtship and leaves her briefly in the lurch. McFarland (1974) has suggested that he may be like a juggler who is trying to keep two balls in the air at once, alternately giving his attention to leading the female and to checking the continued readiness of his nest for her visit.

Courtship interactions of the related ten-spined stickleback can be depicted by a sequence of nine steps, beginning with a female's response to a male's initiative (figure 14.1). However, Morris also shows that each action by one fish may elicit or appear as a response to more than one action by the other fish. Further, as in the fritillary, one partner may be more ready for sexual behavior than is the other, in which case the chain may be foreshortened. Yet, with no action coming more than two stages away from the response with which it is most frequently followed, the nine-stage plan of the sequence remains sufficiently apparent that it is a distinctive characteristic of the species.

Calling such sequences "reaction chains" implies that the selection of each act by each participant is determined only by the preceding act of the other participant. But the acts of each participant affect the other's behavior only by increasing the probability of certain responses, and other stimuli, both external and internal to each participant, are important (Hinde 1970). Thus formalized courting interactions are fluid, and their sequences can be described as following orderly but flexible plans that can be mapped less well as chains than as flow diagrams (see figures 14.2 through 14.5).

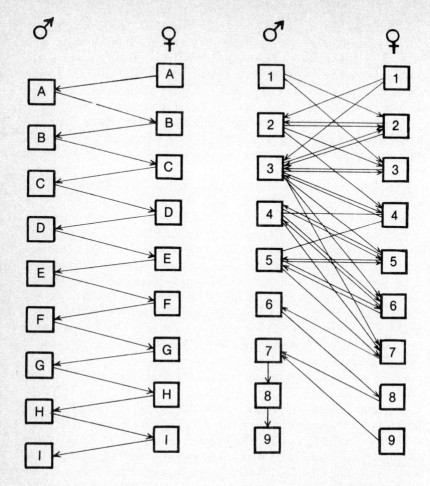

Figure 14.1. The nine-stage courtship of the ten-spined stickleback. The sequence of actions shown by a male and a female of this species as they court is orderly, but flexible. It is an oversimplification to interpret the interaction as a "reaction chain" (on the left) in which each action by one sex follows only one action of the opposite sex. What actually happens is shown on the right: a number of the actions in the sequence can follow or be followed by several actions on the part of the recipient fish. (From Morris 1956b.)

In the typically fluid interactional sequences of at least birds and mammals, components are used as is opportune. Although particular kinds of formalized interactions can be recognized, this is usually done initially on the basis of a combination of patterns that they share and of common circumstances of performance, or sometimes by apparently common functions. The details of behavior differ from one occurrence to another of the same kind of interaction.

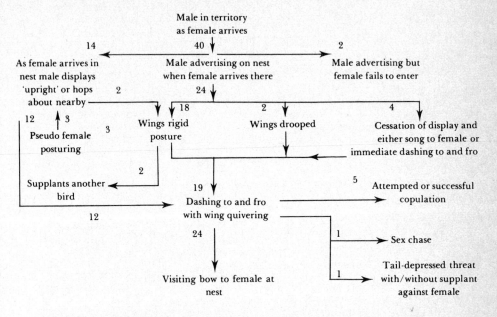

Figure 14.2. Response sequences of a male village weaver to the arrival of a female in his territory. Forty sequences were observed and tallied, and the number of times each step was seen is indicated by the figures printed by each arrow.

Figure 14.3. The displays referred to in figure 14.2 as "male advertising on nest" (left) in which the male hangs below the nest he has built with his head pointing into its entrance, and beats his wings, and the "wings rigid posture" (right) in which the male (at nest or on a twig beside it) suddenly stops beating his wings and holds them extended above his back.

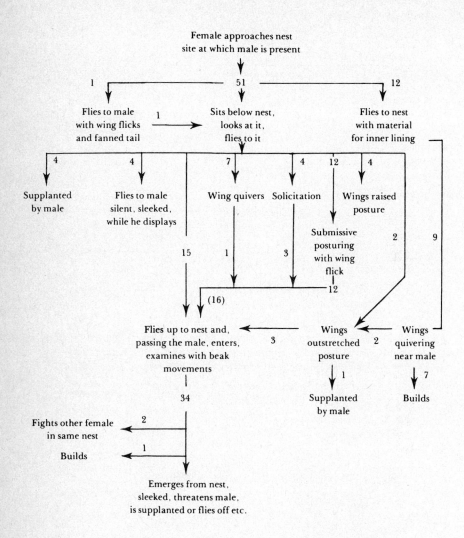

Figure 14.4. The sequence of the female's female's behavior on her arrival in a male's territory. Fifty-one sequences were tallied.

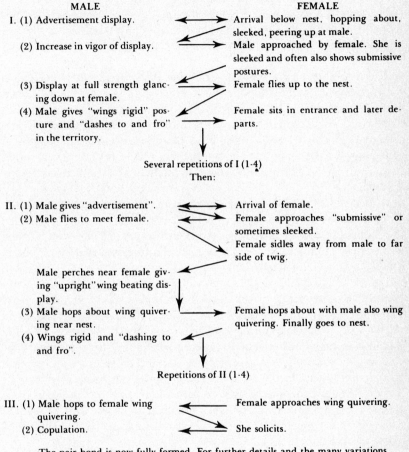

Pair formation in *Ploceus cucullatus* (Summary)

MALE

I. (1) Advertisement display.

    (2) Increase in vigor of display.

    (3) Display at full strength glanc-
ing down at female.

    (4) Male gives "wings rigid" pos-
ture and "dashes to and fro"
in the territory.

FEMALE

Arrival below nest, hopping about,
sleeked, peering up at male.

Male approached by female. She is
sleeked and often also shows submissive
postures.

Female flies up to the nest.

Female sits in entrance and later de-
parts.

Several repetitions of I (1-4)
Then:

II. (1) Male gives "advertisement".
    (2) Male flies to meet female.

Arrival of female.

Female approaches "submissive" or
sometimes sleeked.

Female sidles away from male to far
side of twig.

Male perches near female giving "upright" wing beating display.

    (3) Male hops about wing quiver-
ing near nest.

    (4) Wings rigid and "dashing to
and fro".

Female hops about with male also wing
quivering. Finally goes to nest.

Repetitions of II (1-4)

III. (1) Male hops to female wing
quivering.
    (2) Copulation.

Female approaches wing quivering.

She solicits.

The pair bond is now fully formed. For further details and the many variations,
see the text.

Figure 14.5. An idealized summary of the interactions of male and female
weavers during the pair-formation sequences detailed in figures 14.2 and 14.4.
(From Crook 1963.)

Among the analyses that most clearly reveal the variability which occurs within a single kind of interaction are those of the T-sequences performed by courting golden jackals (Golani and Mendelssohn 1971; Golani 1973, in press). T-sequences can be recognized by a number of common elements of form, including a mutually assumed position in which one jackal stands at right angles to the side of its mate (the two individuals thus creating a T configuration), and by an apparent commonality of function: they begin about four months before the time at which a pair copulates, and gradually develop into the precopulatory interaction.

The flexibility of the basic organization of T-sequences is evident in the extensive differences among pairs in the ways they perform (Golani and Mendelssohn 1971). Each pair has its preferred subroutines. For example, in some only the male will form the top of the T, in others only the female, in others both. Some pairs incorporate a stylized chasing of the female by the male into the interaction, others do not. Even within the performances of a pair, however, the details of behavior are greatly variable.

Golani (1973) made records of many positions each jackal takes in a T-sequence—such as positions of ears, tail, and body—using a checklist of 72 possible acts to categorize stance once every second from a frame of cine film. He found that extremely few simultaneous configurations in the behavior of either individual repeat from interaction to interaction. However, many of the positions classified by this technique were artificially frozen segments of the movements being made by body parts (their "trajectories") as they moved into fixed positions (Golani in press). This was found by analyzing movements minutely with the Eshkol-Wachman movement notation procedure. The movements, as well as the positions, are shown to be variable—for example, as each participant adjusts to the other in taking up its positions. It is not yet known, though, to what extent sequences of acts vary within these interactions, as neither of Golani's procedures deals with sequential ordering (other than within the sequences making up the trajectories of single movements).

With all the variability, however, regularity was observed at a gross level in the general kinds of positions taken and movements made by jackal mates with respect to each other—for instance, in their mutual adoption of the T-formations which were used to name the interactions. What is regular in the interactions, in fact, shifts seasonally, as a sample of a few patterns shows: early in the season a springy form of trotting, sniffing the ground, urinating, and scraping are common; later, circling of mates around each other, adopting the T-formation, putting the forelegs or head on the mate's back, and other patterns are superim-

posed on the first set; in the third phase the female regularly begins to mount the male, and he then follows her, licking her vulva; the acts of the former phases soon drop out and the male starts to mount the female; within a few days of this point, they copulate.

The variability of positions, movements, and behavioral preferences of individuals and pairs measured by Golani reflects the diversity of options available to each mate as a pair of jackals performs a T-sequence interaction. The existence of interactional rules is evident in the cooperation necessary for the mates both to be able to adopt parallel, antiparallel, T, and other mutually determined spatial arrangements, and to develop the seasonal progression of these interactions toward the final precopulatory exchanges. As Golani (in press) expresses it, the motor choices shown by each participant in a T-sequence are "embedded" within a "broader context of shared motor routes" by which the mates achieve positions relative to each other. An example of a specific kind of movement, often made by the male but requiring his mate's cooperation, is the tracing of a horizontal circle at a specific distance around the female's head while he is continuously changing his location. Despite their variability in detail, T-sequences are orchestrated and bounded by rules for cooperative behavior.

SYNCHRONY IN COURTSHIP INTERACTIONS   Synchrony of the actions of each participant with each other is a prominent feature of the plan of formalized interactions in courting and other encounters in many species.

Individual participants who synchronize may perform the same or different acts. For instance, in the complicated displaying parties of courting male pond ducks such as mallards synchrony can be very precise, several males "exploding" into display within one-quarter of a second or less. Weidmann and Darley (1971) showed that before groups synchronously begin down-up, grunt-whistle, or other displays one or more individuals asynchronously begin two other displays: first the introductory shake that apparently indicates readiness for group display, and then head-flicks when the acts performed in group display are imminent. These individual preliminaries probably permit watching males to pick up the beat of an initiator's performance (in the terms used by Kendon and Ferber 1973; and see below) and fall into step.

How male ducks competing for the attention of a female may profit by synchronizing their displays with each other is not yet known, but the combined performance may provide a "super-normal stimulus" to a female, increasing the likelihood that she will become responsive. On the other hand, various functions can be postulated when competing males of some species of fireflies flash in synchrony with each other. Their

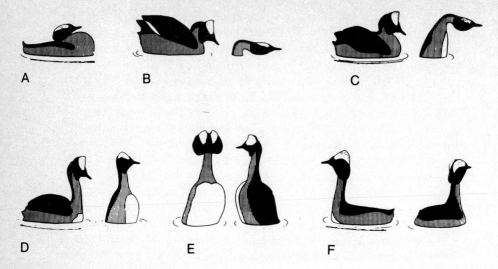

Figure 14.6. Synchrony in a courting performance of horned grebes. A court-ing horned grebe *Podiceps auritus* initiates a "discovery ceremony" by adopt-ing the striking "bouncy posture" (*A*) in which it pulls back its head and puffs out and raises its breast. Suddenly it dives. It postures and dives several times, then emerges from a dive facing away from the bird (now in a cat dis-play, *B*) to which it has been displaying. Slowly it rises in the ghostly penguin display (*C*) nearly to its full height (*D*) and turns toward the second bird— which has been rising too (*E*). Now in synchrony, the birds then simultane-ously swim, both standing erect, in what is known as a penguin dance per-formance. After several penguin dances they terminate the interaction by both formally turning away (*F*) and swimming slightly apart. (From Storer 1969.)

combined flashing may provide greater stimulation to females, but in the darkness in which mates are sought it could also have other effects. For instance, Otte and Smiley (ms.) have suggested that it may assist females in locating males of the correct species if they identify these primarily by the duration of interflash intervals; synchrony would lessen the confusion otherwise generated by numerous males flashing asyn-chronously as they fly. Similarly, when populations are dense synchro-nous flashing by males may enable each to recognize the answering flashes of a conspecific female; these are distinguished from those of females of other species through the duration of the delay between her flash and that of the male to whom she is responding.

Synchrony appears in parts of the prolonged and complexly formal-ized courtship interactions of many species of acquatic birds called grebes. Indeed, it is one feature of the spectacular joint performances that attracted Huxley and other pioneers to the great crested grebe (see chapter 1). The synchronizing birds are not rivals, but mates.

Storer (1969) has provided a very detailed description of the "discovery ceremony" of the horned grebe *Podiceps auritus,* showing the asynchronous acts with which a pair begin this performance and the exact point at which they fall into synchrony (figure 14.6). One member of the pair may begin diving or displaying, and its mate may adopt the cat attitude display (similar to that of the great crested grebe shown in chapter 1, figure 1.1). In response to the cat attitude display, the first grebe then does a series of "bouncy dive" displays and, moving close to the mate, emerges from one in the "ghostly penguin" display. As it slowly rises in the water its back is to its mate; when nearly at full height it turns slowly toward the mate which now raises its head from the "cat" display, lifting it high. At this point the two individuals fall into synchrony, swim together and rise nearly vertically into the penguin dance display. Synchronous penguin dances then alternate about ten times with a "habit preening" display (which may or may not break synchrony) before the formal termination to the interaction, a slow, mutual, "ceremonial turning away" that is followed either by swimming slightly apart and relaxing, or by yet other displaying.

Greetings

Greetings, in which one or more individuals display as they come into each other's presence, are useful events in which to study formalized interactions. Prominent during pair formation, they are also a part of the everyday life of at least those species of birds and mammals with persisting social bonds. In any potentially agonistic circumstance, individuals who join or are joined by others must be at least slightly ready to attack or escape; the exchange of a greeting lessens the likelihood that the agonistic potential will convert into action. Each participant must usually reassure and appease the other, and greetings are formalized interactions incorporating reciprocal, simultaneous, participant acts.

Like grebe performances (which are used both in courting and in greeting), the greetings performed by pairs of gulls and skuas (summarized by Tinbergen 1959b; Moynihan 1962a) are among the most elaborate known for birds, and they are formally structured. The procedures differ somewhat among species, but in many the male, or sometimes both participants, may begin by performing the long call—in effect, this is a challenge. Both mates then align in parallel and utter mew calls, and then perform choking in parallel. Next, both may assume parallel upright postures and proceed to less agonistic displays related to or involving uprights, including a head flagging display in which they turn their heads away from each other.

Tinbergen has emphasized the significance of this sequence in pair formation by referring to it as a single "major display" built of ele-

mentary displays (1959b:11) in "rigidly fixed" succession (p. 43). That the order is not rigidly fixed, however, is apparent in the detailed descriptions available for many species. For instance, Moynihan (1962a:82–86) notes that greeting interactions of the gray gull *Larus modestus* "tend to be rather different in different circumstances, partly depending upon which sex is approaching the other and how the approach is made"; variations in the order, or omission of some displays, are more common than the idealized sequence. The general rule seems to be that gulls' greeting interactions begin with displays conveying relatively high probabilities of agonistic behavior and proceed toward displays indicating lowered probabilities of such acts, with freedom for each participant to perform displays appropriate to its state. The participants largely adhere to a cooperative rule, each accommodating by tending to perform the same display as its mate most of the time, shifting displays when the mate does. As a display sequence develops and is performed mutually, the likelihood of either individual attacking or fleeing decreases.

Perhaps because the circumstances can lead to agonistic behavior, the orientations of the participants in most greetings appear to be formally limited; they orient in parallel or obliquely, but do not align directly toward each other. In addition, they give other displays that may carry information about the probability of agonistic behavior. The dramatic triumph ceremony of geese is a good illustration. It is often performed when a victorious gander returns to his family after driving off intruders (Heinroth 1911; Lorenz 1966:150–185). The return of a graylag gander, for instance, leads to a "cackling" greeting with his mate or family (Fischer 1965). The geese orient obliquely to one another, each in the posture that the gander employed when he went on the attack but not in the direct alignment he adopted then. Instead of fleeing and dispersing, as the intruders fled, the family remains intact and quiets down. In other species the displays used with the oblique orientation may not carry agonistic messages. Paired eastern kingbirds, for instance, greet each other with the kit-ter vocalization (see chapter 3) and a wing flutter display that encode messages about locomotion and the general set of unspecified alternatives (W. John Smith 1966). These are performed even when greeting in very agonistic circumstances, such as during disputes between two territorial males who are alternately fighting and perching 2 or 3 meters apart to threaten toward each other. Sometimes the mate of each male will be perched some distance behind him, and he may turn and exchange the usual kit-ter and wing fluttering greeting with her at intervals without leaving the dispute.

Still other messages may be carried by the displays performed in formalized greetings. In at least a few species, for example, greetings

provide information about readiness to engage in sexual behavior. In polygamous weaverbirds males greet with their first mates and with the females that they are subsequently courting using essentially similar postures, and after greeting may copulate even with a mate who has by then hatched her eggs (Crook 1964). In monogamous weavers, however, greetings are simpler and apparently carry no information about sexual behavior.

Related Interactions

"Courtship" and "greeting" events overlap. The courtship category is so broadly defined, in fact, that it has been applied both to precopulatory interactions and to all manner of other interactions in which mates or pairing birds appease and reassure each other. Greetings are appeasing-reassuring interactions done as individuals (mates or not) join each other, and are related to threatening-appeasing interactions done when one individual dominates and the other submits. Other appeasing-reassuring and threatening-appeasing interactions occur when an agonistic event disturbs individuals who are already together. Of course, criteria for "being together" can become quite arbitrary. For example, two barnyard hens living in the same flock may perform a brief formalized interaction when they near one another: the dominant thrusts with her head, and the submissive makes bowing movements (Maier 1964). When the flock is compact this action is simply a threatening-appeasing exchange; at other times it is threatening and appeasing done as greeting—in any case it permits the hens to remain within the same group. A range of related circumstances is possible and the single kind of formalized interaction remains appropriate, just as a single display can be performed in a range of circumstances.

Most species do not have different displays and formalized interactions to use in reducing the agonistic potential of events that arise in different circumstances, although they may have differing performances for events of greater or lesser risk. For instance, the "discovery ceremony" of grebes previously discussed as a courtship activity is not limited to the period of pair formation. It continues to serve when mated birds greet each other after temporary absences, and even for appeasing and reassuring apart from greetings. Grebes also have a formalized interaction called head shaking that is used similarly, although less often than the discovery ceremony in greetings and more often in agonistic events that arise when mates are already together. The two kinds of formalized interactions sometimes intergrade, and presumably have similar messages and overlapping functions. Head shaking to some extent replaces discovery ceremonies after pairing, and even in species

with less elaborate formalized interactions there is a distinct quantitative change once a pair bond is firmly established. An example of yet another formalized interaction that grebes employ in greeting and other potentially agonistic circumstances is shown in figure 14.7.

Apprehensive behavior lessens with mutual familiarity, and greeting exchanges become less frequent. That their occurrence depends on the likelihood of agonistic behavior is easily demonstrated by simple experiments. For example, paired eastern kingbirds greet each other infrequently by the time that they have a nest, but a brief resurgence of greeting can be elicited by making one individual temporarily aggressive. If a stuffed owl is placed in the vicinity of the nest the male will attack it, and then will often turn on his mate when she appears behind him and attack her; she involves him in a greeting exchange of displays, which forestalls the attack (W. John Smith 1966).

The behavior of greeting overlaps with yet another category of formalized interactions: "duetting." Duets are combined vocal performances of mates, one leading and the other joining in with either the same vocalization, creating polyphony, or a different one, yielding antiphony (Thorpe 1972). If other individuals, such as offspring, also join the duet becomes a trio, quartet, or "communal" performance. The functions of duetting are mentioned briefly in chapter 6 under the nonbehavioral message of identification; they help to reaffirm the adherence of each mate to their bonded relationship, keep the mates' physiological states synchronized with each other, coordinate their activities, help the individuals to keep track of each other when in dense vegetation or flocks, and aid other comparable tasks depending on the circumstances.

The participants in a duet call and answer, but may or may not be together, joining each other, or even within sight of each other. Duetting events are often not agonistic: control of attack and escape behavior is often less important than a reaffirmation of continued readiness to associate. Yet the duetting performances of many species are performed in a range of circumstances; in some they serve as greetings (for diverse tyrannid flycatchers, see W. John Smith 1971a; research in progress), and even as challenges to pairs on neighboring territories.

Intraspecific chorusing behavior is sometimes organized, and is then similar to duetting in that participating individuals call simultaneously or in a fixed order with respect to their nearby neighbors (see chapter 12). Fixed orders are not always formalized, however; they can result from ad hoc attempts to avoid interference from a neighbor's vocalizations.

Chorusing is often an adaptation for attracting mates, and the chorusing individuals are in competition with each other. Choruses can also

Figure 14.7. The diving ceremony of the New Zealand dabchick. This is an example of a simple, mutually performed "ceremony" that can be described as a formalized interaction. Note the coordinated moves of the two participants; for example, in the sequence A through F the bird on the right performs a series of movements with its bill such as might occur in an attack while its mate holds its head back; in G the mate in turn begins this or another sequence (with its bill open, however) and the first bird dives, emerging at a site anticipated by its mate. The performance is repeated several times, the birds sometimes alternating the parts they play. All species of grebes engage in comparable performances in various circumstances that can function in either appeasement or greeting. (From Storer 1971.)

result from the countersinging of territorial rivals (for example, in the synchronous singing of neighboring titi monkeys; Moynihan 1966a), in which case the interactions have been formalized for the function of challenging.

Challenging

Competing individuals who have arrived at a mutual understanding about, for example, the location of a territorial boundary between their claims, may limit their further rivalry by disputing each other primarily in formalized challenging interactions.

Neighbors in gannet colonies, for instance, dispute frequently. Once they have established their borders a challenge often is initiated by one bird performing a display called menacing: the communicator opens its beak and thrusts it at its opponent while twisting its head sideways, then withdraws its beak. The formalized interaction begins when the neighbor joins in and the two synchronize their display thrusts. Although they are near enough to bite each other, they rarely do so in these encounters— even though every gannet menaces a neighbor more than once an hour throughout the daylight hours of the nesting season (J. B. Nelson 1965).

Formalized interactions also occur during the process of establishing territorial boundaries in many species that are less aggressive than are gannets. In mockingbirds, for instance, two challengers will "dance" together. Each stands very upright and thinned, facing the other, and moves whenever the other moves. When one jumps forward the other may jump either forward or to the side, and they jockey for position until each is hopping mainly to the side, along their boundary (Hailman 1960).

As opponents act in unison along boundaries in other cases, they may fall into bouts of displaying in parallel. Neighboring eastern kingbird males may display while flying in parallel (W. John Smith 1966), and savannah sparrows (Potter 1972), willow ptarmigan (Moss 1972), or arctic ground squirrels (Carl 1971) will walk in parallel along a common boundary. Lack of willingness to cross the fixed boundary may be all that determines their parallel orientations, or their interactions may be formalized to include a rule that dictates the performance; it is not yet possible to select between these two possible explanations.

The very complex mutual challenging interactions of male wildebeest who hold neighboring territories are mentioned at the beginning of this chapter. Many territorial vertebrates have essentially comparable, if somewhat less complex, performances. For instance, the dominant male black-tailed prairie dogs of adjacent coterie groups will challenge each other along their boundaries by approaching head-to-head while tooth

chattering. One will then face about and, with tail lifted and spread, present anal glands to his opponent. The opponent usually edges cautiously forward and sniffs, then may turn about-face and present while the first sniffs (W. John Smith et al. 1973; King 1955). This little ceremony with its alternation of turn-taking may interrupt the head-to-head encounter several times over a number of minutes, neither animal taking advantage by attacking the one with its back turned.

Comparable turn-taking is seen as two fighting fish *Betta splendens* first establish their relationships, each carefully matching the previous displays of its opponent in an escalating series until one individual falls behind; very shortly afterward it is defeated by the other fish (Simpson 1968, 1973a). Rules for turn-taking, like rules for behaving synchronously, appear to be widespread in formalized interaction and, of course, are necessary to mesh the contributions of the participants in an orderly way.

Examples from the Study of Human Behavior

In much of our social interaction our circumstances constrain us to do "what is expected of us"—to accept the limits of a standardized part. (The term "part" here takes the sense of the part an actor adopts in a play, although a play without a rigidly fixed script; it is also called a "line" by Goffman 1967.) Coherent interactions, of humans or any animals, necessarily require that each participant plays a part appropriate to the kind of interaction and, unless the tenor of the interaction changes, remains consistently within the limits of that part. Joint adherence by all participants to their requisite parts, each accommodating to the others' performances, yields the formal structure of the interaction and provides what Goffman (1967:19) calls the "ritual state" within which social encounters run smoothly.

The notion that many human interactions are formally structured is the basis of Goffman's concept of an "interchange," which was described above as effectively identical with the ethological concept of formalized interaction. Goffman developed the concept of interchange solely within the framework of human interpersonal rituals involved in "face-work": the business of sustaining one's own part in any encounter and of helping other participants to sustain theirs.

"Among adults in our society almost every kind of transaction, including every coming together into a moment of talk, is opened and closed by ritual" (Goffman 1971:139). The ritual behavior typically begins when one individual signifies that he is involved with and somehow allied to another individual. The second individual must then indicate receipt and appreciation of the signal and acceptance of the implied relationship with the first person. This "little ceremony" is a

simple form of what Goffman calls the "supportive interchange." In line with the accommodative organization of ritual order, he sees the other fundamental form of interchange as being "remedial": when an infraction of social rights and obligations occurs order is restored through a patterned interchange of several moves. For example, a participant may challenge the misconduct of another, who then redefines the offensive act, redefines his part in the interaction, provides compensation, or punishes himself; this offering is accepted, and the offender indicates gratitude for the acceptance (Goffman 1967).

Supportive and remedial interchanges are so basic that Goffman asserts that almost all brief human encounters consist entirely of either one or the other; and they are also the tools for initiating and closing more extended conversations. In the latter case they become "ritual brackets" marking transitions first to the state of increased access of the individuals to one another, and subsequently to the renewal of the state of lesser access (Goffman 1971:79). Supportive and remedial interchanges also function in many other ways: the supportive can be used at various junctures to provide indications of reassurance, or "identificatory sympathy," and the like. The remedial can even operate as a form of closure to an encounter: an individual apologizes, as it were, for reducing the other person's access to him. Both supportive and remedial interchanges have forms specific to different occasions, all as highly conventionalized and automatically performed as are the displays which provide some of their component acts.

Greetings

Changes in the level of access individuals have to each other are marked by greetings and goodbyes, which Goffman (1971) groups within the concept of "access rituals." Human greetings occur when there is an increase of access, with the exception of "passing greetings"; a passing greeting can constitute the whole of the interaction in which it occurs, and is described by Goffman as "almost like the ritualization of a ritual" (1971:79). Greetings take on a variety of different forms in different circumstances, but a detailed observational study by Kendon and Ferber (1973) supports Goffman's contention that they are rituals—formalized interactions with rule-bound structures.

Kendon and Ferber demonstrated an impressive organization of minute details in mutual organization, timing, and sequencing of greeting events recorded on cine film. They found that greetings in several social circumstances studied in the northeastern United States are usually based on two successive exchanges of "salutation displays," a feature most easily seen when the individuals are initially some distance apart.

Shortly after each individual orients to the other they exchange a

"distance salutation," in which one displays, commonly with a head toss, perhaps with a wave and vocalization. The head toss is compounded of various displays such as raising then lowering the eybrows and smiling, and appears from the exploratory work of Eibl-Eibesfeldt (1968, 1971) to be commonly performed by diverse ethnic groups around the world in one form of greeting at a distance.

Following their distance salutations, the individuals may choose not to interact further, or may approach, each leaving his initial site. While approaching they shift gazes, and a quick but distinct aversion of gaze commonly comes just before they begin what Kendon and Ferber call the "close salutation" (figures 14.8 and 14.9). Smiling and vocalizing are meanwhile very likely, and one of a small number of distinctive head sets is assumed as the participants continue to approach. When they arrive at the formalized close salutation phase both halt face-to-face, usually within 2 meters of each other, and then select from a large range of highly formalized acts such as handshaking, embracing, and exchanging remarks patterned on a verbal exchange of information. Although often perfunctory, this verbal exchange can be prolonged and incorporate true speech exchanges; in fact, there are specialized social circumstances in which it is formally prolonged to extreme lengths.

Even if conversation then continues, the greeting is formally closed by the participants moving apart and changing their orientations to each other. This formal closing of one kind of interaction, the greeting, before any other behavior is begun is a widespread characteristic of human behavior, and is not unknown in other species. For instance, as described above, paired horned grebes perform a rigid "ceremonial turning away" (Storer 1969) as they terminate greetings based on both the discovery ceremony and vigorous versions of the head shaking ceremony.

Formalization of the dyadic pattern of greetings does not require them to assume a rigid, nearly identical form. Even within the limited ethnic sample of the Kendon and Ferber study, greetings were found to be extremely variable interactions within the general structure outlined above. Much of the variation appeared to be accounted for by differences in the degree of familiarity among the participants, their status and other established relationships, their sex, the kind of occasion, time elapsed since last meeting, and similar social variables. Formalization provides the participants with a flexible interactional plan for greeting, not a simple lockstep sequence.

Courtship

Although a central topic of ethological research on nonhuman species, the courting behavior of humans has not yet received much detailed ethological study. An exception, focusing on one kind of formalized

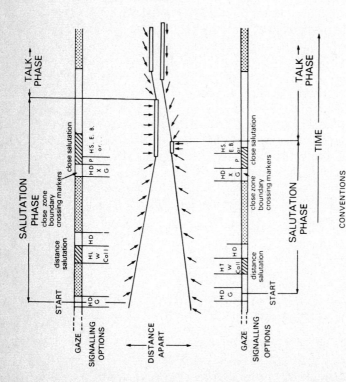
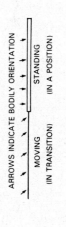

Figure 14.8. Schematic diagram of a typical human greeting, as observed in the Northeastern United States. Greetings with the general characteristics shown here are common initiations to encounters that subsequently become conversational interactions. The greeting, or "salutation phase" of the encounter, develops as the two participants approach each other and take up facing positions. It terminates as they move from these positions into a more nearly side-by-side orientation and begin the "talk phase" of the encounter. Each phase begins with a change in the spatial relationships of the two participants.

HD = Head dip, a sharp, downward tilting of the head and neck, usually with gaze averted by briefly closing the eyes or turning the head

HL = Head lower, in which the head is tilted forward and held there briefly, while the individual keeps looking at the other participant

HT = Head toss, a rapid raising and then lowering of the head; eyebrows may raise and lower as the head does, and a call may be given with the lowering

W = Wave, in which the hand and usually the arm are raised, and may be flapped or wagged

G = Groom, involving various manual adjustments of clothing, touching hair, or patting, rubbing, or scratching self

X = Body cross, a movement of the hand(s) or arm(s) across the body

P = Palm presentation, an orientation of the palm of the hand out toward the other participant, sometimes in preparation for a handshake

HS = Handshake

E = Embrace

B = Bow

. . . = Other acts that may be performed in close salutations

(Based on information provided by A. Kendon personal communication)

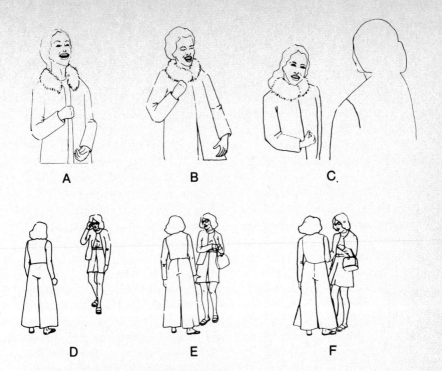

Figure 14.9. Examples of greeting behavior by a hostess and guests at a lawn party. *A*. The hostess, having sighted an approaching guest, does a head toss display with raised eyebrows and an open mouth smile, part of her distance salutation. *B*. She then terminates the distance salutation with a head dip, shutting her eyes. *C*. The guest is now completing her approach, and the hostess is again looking at her. She has tilted her head to one side, an example of one of the distinctive head-sets characteristic of this phase of a greeting. *D*. In a different event, the hostess stands after a distance salutation as an approaching guest touches her hair in a grooming movement. *E*. Hostess and guest perform close zone boundary crossing markers, the guest tilting her head and holding her right arm across her body in a body cross, while the hostess does a palm presentation. *F*. The hostess and guest shake hands as part of a close salutation. These line drawings were traced from frames of a cine film record of the events. (From Kendon and Ferber 1973.)

interaction that is a part of courtship, is Kendon's study of a kissing round (in press 2). The study is based on 4 minutes of cine film of a couple sitting on a park bench; the subjects were unaware of the camera, which was about 90 meters from them. The recording technique has its limitations; for example, the couple's faces can be seen in the film record only when they turn to face each other and move their faces close together, and their vocalizations are unrecorded. Nonetheless, it provided definite indications of interactional formalization.

The event can be divided into five phases. In the first, third, and fifth

phases the couple face, approach, make facial contact, and kiss; the second and fourth phases are brief transitions in which the two persons are disengaged. Throughout, the woman's face shows much more change than does the man's and her expressions appear to regulate who takes the initiative at each juncture. For instance, when she smiles with her mouth closed she is receptive to being approached and kissed, but when her smile shows her teeth he does not try to kiss her. With the latter form of smile, she takes the initiative and may approach him or disengage from him. (At the end of the third and fifth phases she also protrudes her tongue on withdrawing from him, performing the display described in chapter 5 as providing a message of depressed readiness to interact.) The woman's control of initiative and the man's cooperation permit the interaction to be orderly and flexible.

The second and fourth phases of the event, the transitional phases, appear to be "negotiations," and are characterized by pre-enactments of the initial portions of the phases that follow. In the second phase, for instance, the man turns toward the woman, who then turns toward him with a closed-mouth smile; he then turns away, and after she turns more fully toward him she too turns away. Slightly over a second later the third phase begins as he turns toward her and waits until she turns to him: that is, they begin a period of involvement with the sequence of movements that was pre-enacted in the preceding transitional phase. In the fourth phase the man again turns toward the woman, but instead of turning she leans forward, smiling with her teeth showing; when the fifth phase begins, the woman takes the initiative, as pre-enacted in this brief exchange. Kendon points out that these transitional phases are where change occurs within the event, and change thus appears to require an interactional rule of two steps: a pre-enactment, and then initiation of the new sequence itself. This may be a reflection of a more widely used interactional rule, a point discussed further below.

Conversing

Conversation is another human interchange with a formal structure. Conversing depends on cooperation to be begun, sustained, and terminated.

A typical way to begin conversing is with a formalized "summons-answer sequence" (Schegloff 1968). A summons can be an address ("Mr. Jones?" or "waiter") or a courtesy phrase ("pardon me"), or even a tactual display (a tap on the shoulder) or a visible display (for example, in a classroom, raising a hand). The answer is likely to be an open-ended question ("yes?" or "what?") that does not presage the content of the summoner's reply. A summons-answer sequence commits

each participant to follow through: the summoner will speak again, and the answerer will listen.

For a conversation to be further sustained, the participants need to develop a "working consensus" (Goffman 1959) on the topic, on who will guide its development, the acceptable level for expressing emotionality, and several other features. Each must assess and accommodate to the other's style of interacting so that they may interrelate their performances.

Participants set the stage for a conversation by jointly establishing what Argyle and Kendon (1967) have called "standing features" of orientations—distances apart and postures that facilitate their tasks and serve notice of each participant's agreement to continue. (These are "standing" features in that they are relevant throughout the conversational interaction; they do need to be actively maintained, however; see Kendon ms.) Relatively little is known about the extent to which some of these standing features have been formalized. For instance, much has been postulated about limitations on the distances apart and angular relationships of participants, but little measurement has been done in naturalistic circumstances. However, Kendon's studies of cine films (for example, in his work on greetings, see above) have enabled him to recognize various spatial "arrangements" that he groups within a concept of F-formations.

A F-formation arises when two or more people orient with equal, direct, and exclusive access to the space between them (Kendon in press 1). Many different spatial arrangements are possible: for example, vis-à-vis, L-shaped arrangements, side-by-side, and (with more than two participants) semicircular or triangular. The participants cooperate to maintain their arrangement, each compensating for positional deviations of the others. Outsiders also cooperate, leaving a buffer zone around the formation. For an outsider to become a member takes cooperation that appears to be governed by interactional rules. One procedure is for him to approach, stop a little distance away and wait until a member looks at him; he and this member then exchange gestures or utterances, then the member makes a spatial-orientational move (such as stepping back a bit) with which the outsider coordinates in a move that yields him access to the shared space of the formation's members. Departing from a F-formation also requires formal moves. For instance, the departing individual may first move slightly away and then step back in, then he and the others exchange parting remarks and he turns and walks away— the procedure in such a case having a "pre-enactment" feature of the sort described previously for changes in the activity of a kissing round.

A particularly detailed analysis of the interactional moves by which a sample conversation was started has been provided by Kendon (1973),

based on a frame-by-frame study of a cine film record. Five individuals were seated, awaiting a sixth who was to chair an informal discussion. He sat down, made some introductory remarks, then looked at each of the others in turn, receiving from each a brief headnod, as if confirming that they gave him leave to proceed. He did not immediately proceed, but lowered his head in a typical gaze-averting move of an individual who is formulating what he is about to say (as Kendon had demonstrated in earlier work), although in this case perhaps also a move indicating that he was receptive to overtures. He leaned forward, then reached for a cup on a table. As he touched the cup, and before picking it up or leaning back, he rotated his head to look directly at a participant. At *precisely* the moment he began to rotate his head she began to turn to look at him, so that they began to form the axis of interaction simultaneously as if she had known in advance that she would be the target of his next address. Not until their faces were oriented at one another did the chairman say anything.

Checking back on the film, Kendon found that this member had slightly raised her eyebrows, then lowered her head and looked away from the chairman when he looked at her while scanning the group. As he had leaned toward his cup she had again turned her head away—but she tilted her head forward as he began to reach. By this time she was sharing his rhythm of movement, indicating that she was attentive and thus increasing the likelihood that he would address her. And as he did select her she was already moving in synchrony with him, just as we have seen that ducks and grebes come into synchrony for some kinds of interactions. From this and other work (1970; Kendon and Ferber 1973) Kendon has proposed the general hypothesis that moving into synchrony with another person is one interactional formalization that communicates readiness to interact without risking a verbal request and its entailed commitment.

A speaker and listener who establish and maintain an axis of interaction orient jointly to one another, and as each repeatedly scans the other their eyes briefly meet. The axial listener often adopts a posture similar to (or "congruent" with; see Scheflen 1964) that of the speaker and employs headnods, changes in facial expressions, and other movements that appear to be formalized signals of continued attention at appropriate points during the flow of speech. Both speaker and listener may fall into what Condon and Ogston (1967) called "interactional synchrony": holding positions, then shifting simultaneously in a movement-mirroring fashion. They manage the allotment of speaking and listening activities by "turn-taking signals" (Duncan 1972): the speaker can signal readiness to yield by intonation (such as sustaining pitch at a terminal junction), drawling in a terminal clause, relaxing a tensed hand, uttering

a formalized remark ("or something" or "you know") after trailing off speech or decreasing loudness. Alternatively, he can attempt to suppress temporarily a turn by his listener by continuously gesticulating. These signals are formalizations at the display level, but the joint adherence of speaker and listener to the interactional format in which these displays are relevant, their playing of their proper parts and not forcing the other to yield, indicates formalization at the interactional level.

(Conversing is peculiar to humans, of course, because it requires speech, but other animals have formalized interactions that require the turn-taking that is an essential organizational feature of conversations. Examples described previously include some of the chorusing of frogs or birds and the challenges of prairie dogs; another is the social play that is well developed in many species of mammals. Yet another is a "dance" performance of birds called blue-backed manakins *Chiroxiphia* species, in which males perch side by side and each jumps up with a call in turn. If a female is present they orient toward her and the performance becomes a revolving cartwheel, each in turn jumping and fluttering backwards away from her while the perched one hitches forward toward her, to jump in turn while the other alights in the place it vacated; see Sick 1959, 1967; D. W. Snow 1963.)

Turn-taking although a rule of fundamental importance in organizing conversations, is not inviolable. In fact, rules exist to violate systematically its one speaker at a time feature by arranging for precise overlaps—for example, when a listener joins in with the same thing the speaker is saying (a special case of in-unison synchrony). By doing so, the listener proves that he knows what the speaker knows, rather than just claiming so—which is all he can do if he lets the speaker finish without overlap (Jefferson 1973). This procedure for overlapping is a formal device, and a speaker may even seek to elicit it. Jefferson explores other formalized violations of turn-taking, such as those used to mark interruptions as being deletable or to indicate that the listener will repeat his attempt to speak. Potential deletability can be marked by a speaker placing a term after he has completed an utterance (often this is an address term like "ma'am," or the listener's name) or by a listener stating an interruption completely while the speaker continues (this will not become part of their interaction unless acknowledged by the speaker). On the other hand, a listener who interrupts with only the start of an utterance signals that he does not consider the speaker's talk to be appropriate, and will restart and complete his utterance (Jefferson calls this device a "restart format").

Turn-taking is an example of the formalized "dynamic features" ("dynamic" as distinct from "standing" features, see above) that Argyle and Kendon regarded as structuring the flow of a conversational interac-

tion. These features are based upon utterances, movements and patterns of looking within the conversation. Such acts appear to indicate the direction of the participants' attention and its shifts, provide indications of taking turns as speaker or listener, designate who gets which part at which juncture, and make clear the participants' continuing abidance by the working agreement with which they are managing the interaction (Argyle and Kendon 1967). These devices have been variously described, for example as "regulator" acts by Ekman and Friesen (1969a), and as the "directional track" of the interaction by Goffman (1974)—that is, a set of features that are not part of the content or "main track" of the interaction but have a "framing effect" by giving structure to the event.

Part of the interplay between speakers and listeners during a conversation is reflected in the adjustments made by speakers to feedback signals from listeners, and even to the absence of appropriate feedback. Each sentence can be systematically modified as it is spoken, with additions, deletions, and changes of meaning all fit into it in accordance with changes in the interaction between speaker and listener (C. Goodwin 1975).

In natural conversations, a speaker apparently must speak to some specific listeners. Lacking or losing evidence of an attentive listener (available from formalized clues such as gaze direction and head nodding) a speaker will stop speaking or adjust the sentence that is being produced. He may "recycle" part of it, effectively repeating a section for which his listener was momentarily disattending, or add a new segment to a sentence if a recipient who appeared inattentive begins to provide the signals of attentiveness. For example, in "How are you feeling? . . . these days," the new segment is tacked on to extend the sentence as the proposed recipient begins to orient appropriately (C. Goodwin 1975). A speaker can also cause a sentence to mark time, by rephrasing one segment over and again as apposite interjections (". . . Teema, Carrie, and Clara, and myself. The four of us. The four children . . .," also from Goodwin 1975) until a recipient's attention is regained. If a speaker addresses one member of a group but finds that another is attentive instead, a sentence may be constructed with an information content that is augmented or altered partway through to fit differences in the knowledge the two recipients can be supposed to have. Further, Goodwin finds evidence that many solutions to such problems of speakers are standardized—for example, solutions that change declarative sentences to questions (feigning forgetfulness), or that interject segments to give listeners time to get into alignment.

The termination of conversations, and in fact of most human interactions if the participants can expect to be separated more than momen-

tarily, requires formalized interactional devices. A conversation usually cannot, for instance, be terminated simply by one of its participants becoming silent; because of the turn-taking feature of this sort of interaction, silence can be interpreted as the use of a turn (Schegloff and Sacks 1973). Clear-cut endings have to be negotiated, and there are formal ways to do this.

In the terminating phase of a conversation a number of kinds of remarks become more frequent than they are in earlier phases. For example, there may be more statements that refer to and summarize what has been discussed, and more references to the goal or goals of the interaction or the passage of time and the existence of other demands on the participants. Among friends there may then be affirmations that their relationships will continue, and strangers may wish each other well. Measures of the occurrence of such remarks have been made by Kessler (1974) in experiments based on a conceptual framework outlined by Albert and Kessler (in press). Other studies have shown that there are utterances which can be used to introduce an exchange that terminates a conversation. These utterances (for example "We-ell . . . ," "O.K. . . . ," "So-oo . . .") are typically given a downward intonation and are used as the entire turn of one speaker. The other speaker can accept them as either preclosing moves or as points at which to launch the conversation onto a new topic. If they are accepted as the former, this is usually shown by also passing up a turn: for example, by replying with "O.K." to the other person's "We-ell. . . ." Then a two-part terminal section of the conversation can be carried out, such as by each participant in turn saying "goodbye" (Schegloff and Sacks 1973).

Formalized interactional segments of two parts seem to be found when quite diverse changes must be negotiated. They have been found in Kendon's studies of the kissing round, and of the F-formation (in the latter case based on primarily films of conversations), in the Schegloff and Sacks study of procedures for terminating conversations, and in Jefferson's restart format. In each case, a tentative move is performed to set the stage for the move that changes the course of the interaction (accepting termination as a change in course). This two-part sort of structure may be among the more widespread and easily detected patterns of formalized interaction.

Repertoire

Humans appear to have extensive repertoires of formalized interactions and routines, subroutines, parts to play, and alternatives that fit widely varying circumstances. Any individual's repertoire is limited, but we lack adequate descriptions of its limits, size, and complexity. Still, its

extent appears to be much greater than in other species, although managing interactions is a necessary part of the social behavior of all.

If people employ more formalized interactional structures than do other species it may be partly because the interactions available to us can be more complex and less predictable than those available to other species. (Cause and effect are interwoven here, of course. That is, the complexity and variability of human interactions exist in part because of the efficient tools we have developed for managing them.) Further, a problem that humans, at least in large societies, face more often than do members of most other vertebrate species is that of interacting with individuals they do not know. Possibly some of our development of formalized interactions has been to provide means of dealing with the problems of anonymous interactions; by having standardized parts to adopt, we can become predictable to one another.

Still, the use of numerous and diverse formalized techniques for managing interactions can create problems. For instance, Scheflen (1967) observed that each person has available a repertoire of interactional parts to fit many programs, and may confuse some of these parts if distracted by outside sources, or if the situation is ambiguous. Scheflen postulated that during an interaction each participant must continuously assess his role, the kind and pace of the interaction, the adherence of other participants to the same program, and alternative behavioral possibilities. These assessments constantly require recalibration of each individual's behavior, or the program may falter and the interacting group break up.

Properties of Formalized Interactions

Formalized interactions are joint productions of two or more participants and are characterized by rules that constrain the kinds of behavior open to the players. But there are many ways to interact, and not all are formalized, or formalized to the same extent. If we are to learn to recognize the phenomenon consistently, we must determine which properties characterize interactional formalization. We can start by describing a number of features that seem typical of those formalized interactions with which we are now familiar.

First, the *sequence* of moves by the participants is predictable, although the limits of predictability vary among species and among different kinds of formalized interactions.

Second, in sequencing and in other characteristics, most formalized interactions permit their participants rule-bound *flexibility* with which to accommodate to both shifting circumstances and differences in individual styles, temperaments, and the like. Sequences can incorporate junc-

tures at which multiple options are possible; the acts selected to be used in making an interactional move can differ from event to event; and pacing, duration, and similar features can vary.

Third, specialized and mutually maintained *orientations* of the participants appear to be important in formalized interactions. Particular orientations continue or recur throughout each event: the players align in parallel, antiparallel, at right angles, facing, shoulder-to-shoulder, and so on, and cooperate to sustain or permit these spatial arrangements, often while adopting special postures and observing fixed interindividual distances.

Fourth, the *timing* of the participants' acts is variously specialized. Usually there is considerable synchrony, each participant changing behavior when the other participant does, or changing in synchrony with the participant who is momentarily at the focus of a group of more than two individuals. Participants often alter their behavior in the same or a complementary way. Where a formalized interaction allots its participants alternating, reciprocal parts (such as in the speaker and listener parts of a conversation), specialized timing devices determine turn-taking, and specialized ways are used to seize or control the initiative at such junctures. Other devices that are used to change the direction of an event often require carefully timed cooperation in a two-step formula (described for junctures in the kissing round studied by Kendon, and for interrupting or terminating a conversation, or joining or leaving the F-formation of a conversing group): first, pre-enactment; second, the termination moves or the initiation of a new direction.

The characteristic sequences, orientations, and features of timing, as well as the programmed flexibility of these interactions are specialized to be informative, permitting participants to anticipate each other's behavior. This is formalization, the fundamental characteristic that is shared by the categories of formalized interactions and display behavior. There are also other important similarities between these two behavioral classes. Like display units, each kind of formalized interaction is performed in diverse circumstances and its functioning depends on sources of information that are contextual to it (who the players are, where and when they interact). Further, although counts have not yet been made, current indications are that these are limited numbers of formalized interactions in the repertoire of each species.

15

A Prologue to the Study of Communicating

Studying the behavior of communicating is necessary for understanding the mechanisms of social behavior, whether those mechanisms involved in the moment-by-moment business of interacting or those enabling the more persistent patterns of social organization. This is because individuals who participate in social interactions and who have enduring relationships with one another cannot do without adequate information. Much of this information is shared through behavior that is specialized to be informative.

Despite its importance, the study of the behavior of communicating is not yet highly developed. The work described in this book is more a prologue than a definitive exercise. To draw a simple schematic diagram of an event in which communication occurs, as in figure 1.2, may foster a false sense of understanding by suggesting that if we can lay open the event as an array of parts we are also able to relate these parts to one another and explain the details of their interdependence. Not so—or not yet so. The parts are much more readily put together by participants than by observers; some parts can scarcely be detected by an observer, and if detected their significance cannot be evaluated.

Communicating is not just a matter of one individual providing a signal and another individual responding to the information that has been made available. Many classes of signals, including displays, badges, constructions, and others, have evolved both special features of form and the capacity to provide particular kinds of information because they are predictably correlated with limited classes of referents. Patterns of response have become based on the ability of animals to detect and recognize specialized signals by means of appropriate receptor structures and neural mechanisms, and on mechanisms for relating the advent of a signal to information from many other sources.

Numerous and diverse relationships among informative contributions from different sources are often relevant in communicating. We are only beginning to appreciate the subtleties and complexities of the acts, relationships, and processes that are involved.

The study of communicating necessarily turns on the role of information. First, information is what is shared among individuals by the process of communicating. Second, this information, by rendering events more predictable, gives advantages to users of the process—it enables recipients to make appropriate choices and communicators to influence those choices. This role suggests several questions that must be addressed, and each question leads to a host of others.

Several fundamental questions have been considered within this book:

What are the sources, both formalized and not, of information that is relevant as two or more individuals communicate?

What kinds of information are made available—and, in the case of formalized sources, why these instead of other kinds?

How does signaling behavior come to be part of a species' repertoire, and how is it specialized for its tasks?

What is accomplished by making information available? How do recipients respond, and how do they or the signalers (or both) profit from the information that is shared?

These questions arise from an analytic perspective that focuses on interactional behavior. Their answers will contribute directly to our understanding of the means by which participants manage orderly encounters. The preceding chapters contain an examination of our progress in seeking answers, and some of the principal points are briefly reviewed below.

Sources of Information

Everything that is sensible is informative. It is not necessarily relevant at all times—or even at any time—but it is capable of providing information to a sensing individual. Yet all species have developed behavior patterns (displays and formalized interactions) and other devices (badges, odors, constructions, and the like) that are specialized as sources of information. These patterns and other devices often provide information that is relevant in interactions and that is less available from other sources. Each species has repertoires of such specializations fit for reception by diverse sensory modalities.

With the exception of the combinatorial possibilities of language, and perhaps also the "songs" of some vertebrate animals, all repertoires of

specialized (formalized) signals appear to be limited to a rather small number of units. Communication does not depend solely on these units, however. Probably all species use the information made available by their formalized behavior and accoutrements along with information obtained from ordinary sources: information from the locus and time of an event; from known relationships, histories, and idiosyncracies of individuals; and from all of the behavior of these individuals as they move, forage, fight, wait, and engage in other acts.

Kinds of Information

To learn what kinds of information are provided by any source, it is necessary to determine what can be known or predicted when that source is available. With what does the performance of a display correlate? Who wears badges (for example, crests or antlers), and when? Who provides particular scents, and when?

Displays make available information (that is, provide messages) about the identities and sometimes the locations of the individuals performing them, and about the behavior in which these individuals may engage. Referents external to the displaying individuals may exist, but have not been surely demonstrated.

The information made available about the behavior that a communicator may select refers to remarkably few kinds of activities, at least in the cases of birds, mammals, and other animals in which either the messages of displays have been studied or sufficient description has been published to permit reasonable inferences about messages. A very few of these behavioral selections—such as attacking and escaping—are narrowly defined; they are typically acts used in circumstances in which inappropriate responses are likely to be especially costly. At the other extreme, some messages refer to such broadly inclusive categories of behavioral selections as locomotion and interaction, and even to behavior apparently specified only as including undefined alternatives that are incompatible with the behavior specified by other messages. There are other broad but less inclusive messages, but most truly narrow behavioral selection messages are not widespread among animals, and many species make use of very few of them.

That the displays of most species yet studied make available only the same small number of behavioral selection messages is initially surprising. However, it appears to be the result of a necessary economy. Because any species has only a small number of displays available in its repertoire, a need exists for each display to be performed in as many circumstances as is practical. For a few displays used in crucial events the number of circumstances can be small, but this is wasteful. The

majority of displays serve best if used more widely. That the displays of so many species deal in such similar messages may simply imply that few messages are of really wide applicability.

In addition to messages that indicate what behavior a communicator may perform, other messages provide an indication of how the individual will act: how likely it is to pick each selection, and how vigorously it will perform, how it will orient its performance, and so forth. This can be described as behavioral information supplemental to the information about behavioral selections, although the importance of this supplemental information is in no way secondary.

Badges, scents, and other nonbehavioral sources of information also provide information that identifies communicators. And they usually give some indications of the behavior that can be expected—often agonistic or sexual behavior. In many cases these sources and behavioral specializations are combined—for example, in raising a crest or hackles—the structure emphasizing the performance.

The generalizations that are now made about messages need further testing, as the base of our knowledge is still fragmentary. Much additional study explicitly devoted to the messages of the formalizations of diverse species is badly needed. This research provides one of the big tasks facing comparative ethology. The challenge is exciting because, far from being a matter on filling in gaps, such research promises to expand our conceptual horizons in ways not yet seen.

The Origins and Molding of Signaling Behavior

Without a fossil record of behavior, we shall never know the evolutionary origins of particular displays. Nonetheless, by comparing the forms that visible displays have evolved we can infer that many kinds of acts with direct functions, like the movements employed in beginning locomotory activities, could readily serve as the displays' precursors. Comparable precursors probably exist for all kinds of displays.

During their evolution, signal acts become differentiated from their precursors as they are increasingly formalized—that is, modified specifically to serve as useful sources of information. They may become stereotyped, exaggerated, simplified, iterated, or have their performances extended or restricted in time or space. However they are altered, displays must remain feasible and practical for their performers, suited to the latters' physical equipment and environmental limitations. The forms evolved by display acts are influenced by characteristics of the climates, landforms, and vegetation within which they must be effective. Forms are also influenced by the characteristics of animals present when displays are performed. Not just animals of the per-

former's species, but also their predators, their prey, the species that are sources of noise which interferes with transmission and reception, and even species with which the communicators form mixed social groups for foraging or the evasion of predators.

Under these various influences the forms of displays diverge, converge, and become more or less readily detectable and locatable. Numerous pressures mold each display, and usually some of the pressures are at odds with each other. As a result, each display's form is a compromise, with complex evolutionary roots.

The Effects of Making Information Available

The performance of formalized acts may have significant effects on interactional behavior, because the information that is provided makes the communicator's behavior more predictable than it might be otherwise. But it is essential to realize that what is contributed by this means is only a part of the information that is used by recipient individuals. Most interactional events are complex. Relevant information may be obtained from a host of sources contextual to those that are formalized.

It is not easy to analyze the ways in which information from different sources is employed by the recipient of a formalized act in selecting its behavior. This is partly because an observer may be unable to detect some of the sources, and partly because recipients may not appear to respond, or may respond and continue responding in the sense that much of their later behavior becomes modified by the information. Because recipients regularly store information, its consequences ramify and become dependent on the continued aquisition of further information. Further, some information is stored genetically, and appears as predispositions to responses.

Functions, the adaptive consequences of sharing information and responding to it, depend on responses. No display that is used in more than one kind of circumstance has a single function, any more than it generates a single kind of response. And the functions of a formalized source of information are difficult to analyze both because of the problems in analyzing responses, and because it is usually difficult to demonstrate that a particular action is adaptive—that is, that it enhances "fitness," increasing the representation of the responsible genotype in future generations.

Observations in natural circumstances usually provide some indications of the ways recipients behave after displays have been performed, and usually suggest how such behavior may be functional for the individuals concerned (usually both the communicator and the recipients, but only for the communicator in those evolutionarily unstable cases in

which the recipient is misled by the form of an information source). Some observations can be made in circumstances that amount to "natural experiments": circumstances in which specific sources of information are absent, constant, or vary in obvious and simple ways. In the final analysis, however, experimental intervention is usually necessary for decisive tests of postulated responses and functions. This intervention must do minimal and definable damage to the social fabric of the events being tested, not grossly violating the participants' expectations, or its results become ambiguous; devising appropriate experiments has proved extremely difficult.

Analyses of responses and functions, while necessary for an understanding of communicating, thus encounter formidable problems because each kind of formalized act that is used in more than one kind of circumstance leads to a variety of responses and functions. Mapping the processes that are involved should begin with simple cases, and must assess the ways in which animals deal with information from multiple sources. As yet, we are not in a position from which a comprehensive categorization of responses and functions can be established to guide research.

Where Do We Stand?

The prologue that our current knowledge, postulates, and theories provide to the study of communicating is substantial though somewhat fragmented. In addition to a considerable descriptive base, we have two broad, complementary perspectives available to guide our research. Further, the behavior performed in communicating and the informative matrices within which it operates are becoming more readily defined. As a result, it is becoming easier to phrase precise questions about the forms, messages, and uses of different sources of information.

The two perspectives now used in studying the behavior of communicating arise from quite different theoretical concerns. The traditional ethological perspective asks what factors lead an individual to signal: what information is it processing that causes it to behave as it does? This derives from an organismic perspective, used for studying the mechanisms that control the behavior of an individual organism. The other view asks not what information a communicator uses in selecting its behavior, but what information it makes available to other individuals that may influence their choices, and how do these individuals respond? This is based on an interactional perspective, explicitly formulated to further our understanding of the mechanisms controlling social behavior. The two perspectives are complementary: the first takes the behavior of an individual as its superordinate unit, the unit where

integration is seen; and the second takes this unit of behavior as a component that contributes to the integration of interactions.

Because the behavior of communicating allows information to be shared, and because communicating is adaptive only when recipients respond, an interactional perspective is particularly appropriate for its analysis. Within this perspective, three levels of abstraction are useful in suggesting specific analytic techniques for specific tasks. At the most abstract level, sources of information are analyzed and compared as physical entities; at the next analytic level their relations to referents are determined; at the least abstract but technically most difficult level, the effects of signal use are studied. Taken from the field of semiotics, these are termed the syntactic, semantic, and pragmatic levels, respectively.

Research from both organismic and interactional perspectives is facilitated by our broadened understanding of the behavior that has been specialized by the processes of formalization. This behavior can now be described in terms of unit acts or patterns, rules for behaving, or combinations of units and rules. Which is done depends on the kinds of postulates to be tested and on the complexity of the behavior being studied.

In a sense, formalization always produces rules for behaving, rules that impose patterns, with various degrees of rigidity, on larger or smaller chunks of behavior, and tie the formalized products to particular referents. At the level of complexity in which the chunks are display acts—the behavior central to much ethological research on communicating—the results of formalization often appear to be more or less fixed unit structures that vary only in limited ways. Yet they do vary, and some displays are notably inconstant in form; some even intergrade with one another. Because they vary, displays cannot be analyzed simply on the basis of form. However, most display units can be recognized by a combination of form and message criteria because they do not vary capriciously, but according to rules by which their shifting forms map shifts in messages.

Other rules exist for combining some displays into compounds, and the compounds may be fixed or flexible. The products of formalization in such cases are more complex than are single displays. Further, additional sets of rules govern combining the behavior of two or more participant individuals (both their displays and other acts) into cooperative patterns, formalized interactions. Many such patterns can incorporate great flexibility, with the result that the behavioral details of formalized interactions can be enormously variable.

The study of the contributions made by different classes of formalized information sources to communicating is one important avenue of research. It is necessary, however, to study the capacities of communicat-

ing individuals to base responses on information from various sources, taking some information as focal and other as contextual. The importance of contextual features in all communicative events is now widely accepted by ethologists, although much systematic study of the relative importance of different sources in specific kinds of events remains to be done. The issue has opened our eyes to the diverse kinds of contributions that are made by both formalized and ordinary sources of information, and to the great complexity inherent in the process of communication.

The wider sphere within which communication is significant sets the eventual goals of the study of communicating as it is treated in this book. Thus the goal is not just to know and understand the forms of specialized signals, or to list and explain the evolution of the kinds of information they make available, or the kinds of responses they facilitate in different circumstances. It is to understand, on the basis of these and other findings, how the behavior of communicating contributes to the management of interactions and the orderliness of relationships among individuals. This understanding is key in any attempt to understand the mechanisms of social behavior.

REFERENCES

INDEX

References

Adler, N. T. 1969. Effects of the male's copulatory behavior on the successful pregnancy of the female rat. *Journal of Comparative and Physiological Psychology* 69:613–622.

Adler, N., and J. A. Hogan. 1963. Classical conditioning and punishment of an instinctive response in *Betta splendens*. *Animal Behavior* 11:351–354.

Ainley, D. G. 1974a. Displays of Adelie penguins: A re-interpretation. In B. Stonehouse, ed., *The biology of penguins*. London: Macmillan.

———. 1974b The comfort behaviour of Adelie and other penguins. *Behaviour* 50:16–51.

Albert, S., and S. Kessler. in press. Six processes for ending social encounters: The conceptual archeology of a temporal process. *Journal of the Theory of Social Behavior*.

Alexander, R. D. 1957. The taxonomy of field crickets of the eastern United States (Orthoptera: Gryllidae: *Acheta*). *Annals of the Entomological Society of America* 50:584–602.

Alexander, R. D. 1967. Acoustical communication in arthropods. *Annual Review of Entomology* 12:495–526.

———. 1968. Arthropods. In T. A. Sebeok, ed., *Animal communication*. Bloomington: University of Indiana Press.

Altmann, S. A. 1962. Social behavior of anthropoid primates: analysis of recent concepts. In E. L. Bliss, ed., *Roots of behavior*. New York: Harper and Brothers.

———. 1965. Sociobiology of rhesus monkeys. II. Stochastics of social communication. *Journal of Theoretical Biology* 8:490–552.

———. 1967. *Social communication among primates*. Chicago: University of Chicago Press.

Ambrose, J. A. 1966. Ritualization in the human infant-mother bond. *Philosophical Transactions of the Royal Society of Britain* 251:359–362.

Anderson, P. W. 1972. More is different. *Science* 177:393–396.

Andersson, M. 1971. Breeding behaviour of the long-tailed skua *Stercorarius longicaudus* (Vieillot). *Ornis Scandinavica* 2:35–54.

————. ms. Social behavior and communication in the great skua.

Andrew, R. J. 1956. Some remarks on behaviour in conflict situations, with special reference to *Emberiza* species. *British Journal of Animal Behaviour* 4:41–45.

————. 1957. A comparative study of the calls of *Emberiza* species (buntings). *Ibis* 99:27–42.

————. 1961a. The motivational organisation controlling the mobbing calls of the blackbird (*Turdus merula*): I, II, III, and IV. *Behaviour* 17:224–246; 288–321; 18:25–43; 161–176.

————. 1961b. The displays given by passerines in courtship and reproductive fighting: A review. *Ibis* 103a:549–579.

————. 1962. Evolution of intelligence and vocal mimicking. *Science* 137:585–589.

————. 1963. Evolution of facial expressions. *Science* 142:1034–1041.

————. 1964. Vocalization in chicks, and the concept of "stimulus contrast." *Animal Behaviour* 12:64–76.

————. 1969a. Animal communication. (Review.) *Science* 164:693–694.

————. 1969b. Signals and responses. *Science* 164:693–694.

————. 1972. The information potentially available in mammal displays. In R. A. Hinde, ed., *Non-verbal communication.* New York: Cambridge University Press.

Argyle, M., and J. Dean. 1965. Eye contact, distance, and affiliation. *Sociometry* 28:289–304.

Argyle, M., and A. Kendon. 1967. The experimental analysis of social performance. In L. Berkowitz, ed., *Advances in experimental social psychology* 3:55–98.

Arglye, M., M. Lalljee, and M. Cook. 1968. The effects of visibility in interaction in a dyad. *Human Relations* 21:3–17.

Armstrong, E. A. 1947. *Bird display and behaviour,* 2nd ed. London: Lindsay Drummond.

————. 1965. *Bird display and behaviour.* New York: Dover.

Ashmole, N. P. 1963. The biology of the Wideawake or sooty tern *Sterna fuscata* on Ascension Island. *Ibis* 103b:297–364.

Austin, J. L. 1962. *How to do things with words.* New York: Oxford University Press.

Ayala, F. J., and T. Dobzhansky. 1974. *Studies in the philosophy of biology. Reduction and related problems.* Berkeley: University of California Press.

Baerends, G. P., R. Brouwer, and H. Tj. Waterbolk. 1955. Ethological studies on *Lebistes reticulatus* (Peters). I. An analysis of the male courtship pattern. *Behaviour* 8:249–334.

Baerends, G. P., and N. A. van der Cingel. 1962. The snap display in the common heron. *Symposium of the Zoological Society of London* 8:7–24.

Baker, M. C. 1974. Genetic structure of two populations of white-crowned sparrows with different song dialects. *Condor* 76:351–356.

————. 1975. Song dialects and genetic differences in white-crowned sparrows (*Zonotrichia leucophrys*). *Evolution* 29:226–241.

Balda, R. G., and G. C. Bateman. 1971. Flocking and annual cycle of the piñon jay, *Gymnorhinus cyanocephalus*. *Condor* 73:287–302.

Baptista, L. F. 1974. The effects of songs of wintering white-crowned sparrows on song development in sedentary population of the species. *Zeitschrift für Tierpsychologie* 34:147–171.

Barfield, R. J., and L. A. Geyer. 1972. Sexual behavior: Ultrasonic postejaculatory song of the male rat. *Science* 176:1349–1350.

Barham, E. G., W. B. Huckabay, R. Gowdy, and B. Burns. 1969. Microvolt electric signals from fishes and the environment. *Science* 164:965–968.

Barlow, George W. 1962. Ethology of the Asian teleost, *Badis badis*. IV. Sexual behavior. *Copeia* 2:346–360.

———. 1963. Ethology of the Asian teleost, *Badis badis*. II. Motivation and signal value of the color patterns. *Animal Behaviour* 11:97–105.

———. 1968. Ethological units of behavior. In D. Ingle, ed., *The central nervous system and fish behavior*. Chicago: University of Chicago Press.

———. 1972. The attitude of fish eye-lines in relation to body shape and to stripes and bars. *Copeia* 1972:4–12.

———. 1974. Extraspecific imposition of social grouping among surgeonfishes (Pisces: Acanthuridae). *Journal of Zoology of London* 174:333–340.

Barlow, G. W., and R. F. Green. 1969. Effect of relative size of mate on color patterns in mouthbreeding cichlid fish, *Tilapia melanotheron*. *Communications in Behavioral Biology* 4:71–78.

Barth, R. H., Jr., 1964. The mating behavior of *Byrsotria fumigata* (Gúerin) (Blattidae, Blaberinae). *Behaviour* 23:1–30.

Bartholomew, G. A., and N. E. Collias. 1962. The role of vocalization in the social behaviour of the northern elephant seal. *Animal Behaviour* 10:7–14.

Bartholomew, R. M. 1967. A study of the winter activity of bobwhites through the use of radio telemetry. *Occasional Papers Adams Center Ecological Studies* 17:1–25.

Bartnik, V. G. 1970. Reproductive isolation between two sympatric dace, *Rhinichthys atratulus* and *R. cataractae,* in Manitoba. *Journal of the Fisheries Research Board of Canada* 27:2125–2141.

Bastock, Margaret. 1967. *Courtship: An ethological study*. Chicago: Aldine.

Bastock, M., and A. Manning. 1955. The courtship of *Drosophila melanogaster*. *Behaviour* 8:85–111.

Bastock, M., D. Morris, and M. Moynihan. 1953. Some comments on conflict and thwarting in animals. *Behaviour* 6:66–84.

Bateson, G. 1955. A theory of play and fantasy. *Psychiatric Research Reports* 2:39–51.

Beer, C. G. 1963, 1964. Ethology—the zoologist's approach to behaviour. *Tuatara* 11:170–177, and 12:16–39.

———. 1970. Individual recognition of voice in the social behavior of birds. *Advances in the Study of Behavior* 3:27–74.

———. 1973. A view of birds. *Minnesota Symposium on Child Psychology*. 7:47–86.

———. 1975a. Was professor Lehrman an ethologist? *Animal Behaviour* 23:957–964.

———. 1975b. Multiple functions and gull displays. In G. Baerends, C. Beer, and A. Manning, eds., *Function and evolution in behaviour*. Oxford: Clarendon Press.

Bekoff, M. 1972. The developmnet of social interaction, play, and metacommunication in mammals: An ethological perspective. *Quarterly Review of Biology* 47:412–434.

———. 1974. Social play and play-soliciting by infant canids. *American Zoologist* 14:323–340.

———. 1976. The communication of play intention: Are play signals functional? *Semiotica*.

Bell, S. M., and M. D. Salter Ainsworth. 1973. Infant crying and maternal responsiveness. In F. Rebelsky and L. Dorman, eds., *Child development and behavior*, 2nd ed. New York: Knopf.

Bennet-Clark, H. C. 1971. Acoustics of insect song. *Nature* 234:255–259.

Bennet-Clark, H. C., and A. W. Ewing. 1969. Pulse interval as a critical parameter in the courtship song of *Drosophila melanogaster*. *Animal Behaviour* 17:755–759.

Bernstein, I. S. 1968. The lutong of Kuala Selangor. *Behaviour* 32:1–16.

———. 1970. Primate status hierarchies. In L. A. Rosenblum, ed., *Primate behavior*. New York: Academic Press.

Bernstein, I. S., and L. G. Sharpe. 1966. Social roles in a rhesus monkey group. *Behaviour* 26:91–104.

Bertram, B. 1970. The vocal behaviour of the Indian hill mynah, *Gracula religiosa*. *Animal Behavior Monographs* 3:81–192.

Biederman, I. 1972. Perceiving real-world scenes. *Science* 177:77–80.

Birdwhistell, R. 1970. *Kinesics and context*. Philadelphia: University of Pennsylvania Press.

Black-Cleworth, Patricia. 1970. The role of electrical discharges in the non-reproductive social behaviour of *Gymnotus carapo* (Gymnotidae, Pisces). *Animal Behaviour Monographs* 3:1–77.

Blair, W. F. 1964. Isolating mechanisms and interspecies interactions in anuran amphibians. *Quarterly Review of Biology* 39:334–344.

———. 1968. Amphibians and reptiles. In T. A. Seboek, ed., *Animal communication*. Bloomington: University of Indiana Press.

Blest, A. D. 1957. The function of eye-spot patterns in Lepidoptera. *Behaviour* 11:209–256.

———. 1961. The concept of ritualisation. In W. H. Thorpe and O. L. Zangwill, eds., *Current problems in animal behaviour*. Cambridge, England: Cambridge University Press.

———. 1964. Ritualisation. In A. L. Thomson, ed., *A new dictionary of birds*. New York: McGraw-Hill.

Blum, M. S. 1969. Alarm pheromones. *Annual Review of Entomology* 14:57–80.

Blurton Jones, N. G. 1960. Experiments on the causation of the threat pos-

tures of Canada geese. Wildfowl Trust, 11th *Annual Report* 1958–1959, pp. 46–52.

—————. 1968. Observations and experiments on causation of threat displays of the great tit (*Parus major*). *Animal Behaviour Monographs* 1:75–158.

—————. 1972a. Characteristics of ethological studies of human behaviour. In N. Blurton Jones, ed., *Ethological studies of child behaviour*. New York: Cambridge University Press.

—————. 1972b. Non-verbal communication in children. In R. A. Hinde, ed., *Non-verbal communication*. New York: Cambridge University Press.

Boase, H. 1925. Courtship of the teal. *British Birds* 19:162–164.

Bock, W. 1959. Preadaptation and multiple evolutionary pathways. *Evolution* 13:194–211.

Bogert, C. M. 1960. The influence of sound on the behavior of amphibians and reptiles. In W. E. Lanyon and W. N. Tavolga, eds., *Animal sounds and communication*. American Institute of Biological Sciences, Publ. 7.

Bolt, R. H., F. S. Cooper, E. E. David, Jr., P. B. Denes, J. M. Pickett, and K. N. Stevens. 1969. Identification of a speaker by speech spectrograms. *Science* 166:338–343.

Bossert, W. H., and E. O. Wilson. 1963. The analysis of olfactory communication among animals. *Journal of Theoretical Biology* 5:443–469.

Bourlière, F. 1955. The natural history of mammals. London: Harrap.

Bowers, J. M., and B. K. Alexander. 1967. Mice: Individual recognition by olfactory clues. *Science* 158:1208–1210.

Brand, J. M., R. M. Duffield, J. G. MacConnell, M. S. Blum, and H. M. Fales. 1973. Caste-specific compounds in male carpenter ants. *Science* 179:388–389.

Brannigan, C. R., and D. A. Humphries. 1972. Human non-verbal behaviour, a means of communication. In N. Blurton Jones, ed., *Ethological studies of child behaviour*. Cambridge, England: Cambridge University Press.

Brémond, J.-C. 1968a. Valeur spécifique de la syntaxe dans le signal de défense territoriale du troglodyte (*Troglodytes troglodytes*). *Behaviour* 30:66–75.

—————. 1968b. Recherches sur la sémantique et les éléments vecteurs d'information dans le signaux acoustiques du Rouge-gorge (*Erithacus rubecula* L.). *La Terre et la Vie* 2:109–220.

Brockway, Barbara F. 1964. Ethological studies of the budgerigar (*Melopsittacus undulatus*): Non-reproductive behavior. *Behaviour* 22:193–222.

Bronson, F. H. 1971. Rodent pheromones. *Biology of Reproduction* 4:344–357.

Brooks, R. J., and E. M. Banks. 1973. Behavioral biology of the collared lemming [*Dicrostonyx groenlandicus* (Traill)]: An analysis of acoustic communication. *Animal Behaviour Monographs* 6:1–83.

Brower, L. P. 1963. The evolution of sex-limited mimicry in butterflies. Mimicry symposium. *Proceedings XVI International Congress of Zoology, Washington* 4:173–179.

Brower, L. P., and J. vanZandt Brower. 1972. Parallelism, convergence, di-

vergence, and the new concept of advergence in the evolution of mimicry. *Connecticut Academy of Arts and Science Transactions* 44:59–67.

Brown, J. L. 1964. The integration of agonistic behavior in the Steller's jay *Cyanocitta stelleri* (Gmelin). *University of California Publications in Zoology* 60:223–328.

———. 1974. Alternate routes to sociality in jays—with a theory for the evolution of altruism and communal breeding. *American Zoologist* 14: 63–80.

Brown, R. 1958. *Words and things. An introduction to language.* New York: Free Press.

Brown, R. G. B. 1962. The reactions of gulls (Laridae) to distress calls. *Annal Épiphytes* 13:153–155.

———. 1964. Courtship behaviour in the *Drosophila obscura* group. I. *D. pseudoobscura. Behaviour* 23:61–106.

———. 1965. Courtship behaviour in the *Drosophila obscura* group. II. Comparative studies. *Behaviour* 25:283–323.

———. 1967. Breeding success and population growth in a colony of herring and lesser black-backed gulls *Larus argentatus* and *L. fuscus. Ibis* 109:502–515.

Brown, W. L., and E. O. Wilson. 1956. Character displacement. *Systematic Zoology* 5:49–64.

Bruner, J. S. 1957. On perceptual readiness. *Psychological Review* 64:123–152.

———. 1968. *Processes of cognitive growth: Infancy.* The Heinz Werner Lecture Series, 1968. Worcester, Mass.: Clark University Press.

Buechner, H. K., and R. Schloeth. 1965. Ceremonial mating behavior in Uganda kob (*Adenota kob thomasi* Neumann). *Zeitschrift für Tierpsychologie* 22:209–225.

Burger, J., and C. G. Beer. 1975. Territoriality in the laughing gull (*L. atricilla*). *Behaviour* 55:301–320.

Busnel, R.-G. 1963. *Acoustic behaviour of animals.* New York: Elsevier.

Busnel, R.-G., and J.-C. Brémond, 1961. Étude préliminaire du décodage des informations contenues dans le signal acoustique territorial du Rouge-gorge (*Erithacus rubecula* L.). *Comptes rendus des Séances, Académie des Sciences* 252:608–610.

———. 1962. Recherche du support de l'information dans le signal acoustique de défense territoriale du Rouge-gorge (*Erithacus rubecula* L.). *Comptes rendus des Séances, Académie des Sciences* 254:2236–2238.

Butterfield, P. A. 1970. The pair bond in the zebra finch. In J. H. Crook, ed., *Social behaviour in birds and mammals.* New York: Academic Press.

Cade, W. 1975. Acoustically orienting parasitoids: Fly phonotaxis to cricket song. *Science* 190:1312–1313.

Capranica, R. R. 1965. *The evoked vocal response of the bullfrog: A study of communication by sound.* Research Monographs 33. Cambridge, Mass.: MIT Press.

———. 1968. The vocal repertoire of the bullfrog (*Rana catesbeiana*). *Behaviour* 31:302–325.

Capranica, R. R., L. S. Frishkopf, and E. Nevo. 1973. Encoding of geographic dialects in the auditory system of the cricket frog. *Science* 182: 1272–1275.

Carl, E. A. 1971. Population control in arctic ground squirrels. *Ecology* 52: 395–413.

Carpenter, C. R. 1934. A field study of the behavior and social relations of howling monkeys (*Alouatta palliata*). *Comparative Psychology Monographs* 10(2):1–168.

———. 1940. A field study in Siam of the behavior and social relations of the gibbon. *Comparative Psychology Monographs* 16(5):1–212.

Carrick, R. 1963. Ecological significance of territory in the Australian magpie *Gymnorhina tibicen*. *Proceedings XIIIth International Ornithological Congress*, pp. 740–753.

Carroll, J. B. 1964. Words, meanings, and concepts. *Harvard Educational Review* 34:178–202.

Chadab, R., and C. W. Rettenmeyer. 1975. Mass recruitment by army ants. *Science* 188:1124–1125.

Chalmers, N. R. 1968. The visual and vocal communication of free living mangabeys in Uganda. *Folia primatologica* 9:258–280.

Chance, M. R. A. 1962. An interpretation of some agonistic postures: the role of "cut-off" acts and postures. *Symposium of the Zoological Society of London* 8:71–89.

———. 1967. Attention structure as the basis of primate rank orders. *Man* 2:503–518.

Chapman, F. M. 1938. *Life in an air castle. Nature studies in the tropics.* New York: Appleton-Century.

Chappuis, C. 1971. Un exemple de l'influence du milieu sur les émissions vocales des oiseaux: l'évolution des chants en forêt équatoriale. *La Terre et la Vie* 118:183–202.

Charnov, E. L., and J. R. Krebs. 1975. The evolution of alarm calls: Altruism or manipulation? *American Naturalist* 109:107–112.

Cherry, C. 1957. *On human communication.* New York: Wiley.

———. 1966. *On human communication,* 2nd ed. Cambridge, Mass.: MIT Press.

Chivers, D. J. 1969. On the daily behaviour and spacing of howling monkey groups. *Folia primatologica* 10:48–102.

Chomsky, N. 1965. *Aspects of the theory of syntax.* Cambridge, Mass.: MIT Press.

Church, R. M. 1957. Transmission of learned behavior between rats. *Journal of Abnormal and Social Psychology* 54:163–165.

Clarke, W. D. 1963. Function of bioluminescence in mesopelagic organisms. *Nature* 198(4887):1244–1246.

Cody, M. L. 1969. Convergent characteristics in sympatric populations: A possible relation to interspecific territoriality. *Condor* 71:222–239.

———. 1970. Chilean bird distribution. *Ecology* 51:455–464.

———. 1971. Finch flocks in the Mohave desert. *Theoretical Population Biology* 2:142–158.

———. 1973. Character convergence. *Annual Review of Ecology and Systematics* 4:189–211.

———. 1974. Optimization in ecology. *Science* 183:1156–1164.

Cody, M. L., and J. H. Brown, 1969. Song asynchrony in neighbouring bird species. *Nature* 222:778–780.

———. 1970. Character convergence in Mexican finches. *Evolution* 24:304–310.

Colvin, M. A. 1973. Analysis of acoustic structure and function in ultrasounds of neonatal *Microtus*. *Behaviour* 44:234–263.

Comfort, A. 1971. Likelihood of human pheromones. *Nature* 230:432–433.

Condon, W. S., and W. D. Ogston. 1967. A segmentation of behavior. *Journal of Psychiatric Research* 5:221–235.

Condon, W. S., and L. W. Sander. 1974. Neonate movement is synchronized with adult speech: Interactional participation and language acquisition. *Science* 183:99–101.

Cook, A., O. S. Bamford, J. D. B. Freeman, and D. J. Teidman. 1969. A study of the homing habit of the limpet. *Animal Behaviour* 17:330–339.

Cook, S. B. 1969. Experiments on homing in the limpet *Siphonaria normalis*. *Animal Behaviour* 17:679–682.

Cooke, F., and C. M. McNally. 1975. Mate selection and colour preferences in lesser snow geese. *Behaviour* 53:151–170.

Cooper, K. W. 1955. An instance of delayed communication in solitary wasps. *Nature* 178:601–602.

Coulson, J. C. 1966. The influence of the pair-bond and age on the breeding biology of the kittiwake gull *Rissa tridactyla*. *Journal of Animal Ecology* 35:269–279.

Coutlee, Ellen L. 1968. Comparative breeding behavior of Lesser and Lawrence's goldfinches. *Condor* 70:228–242.

Craig, Wallace. 1908. The voices of pigeons regarded as a means of social control. *American Journal of Sociology* 14:86–100.

Crane, J. 1949. Comparative biology of salticid spiders at Rancho Grande, Venezuela. Part 4. An analysis of display. *Zoologica* 34:159–214.

———. 1957. Basic patterns of display in fiddler crabs (Ocypodidae, Genus *Uca*). *Zoologica* 42:69–82.

———. 1967. Combat and its ritualization in fiddler crabs (Ocypodidae) with special reference to *Uca rapax* (Smith). *Zoologica* 52:50–76.

———. 1975. *Fiddler crabs of the world*. Princeton: Princeton University Press.

Crews, D. 1975. Effects of different components of male courtship behaviour on environmentally induced ovarian recrugescence and mating preferences in the lizard, *Anolis carolinensis*. *Animal Behaviour* 23:349–356.

Crook, J. H. 1962. The adaptive significance of pair formation types in weaver birds. *Symposium of the Zoological Society of London* 8:57–70.

———. 1963. Comparative studies on the reproductive behaviour of two closely related weaver bird species (*Ploceus cucullatus* and *Ploceus nigerrimus*) and their races. *Behaviour* 21:177–232.

———. 1964. The evolution of social organisation and visual communica-

tion in the weaver birds (Ploceinae). *Behaviour,* Suppl. 10, pp. 1–178.

———. 1969. Function and ecological aspects of vocalization in weaver birds. In R. A. Hinde, ed., *Bird vocalizations.* Cambridge, England: Cambridge University Press.

———. 1970. The socio-ecology of primates. In J. H. Crook, ed., *Social behaviour in birds and mammals.* London: Academic Press.

Crook, J. H., and J. S. Gartlan. 1966. Evolution of primate societies. *Nature* 210:1200–1203.

Cullen, E. 1957. Adaptations in the kittiwake to cliff-nesting. *Ibis* 99:275–302.

Cullen, J. M. 1959. Behaviour as a help in taxonomy. *Systematics Association Publication* 3:131–140.

———. 1972. Some principles of animal communication. In R. A. Hinde, ed., *Non-verbal communication.* New York: Cambridge University Press.

Cullen, J. M., and N. P. Ashmole. 1963. The black noddy *Anous tenuirostris* on Ascension Island. 2. *Behaviour. Ibis* 103b:423–446.

Curio, E. 1971. Die akustiche Wirkung von Feindalarmen auf einige Singvögel. *Journal für Ornithologie* 112:365–372.

———. 1975. The functional organization of anti-predator behaviour in the pied flycatcher. A study of avian visual perception. *Animal Behaviour* 23:1–115.

Cutting, J. E., and B. S. Rosner. 1974. Categories and boundaries in speech and music. *Perception and Psychophysics* 16:564–571.

Daanje, A. 1950. On the locomotory movements in birds and the intention movements derived from them. *Behavior* 3:49–98.

Dagg, A. I. 1970. Tactile encounters in a herd of captive giraffes. *Journal of Mammalogy* 51:279–287.

Dane, B., and W. G. van der Kloot. 1964. An analysis of the display of the goldeneye duck (*Bucephala clangula* (L.)). *Behaviour* 22:282–328.

Dane, B., C. Walcott, and W. H. Drury. 1959. The form and duration of the display actions of the goldeneye (*Bucephala clangula*). *Behaviour* 14:265–281.

Darwin, Charles. 1872. *The expression of the emotions in man and animals.* London: Appleton.

David, Charles N., and Robert J. Connover. 1961. Preliminary investigation on the physiology and ecology of luminescence in the copepod *Metridia lucens. Biological Bulletin* 121:92–107.

Davies, S. J. J. F. 1963. Aspects of the behaviour of the magpie goose *Anseranas semipalmata. Ibis* 105:76–98.

Davis, J. M. 1975. Socially induced flight reactions in pigeons. *Animal Behaviour* 23:597–601.

Dawkins, R., and M. Dawkins. 1973. Decisions and the uncertainty of behaviour. *Behaviour* 45:83–103.

DeFleur, M. L. 1970. *Theories of mass communication,* 2nd ed. New York: McKay.

Delacour, J., and E. Mayr. 1945. The family Anatidae. *Wilson Bulletin* 57:3–55.

Delius, J. D. 1963. Das Verhalten der Feldlerche. *Zeitschrift für Tierpsychologie* 20:297–348.

———. 1973. Agonistic behaviour of juvenile gulls, a neuroethological study. *Animal Behaviour* 21:236–246.

Dethier, V. G. 1964. Microscopic brains. *Science* 143:1138–1145.

DeVore, I. 1965. *Primate behavior. Field studies of monkeys and apes.* New York: Holt, Rinehart and Winston.

DeVore, I., and K. R. L. Hall. 1965. Baboon ecology. In I. DeVore, ed., *Primate behavior.* New York: Holt, Rinehart and Winston.

Diamond, J. M., and J. W. Terborgh. 1968. Dual singing by New Guinea birds. *Auk* 85:62–85.

Dichgans, J., R. Held, L. R. Young, and T. Brandt. 1972. Moving visual scenes influence the apparent direction of gravity. *Science* 178:1217–1219.

Diebold, A. Richard. 1968. Anthropological perspectives: Anthropology and the comparative psychology of communicative behavior. In T. A. Sebeok, ed., *Animal communication.* Bloomington: University of Indiana Press.

Dilger, W. C. 1956. Hostile behavior and reproductive isolating mechanisms in the avian genera *Catharus* and *Hylocichla. Auk* 73:313–353.

———. 1960. Agonistic and social behavior of captive redpolls. *Wilson Bulletin* 72:114–132.

———. 1962. The behavior of lovebirds. *Scientific American* 206(1):88–98.

Dingle, H. A. 1969. A statistical and information analysis of aggressive communication in the mantis shrimp *Gonodactylus bredini* Manning. *Animal Behaviour* 17:561–575.

———. 1972. Aggressive behavior in stomatopods and the use of information theory in the analysis of animal communication. In H. A. Winn and B. Olla, eds., *Behavior of marine animals. Current perspectives in research.* Vol. 1: *Invertebrates.* New York: Plenum Press.

Dodson, C. H., R. L. Dressler, H. G. Hills, R. M. Adams, and N. H. Williams. 1969. Biologically active compounds in orchid fragrances. *Science* 164:1243–1249.

Douglas, M. 1966. *Purity and danger: An analysis of concepts of pollution and taboo.* London: Routledge and Kegan Paul.

Driver, P. M., and D. A. Humphries. 1969. The significance of the high-intensity alarm call in captured passerines. *Ibis* 111:243–244.

Drury, W. H., Jr., and W. John Smith. 1968. Defense of feeding areas by adult herring gulls and intrusion by young. *Evolution* 22:193–201.

Duke, J. D. 1968. Lateral eye movement behavior. *Journal of General Psychology* 78:189–195.

Ducan, S., Jr. 1969. Nonverbal communication. *Psychological Bulletin* 72:118–137.

———. 1972. Some signals and rules for taking turns in conversations. *Journal of Personality and Social Psychology* 23:283–292.

Eibl-Eibesfeldt, I. 1959. Der Fisch *Aspidontus taeniatus* als Nachahmer des Putzers *Labroides dimidiatus. Zeitschrift für Tierpsychologie* 16:19–25.

———. 1968. Zur Ethologie menschlichen Grussverhalten I. Beobachtungen

an Balinesen, Papuas und Samoanern nebst vergleichenden Bemerkungen. *Zeitschrift für Tierpsychologie* 25:727–744.

————. 1971. Zur Ethologie Grussverhaltens: II. Das Grussverhalten und einge andere Muster freundlicher Kontaktaufnahme der Waika (*Yanoáma*). *Zeitschrift für Tierpsychologie* 29:196–213.

Eibl-Eibesfeldt, I., and W. Wickler. 1968. Die ethologische Dutung einiger Wächterfiguren auf Bali. *Zeitschrift für Tierpsychologie* 25:719–726.

Eisenberg, J. F., and E. Gould. 1966. The behavior of *Solenodon paradoxus* in captivity with comments on behavior of other Insectivora. *Zoologica* 51:49–58.

Eisenberg, J. F., G. M. McKay, and M. R. Jainudeen. 1971. Reproductive behavior of the Asiatic elephant (*Elephas maximus maximus* L.) *Behaviour* 38:193–225.

Eisenberg, J. F., N. A. Muckenhirn, and R. Rudran. 1972. The relation between ecology and social structure in primates. *Science* 176:863–874.

Ekman, P. 1970. Universal facial expressions of emotion. *California Mental Health Research Digest* 8:151–158.

Ekman, P., and W. V. Friesen. 1969a. The repertoire of nonverbal behavior: Categories, origins, usage, and coding. *Semiotica* 1:49–98.

————. 1969b. Nonverbal leakage and clues to deception. *Psychiatry* 32:88–105.

Ekman, P., E. R. Sorenson, and W. V. Friesen. 1969. Pan-cultural elements in facial displays of emotion. *Science* 164:86–88.

Emlen, John T., Jr. 1960. Introduction. In W. E. Lanyon and W. N. Tavolga, eds., *Animal sounds and communication*. American Institute of Biological Sciences Publication 7, pp. 9–13.

Emlen, S. T. 1971. The role of song in individual recognition in the indigo bunting. *Zeitschrift für Tierpsychologie* 28:241–246.

————. 1972. An experimental analysis of the parameters of bird song eliciting species recognition. *Behaviour* 41:130–171.

Emlen, S. T., and N. J. Demong. 1975. Adaptive significance of synchronized breeding in a colonial bird: A new hypothesis. *Science* 188:1029–1031.

Epstein, A. N. 1967. Oropharyngeal factors in feeding and drinking. Handbook of physiology—alimentary canal. In C. F. Code, ed., *Handbook of physiology: Section on alimentary canal*. Washington, D.C.: American Physiological Society 197–218.

Erickson, C. J. 1973. Mate familiarity and the reproductive behavior of ringed turtle doves. *Auk* 90:780–795.

Erulkar, S. D. 1972. Comparative aspects of spatial localization of sound. *Physiological Reviews* 52:237–360.

Esch, H. 1961. Ueber die Schallerzeugung beim Werbetanz der Honigbiene. *Zeitschrift für vergleichende Physiologie* 45:1–11.

————. 1967. The evolution of bee language. *Scientific American* 216(4): 96–104.

Espmark, Y. 1975. Individual characteristics in the calls of reindeer calves. *Behaviour* 54:50–59.

Estes, R. D. 1969. Territorial behavior of the wildebeest (*Connochaetes taurinus* Burchell, 1823). *Zeitschrift für Tierpsychologie* 26:284–370.

———. 1972. The role of the vomeronasal organ in mammalian reproduction. *Mammalia* 36:315–341.

———. ms. 1. The significance of breeding synchrony in the wildebeest.

———. ms. 2. Sexual convergence versus sexual dimorphism in gregarious ungulates.

Estes, R. D., and R. K. Estes. in press. *The behavior and ecology of African mammals. I. Hoofed mammals.* Cambridge, Mass.: Harvard University Press.

Estes, R. D., and J. Goddard. 1967. Prey selection and hunting behavior of the African wild dog. *Journal of Wildlife Management* 31:52–70.

Ewer, R. F. 1968. *Ethology of mammals.* London: Logos Press.

Ewing, A. W., and H. C. Bennet-Clark. 1968. The courtship songs of *Drosophila. Behaviour* 31:288–301.

Faegri, K., and L. van der Pijl. 1966. *The principles of pollination ecology.* London: Pergamon.

Fagen, R. 1974. Selective and evolutionary aspects of animal play. *American Naturalist* 108:850–858.

Falls, J. B. 1963. Properties of bird song eliciting response from territorial males. *Proceedings XIIIth International Ornithological Congress*, pp. 259–271.

———. 1969. Functions of territorial song in the white-throated sparrow. In R. A. Hinde, ed., *Bird vocalizations.* New York: Cambridge University Press.

Farlow, J. O., and P. Dodson. 1975. The behavioral significance of frill and horn morphology in ceratopsian dinosaurs. *Evolution* 29:353–361.

Ficken, M. S., and R. W. Ficken. 1965. Comparative ethology of the chestnut-sided warbler, yellow warbler and American redstart. *Wilson Bulletin* 77:363–375.

———. 1970. Responses of four warbler species to playback of their two song types. *Auk* 87:296–304.

———. 1973. Effect of number, kind and order of song elements on playback responses of the golden-winged warbler. *Behaviour* 46:114–128.

Ficken, R. W., M. S. Ficken, and J. P. Hailman. 1974. Temporal pattern shifts to avoid acoustic interference in singing birds. *Science* 183:762–763.

Fischer, H. 1965. Das Triumphgeschrei der Graugans (*Anser anser*). *Zeitschrift für Tierpsychologie* 22:247–304.

Fisher, H. I. 1972. Sympatry of Laysan and black-footed albatrosses. *Auk* 89:381–402.

Fisher, J. 1954. Evolution and bird sociality. In J. Huxley, A. C. Hardy, and E. B. Ford, eds., *Evolution as a process.* London: Allen and Unwin.

Fossey, D. 1972. Vocalizations of the mountain gorilla (*Gorilla gorilla beringei*). *Animal Behaviour* 20:36–53.

Free, J. B. 1971. Stimuli eliciting mating behaviour of bumblebee (*Bombus pratorum* L.) males. *Behaviour* 40:55–61.

Free, J. B., and C. G. Butler. 1959. *Bumblebees.* London: Collins.

Free, J. B., and Ingrid H. Williams. 1970. Exposure of the Nasanov gland by honeybee (*Apis mellifera*) collecting water. *Behaviour* 37:286–290.

Friedmann, H. 1955. The honey-guides. *Smithsonian Institution, U.S. National Museum Bulletin* 208:1–292.

———. 1960. The parasitic weaverbirds. *Smithsonian Institution, U.S. National Museum Bulletin* 223:1–196.

Frijda, N. H. 1969. Recognition of emotion. In L. Berkowitz, ed., *Advances in experimental social psychology*. New York: Academic Press, Vol. IV.

Frings, H., and M. Frings. 1968. Other invertebrates. In T. A. Sebeok, ed., *Animal communication*. Bloomington: University of Indiana Press.

Frings, H., M. Frings, B. Cox, and L. Peissner. 1955. Auditory and visual mechanisms in food-finding behavior of the herring gull. *Wilson Bulletin* 67:155–170.

Frishkopf, L. S., R. R. Capranica, and M. H. Goldstein, Jr. 1968. Neural coding in the bullfrog's auditory system—a teleological approach. *Proceedings of the Institute of Electrical and Electronics Engineers* 56:969–980.

Fry, C. H. 1974. Vocal mimesis in nestling greater honeyguides. *Bulletin of the British Ornithological Club* 94(2):58–59.

Gardner, R. A., and B. T. Gardner. 1969. Teaching sign language to a chimpanzee. *Science* 165:664–672.

Gartlan, J. S. 1968. Structure and function in primate society. *Folia primatologica* 8:90–120.

Gelbach, F. R., J. F. Watkins, II, and J. C. Kroll. 1971. Pheromone trail-following studies of typhlopid, leptotypholopid, and colubrid snakes. *Behaviour* 40:282–294.

Geist, V. 1966. The evolutionary significance of mountain sheep horns. *Evolution* 20:558–566.

———. 1968. On the interrelation of external appearance, social behaviour and social structure of mountain sheep. *Zeitschrift für Tierpsychologie* 25:199–215.

———. 1974. On fighting strategies in animal combat. *Nature* 250:354.

Genest, H., and G. Dubost. 1974. Pair-living in the mara (*Dolichotis patagonum* Z.). *Mammalia* 38:155–162.

Gibb, J. 1956. Food, feeding habits and territory of the rock pipit *Anthus spinoletta. Ibis* 98:506–530.

Gill, F. B., and W. E. Lanyon. 1964. Experiments on species discrimination in blue-winged warblers. *Auk* 81:53–64.

Gilliard, E. T. 1956. Bower ornamentation versus plumage characters in bowerbirds. *Auk* 73:450–451.

———. 1963. The evolution of bowerbirds. *Scientific American* 209(2): 38–46.

Goffman, E. 1959. *Presentation of self in everyday life*. New York: Doubleday.

———. 1963. *Behavior in public places*. New York: Free Press.

———. 1967. *Interaction ritual*. Garden City, N.Y.: Anchor Books.

———. 1969. *Strategic interaction*. Philadelphia: University of Pennsylvania Press.

———— 1971. *Relations in public*. New York: Basic Books.

————. 1974. *Frame analysis*. New York: Harper Colophon.

Golani, I. 1973. Non-metric analysis of behavioral interaction sequences in captive jackals (*Canis aureus* L.). *Behaviour* 44:89–112.

————. in press. Homeostatic motor processes in mammalian interactions—a choreography of display. In P. P. G. Bateson and P. H. Klopfer, eds., *Perspectives in ethology*, II. New York: Plenum Press.

Golani, I., and H. Mendelssohn. 1971. Sequences of precopulatory behavior of the jackal (*Canis aureus* L.). *Behaviour* 38:169–192.

Goldman, P. 1973. Song recognition by field sparrows. *Auk* 90:106–113.

Goodenough, W. H. 1965. Rethinking "status" and "role" toward a general model of the cultural organization of social relationships. In Association of Social Anthropologists, Monograph 1, *The relevance of models for social anthropology*, pp. 1–24.

Goodrich, B. S., and R. Mykytowycz. 1972. Individual and sex differences in the chemical composition of pheromone-like substances from the skin glands of the rabbit, *Oryctolagus cuniculus*. *Journal of Mammalogy* 53: 540–548.

Goodwin, C. 1975. The construction of the turn at talk as an interactive process. Paper presented at Conference on Culture and Communication, 15 March 1975, Temple University, Philadelphia; mimeographed.

Goodwin, D. 1953. Observations on voice and behaviour of the red-legged partridge *Alectoris rufa*. *Ibis* 95:581–614.

Gorman, M. L. 1976. A mechanism for individual recognition by odour in *Herpestes auropunctatus* (Carnivora: Viverridae). *Animal Behaviour* 24: 141–145.

Gorman, M. L., D. B. Nedwell, and R. M. Smith. 1974. An analysis of the contents of the anal scent pockets of *Herpestes auropunctatus* (Carnivora: Viverridae). *Journal of Zoology of London* 172:389–399.

Gould, E. 1969. Communication in three genera of shrews (Soricidae): *Suncus, Blarina,* and *Crytotis*. *Communications in Behavioral Biology A,* 3:11–31.

Gould, J. L. 1974. Honey bee communication. *Nature* 252:300–301.

————. 1975. Honey bee recruitment: the dance-language controversy. *Science* 189:685–693.

Gould, J. L., M. Henerey, and M. C. MacLeod. 1970. Communication of direction by the honey bee. *Science* 169:544–554.

Gould, S. J. 1974. The origin and function of "bizarre" structures: Antler size and skull size in the "Irish elk," *Megaloceros giganteus*. *Evolution* 28:191–220.

Grant, E. C. 1968. An ethological description of non-verbal behaviour during interviews. *British Journal of Medical Psychology* 41:177–183.

————. 1969. Human facial expression. *Man* 4:525–536.

Grant, P. R. 1972. Convergent and divergent character displacement. *Biological Journal of the Linnaean Society* 4:39–68.

Green, S. 1975. Communication by a graded system in Japanese monkeys. In L. Rosenblum, ed., *Primate behavior*. New York: Academic Press.

Griffin, D. R. 1958. *Listening in the dark. The acoustic orientation of bats and men.* New Haven: Yale University Press.

———. 1971. The importance of atmospheric attenuation for the echolocation of bats (Chiroptera). *Animal Behaviour* 19:55–61.

———. in press. *The question of animal awareness, evolutionary continuity of mental experience.* New York: Rockefeller University Press.

Grimes, L. G. 1974. Dialects and geographical variation in the song of the Splendid Sunbird *Nectarinia coccinigaster. Ibis* 116:314–329.

Grinnell, Joseph. 1903. Call notes of the bush-tit. *Condor* 5:85–87.

Gurtler, W. 1973. Artisolierende Parameter des Revierrufs der Türkentaube (*Streptopelia decaocto*). *Journal für Ornithologie* 114:305–316.

Guthrie, R. D. 1971. A new theory of mammalian rump patch evolution. *Behaviour* 38:132–145.

Güttinger, H. R. 1973. Kopiervermögen von Rhythmus und Strophenaufbau in der Gesangsentwicklung einiger *Lonchura*-Arten (Estrildidae). *Zeitschrift für Tierpsychologie* 32:374–385.

Gwinner, E., and J. Kneutgen. 1962. Über die biologische Bedeutung der "Zweckdienlichen" Andwendung erlerntner Laute bei Vögeln. *Zeitschrift für Tierpsychologie* 19:692–696.

Hailman, J. P. 1960. Hostile dancing and fall territory of a color-banded mockingbird. *Condor* 62:464–468.

———. 1963. Why is the Galápagos lava gull the color of lava? *Condor* 65:528.

———. 1967. The ontogeny of an instinct. *Behaviour Supplement* 15.

———. 1969. How an instinct is learned. *Scientific American* 221(6):98–106.

———. 1975. Review. Locatable and nonlocatable acoustic signals for barn owls, by M. Konishi. *Bird-Banding* 46:85–86.

Hailman, J. P., and J. J. I. Dzelkalns. 1974. Mallard tail-wagging: Punctuation for for animal communication? *American Naturalist* 108:236–238.

Haldane, J. B. S. 1953. Animal ritual and human language. *Diogenes* 4:3–15.

Haldane, J. B. S., and H. Spurway. 1954. A statistical analysis of communication in "Apis mellifera" and a comparison with communication in other animals. *Insectes Sociaux* 1:247–283.

Hall, John F. 1965. *The psychology of learning.* Philadelphia: Lippincott.

Hall, J. R. 1970. Synchrony and social stimulation in colonies of the black-headed weaver *Ploceus cucullatus* and Vieillot's black weaver *Melanopteryx nigererrimus. Ibis* 112:93–104.

Hall, K. R. L., and I. DeVore. 1965. Baboon social behavior. In I. DeVore, ed., *Primate behavior.* New York: Holt, Rinehart and Winston.

Hamilton, W. D. 1963. The evolution of altruistic behaviour. *American Naturalist* 97:354–356.

———. 1964. The genetical evolution of social behaviour. *Journal of Theoretical Biology* 7:1–52.

———. 1971. Geometry for the selfish herd. *Journal of Theoretical Biology* 31:295–311.

Hamilton, W. J., III. 1962. Evidence concerning the function of nocturnal call notes of migratory birds. *Condor* 64:390–401.

―――. 1967. Social aspects of bird orientation mechanisms. In R. M. Storm, ed., *Animal Orientation*. Corvallis: Oregon State University Press.

Hanby, J. P. 1974. Male-male mounting in Japanese monkeys (*Macaca fuscata*). *Animal Behaviour* 22:836–849.

Hansen, E. W. 1966. The development of maternal and infant behavior in the rhesus monkey. *Behaviour* 27:107–149.

Hanson, F. E., J. F. Case, E. Buck, and J. Buck. 1971. Synchrony and flash entrainment in a New Guinea firefly. *Science* 174:161–164.

Harrison, C. J. O. 1965. Allopreening as agonistic behaviour. *Behaviour* 24:161–209.

Hastings, J. W. 1971. Light to hide by: Ventral luminescence to camouflage the silhouette. *Science* 173:1016–1017.

Hazlett, B. A. 1968. Size relationship and aggressive behavior in the hermit crab *Clibanarius vittatus*. *Zeitschrift für Tierpsychologie* 25:608–614.

―――. 1972. Stereotypy of agonistic movements in the spider crab *Microphrys bicornutus*. *Behaviour* 42:270–278.

Hazlett, B. A., and W. H. Bossert. 1965. A statistical analysis of the aggressive communication systems of some hermit crabs. *Animal Behaviour* 13:357–373.

Hazlett, B. A., and G. F. Estabrook. 1974. Examination of agonistic behavior by character analysis II. Hermit crabs. *Behaviour* 49:88–110.

Heiligenberg, W. 1974. Processes governing behavioral states of readiness. *Advances in the Study of Behavior* 5:175–200.

Heinroth, O. 1911. Beiträge zur Biologie namentlich Ethologie und Psychologie der Anatiden. *Verhalten 5 International Ornithologishe Kongress* Berlin, pp. 333–342.

Helb, H. W. 1973. Analyse der artisolierenden Parameter im Gesang des Fitis (*Phylloscopus t. trochilus*) mit Untersuchungen zur Objektivierung der analytischen Methode. *Journal für Ornithologie* 114:145–206.

Hendry, L. B. 1976. Insect pheromones: Diet related? *Science* 192:143–145.

Hendry, L. B., J. K. Wichmann, D. M. Hindenlang, R. O. Mumma, and M. E. Anderson. 1975. Evidence for origin of insect sex pheromones: Presence in food plants. Science 188:59–63.

Hess, E. H. 1972. "Imprinting" in a natural laboratory. *Scientific American* 227(2):24–31.

Hinde, R. A. 1952. The behaviour of the great tit (*Parus major*) and some other related species. *Behaviour Supplement* 2:1–201.

―――. 1955. A comparative study of the courtship of certain finches (Fringillidae). *Ibis* 97:706–745.

―――. 1956a. A comparative study of the courtship of certain finches (Fringillidae). *Ibis* 98:1–23.

―――. 1956b. The behaviour of certain cardueline F_1 interspecies hybrids. *Behaviour* 9:202–213.

―――. 1956c. The biological significance of the territories of birds. *Ibis* 98:340–369.

————. 1958. The nest-building behaviour of domesticated canaries. *Proceedings of the Zoological Society of London* 131:1–48.

————. 1959. Behaviour and speciation in birds and lower vertebrates. *Biological Reviews* 34:85–128.

————. 1965. Interaction of internal and external factors in the integration of canary reproduction. In F. A. Beach, ed., *Sex and Behavior*. New York: Wiley.

————. 1969. *Bird vocalizations*. Cambridge, England: Cambridge University Press.

————. 1970. *Animal behaviour. A synthesis of ethology and comparative psychology*, 2nd ed. New York: McGraw-Hill.

————. 1972. *Non-verbal communication*. Cambridge, England: Cambridge University Press.

————. 1974. *Biological bases of human social behaviour*. New York: McGraw-Hill.

————. 1975a. The concept of function. In G. Baerends, C. Beer, and A. Manning, eds., *Function and evolution in behaviour*. Oxford: Clarendon Press.

————. 1975b. Interactions, relationships and group structure in non-human primates. *Symposium of the 5th Congress International Primate Society (1974), Nagoya*.

Hinde, R. A., and T. E. Rowell. 1962. Communication by postures and facial expressions in the rhesus monkey (*Macaca mulatta*). *Proceedings of the Zoological Society of London* 138:1–21.

Hinde, R. A., and N. Tinbergen. 1958. The comparative study of species-specific behavior. In A. Roe and G. G. Simpson, eds., *Behavior and evolution*. New Haven: Yale University Press.

Hingston, R. W. G. 1933. *The meaning of animal colour and adornment*. London: Edward Arnold.

Hockett, C. F. 1960. Logical considerations in the study of animal communication. In W. E. Lanyon and W. N. Tavolga, eds., *Animal sounds and communication*. American Institute of Biological Science, Publication 7, pp. 392–430.

Hockett, C. F., and S. A. Altmann. 1968. A note on design features. In T. A. Sebeok, ed., *Animal communication*. Bloomington: University of Indiana Press.

Hodgdon, H. E., and J. S. Larson. 1973. Some sexual differences in behaviour within a colony of marked beavers (*Castor canadensis*). *Animal Behaviour* 21:147–152.

Hoffman, H. S., D. Schiff, J. Adams, and J. L. Searle. 1966. Enhanced distress vocalization through selective reinforcement. *Science* 151:352–354.

Hogan, J. A. 1965. An experimental study of conflict and fear: An analysis of behaviour of young chicks toward a mealworm. I. The behaviour of chicks which do not eat the mealworm. *Behaviour* 25:45–97.

————. 1966. An experimental study of conflict and fear: An analysis of behavior of young chicks toward a mealworm. II. The behaviour of chicks which eat the mealworm. *Behaviour* 27:273–289.

Höhn, E. O. 1969. The phalaropes. *Scientific American* 220(6):104–111.

Hölldobler, B. 1971. Communication between ants and their guests. *Scientific American* 224(3):86–93.

Holmes, R. T. 1973. Social behaviour of breeding western sandpipers *Calidris mauri. Ibis* 115:107–123.

Hooker, B. I. 1968. Birds. In T. A. Sebeok, ed., *Animal communication*. Bloomington: Indiana University Press.

Hooker, T., and B. I. Hooker (Lade). 1969. Duetting. In R. A. Hinde, ed., *Bird vocalizations*. New York: Cambridge Unversity Press.

Hopkins, C. D. 1972. Sex differences in electric signalling in electric fish. *Science* 176:1035–1037.

————. 1974. Electric communication: Functions in the social behavior of *Eigenmannia virescens. Behaviour* 50:270–305.

Horn, H. S. 1968. The adaptive significance of colonial nesting in the Brewer's blackbird (*Euphagus cyanocephalus*). *Ecology* 49:682–694.

Howell, T. R., B. Araya, and W. R. Millie. 1974. Breeding biology of the gray gull, *Larus modestus. University of California Publications in Zoology* 1041:1–57.

Hulet, W. H., and G. Musil. 1968. Intracellular bacteria in the light organ of the deep sea angler fish *Melanocetus murrayi. Copeia* 1968:506–512.

Humphries, D. A., and P. M. Driver. 1967. Erratic display as a device against predators. *Science* 156:1767–1768.

Hunsaker, D. 1962. Ethological isolating mechanisms in the *Scleropus torquatus* group of lizards. *Evolution* 16:62–74.

Hunt, George L., Jr., and Molly W. Hunt. 1976. Exploitation of fluctuating food resources by western gulls. *Auk* 93:301–307.

Huxley, J. 1914. The courtship habits of the great crested grebe (*Podiceps cristatus*); with an addition to the theory of sexual selection. *Proceedings of the Zoological Society of London* 35:491–562.

————. 1923. Courtship activities in the red-throated diver (*Colymbus stellatus* Pontopp.); together with a discussion of the evolution of courtship in birds. *Journal of the Linnean Society of London, Zoology* 53:253–292.

————. 1966. A discussion on ritualization of behaviour in animals and man. Introduction. *Philosophical Transactions of the Royal Society of Britain* 251:249–271.

Hymes, D. 1967. Models of the interaction of language and social setting. *Journal of Social Issues* 23:8–28.

Ickes, R. A., and M. S. Ficken. 1970. An investigation of territorial behavior in the American redstart utilizing recorded songs. *Wilson Bulletin* 82:167–176.

Immelmann, K. 1969. Song development in the zebra finch and other estrildid finches. In R. A. Hinde, ed., *Bird vocalizations*. New York: Cambridge University Press.

Impekoven, Monika. 1971. Calls of very young black-headed gull chicks under different motivational states. *Ibis* 113:91–96.

————. 1973. The response of incubating laughing gulls (*Larus atricilla* L.) to calls of hatching chicks. *Behaviour* 46:94–113.

Isaac, D., and P. Marler. 1963. Ordering of sequences of singing behaviour of mistle thrush in relationship to timing. *Animal Behaviour* 11:179–188.

Itani, Junichiro. 1963. Vocal communication of the wild Japanese monkey. *Primate* 4(2):11–66.

Jacobs, W. 1953. Verhaltensbiologische Studien an Feldheuschrecken. *Zeitschrift für Tierpsychologie Beiheft* 1:1–128.

Jacobson, M. 1965. *Insect sex attractants*. New York: Wiley.

Janzen, D. H. 1971. Euglossine bees as long-distance pollinators of tropical plants. *Science* 171:203–205.

Jay, Phyllis. 1965. The common langur of north India. In I. Devore, ed., *Primate behavior*. New York: Holt, Rinehart and Winston.

Jaynes, Julian. 1969. The historical origins of "ethology" and "comparative psychology." *Animal Behavior* 17:601–606.

Jefferson, G. 1973. A case of precision timing in ordinary conversation: Overlapped tag-positioned address terms in closing sequences. *Semiotica* 9:47–96.

Jehl, J. R., Jr. 1973. Breeding biology and systematic relationships of the Stilt Sandpiper. *Wilson Bulletin* 85:115–147.

Johnsgard, P. A. 1960. Pair-formation mechanisms in *Anas* (Anatidae) and related genera. *Ibis* 102:616–618.

————. 1965. *Handbook of waterfowl behavior*. Ithaca, N.Y.: Cornell University Press.

Johnson, D. L. 1967. Honey bees: Do they use the direction information contained in their dance maneuver? *Science* 155:844–847.

Johnson, N. K. 1963. Biosystematics of sibling species of flycatchers in the *Empidonax hammondii-oberholserii-wrightii* complex. *University of California Publications in Zoology* 66:79–238.

Johnson, R. P. 1973. Scent marking in mammals. *Animal Behaviour* 21:521–535.

Jolly, A. 1972. *The evolution of primate behavior*. New York: Macmillan.

Kahl, M. P. 1971. Social behavior and taxonomic relationships of the storks. *Living Bird* 10:151–170.

Karplus, I., R. Szlep, and M. Tsurnamal. 1972. Associative behavior of the fish *Cryptocentrus cryptocentrus* (Gobiidae) and the pistol shrimp *Alpheus djiboutensis* (Alpheidae) in artificial burrows. *Marine Biology* 15:95–104.

Kendon, A. 1967. Some functions of gaze-direction in social interaction. *Acta Psychologica* 26:22–63.

————. 1970. Movement coordination in social interaction: Some examples described. *Acta Psychologica* 32:100–125.

————. 1972a. Some relationships between body motion and speech. An analysis of an example. In A. W. Siegman and B. Pope, eds., *Studies in dyadic communication*. New York: Pergamon.

————. 1972b. Review. Kinesics and context: Essays on body motion, by R. L. Birdwhistell. *American Journal of Psychology* 85:441–455.

————. 1973. The role of visible behavior in the organization of face-to-face interaction. In M. von Cranach and I. Vine, eds., *Movement and communication in man and chimpanzee*. New York: Academic Press.

————. in press 1. The F-formation system: spatial organization in social encounters. In A. Kendon, ed., *Studies in the behavior of face to face interaction*. Bloomington, Ind.: Research Center for the Language Sciences and Semiotics.

————. in press 2. Some functions of the face in a kissing round. *Semiotica* 15.

Kendon, A., and M. Cook. 1969. The consistency of gaze patterns in social interaction. *British Journal of Psychology* 60:481–494.

Kendon, A., and A. Ferber. 1973. A description of some human greetings. In R. P. Michael and J. H. Crook, eds., *Comparative ecology and behaviour of primates*. New York: Academic Press.

Kennedy, J. S., and D. Marsh. 1974. Pheromone-regulated anemotaxis in flying moths. *Science* 184:999–1001.

Kessel, E. L. 1955. The mating activities of balloon flies. *Systematic Zoology* 4:97–104.

Kessler, S. J. 1974. An empirical study of interpersonal endings. Ph.D. dissertation, City University of New York.

Kilham, L. 1972. Habits of the crimson-crested woodpecker in Panama. *Wilson Bulletin* 84:28–47.

————. 1974. Copulatory behavior of downy woodpeckers. *Wilson Bulletin* 86:23–34.

King, J. A. 1955. Social behavior, social organization, and population dynamics in a black-tailed prairie dog town in the Black Hills of South Dakota. *University of Michigan Contributions of the Laboratory of Vertebrate Biology* 67:1–123.

King, J. R. 1972. Variation in the song of the rufous-collared sparrow, *Zonotrichia capensis*, in northwestern Argentina. *Zeitschrift für Tierpsychologie* 30:344–373.

Kittredge, J. S., and F. T. Takahashi. 1972. The evolution of sex pheromone communication in the Arthropoda. *Journal of Theoretical Biology* 35:467–471.

Kleiman, D. G. 1966. Scant marking in the Canidae. *Symposium of the Zoological Society of London* 18:167–177.

————. 1967. Some aspects of social behavior in the Canidae. *American Zoologist* 7:365–372.

Kleiman, D. G., and J. F. Eisenberg. 1973. Comparison of canid and felid social systems from an evolutionary perspective. *Animal Behaviour* 21:637–659.

Klopfer, Peter H., and J. J. Hatch. 1968. Experimental considerations. In T. A. Seboek, ed., *Animal communication*. Bloomington: University of Indiana Press.

Klun, J. A., O. L. Chapman, K. C. Mattes, P. W. Wojtkowski, M. Beroza, and

P. E. Sonnet. 1973. Insect sex pheromones: Minor amount of opposite geometrical isomer critical to attraction. *Science* 181:661–663.

Konishi, M. 1970. Comparative neurophysiological studies of hearing and vocalizations in songbirds. *Zeitschrift für vergleichende Physiologie* 66: 257–272.

————. 1973. Locatable and nonlocatable acoustic signals for barn owls. *American Naturalist* 107:775–785.

Konishi, M., and F. Nottebohm. 1969. Experimental studies in the ontogeny of avian vocalizations. In R. A. Hinde, ed., *Bird vocalizations*. Cambridge, England: Cambridge University Press.

Koref-Santibañez, Susi. 1972. Courtship behavior in the semispecies *Drosophilia paulistorum*. *Evolution* 26:108–115.

Krebs, J. R. 1971. Territory and breeding density in the great tit, *Parus major* L. *Ecology* 52:2–22.

————. 1974. Colonial nesting and social feeding as strategies for exploiting food resources in the great blue heron (*Ardea herodias*) *Behaviour* 51: 99–134.

Kruijt, J. P. 1964. Ontogeny of social behaviour in Burmese red junglefowl (*Gallus gallus spadiceus*) Bonnaterre. *Behaviour Supplement* 12:1–201.

Kruuk, H. 1964. Predators and anti-predator behaviour of the black-headed gull (*Larus ridibundus* L.). *Behaviour Supplement* 11.

Kummer, Hans. 1968. *Social organization of hamadryas baboons*. Chicago: University of Chicago Press.

————. 1971. *Primate societies. Group techniques of ecological adaptation*. Chicago: Aldine-Atherton.

Labov, W. 1973. The boundaries of words and their meanings. In C.-J. Bailey and R. Shuy, eds., *New ways of analyzing variation in English*. Washington: Georgetown University Press.

Lack, D. 1940. The releaser concept of bird behaviour. *Nature* 145:107–108.

————. 1953. *The life of the robin*. London: Pelican.

————. 1966. *Population studies of birds*. Oxford: Clarendon Press.

Lade, Barbara I. 1968. Birds. In T. A. Sebeok, ed., *Animal communication*. Bloomington: University of Indiana Press.

Lanyon, Wesley E. 1963. Experiments on species discrimination in *Myiarchus* flycatchers. *American Museum Novitates* 2126:1–16.

Lanyon, W. E., and W. N. Tavolga. 1960. *Animal sounds and communication*. Washington, D.C.: American Institute of Biological Sciences, Publ. 7.

Leach, E. R. 1966. Ritualization in man in relation to conceptual and social development. *Philosophical Transactions of the Royal Society of Britain* 251:403–408.

Le Boeuf, B. J., and R. S. Peterson. 1969. Dialects in elephant seals. *Science* 166:1654–1656.

Le Boeuf, B. J., and L. F. Petrinovich. 1974. Dialects of northern elephant seals, *Mirounga angustirostris:* Origin and reliability. *Animal Behaviour* 22:656–663.

Lehrman, D. S. 1965. Interaction between internal and external environments

in the regulation of the reproductive cycle of the ring dove. In F. A. Beach, ed., *Sex and Behavior*. New York: Wiley.

Lein, M. R. 1972. Territorial and courtship songs of birds. *Nature* 237:48–49.

———. 1973. The biological significance of some communication patterns of wood warblers (Parulidae). Ph.D. dissertation, Harvard University.

Lemon, R. E. 1967. The response of cardinals to songs of different dialects. *Animal Behaviour* 15:538–545.

———. 1968. The relation between organization and function of song in cardinals. *Behaviour* 32:158–178.

Lemon, R. E., and C. Chatfield. 1971. Organization of song in cardinals. *Animal Behaviour* 19:1–17.

Lenneberg, Eric H. 1967. *Biological foundations of language*. New York: Wiley.

Leventhal, H., and E. Sharp. 1965. Facial expressions as indicators of distress. In S. S. Tomkins and C. E. Izards, eds., *Affect, cognition, and personality*. New York: Springer.

Lind, H. 1959. Studies on courtship and copulatory behavior in the goldeneye (*Bucephala clangula* (L.)). *Dansk Ornithologisk Forenings Tidsskrift* 53:177–219.

———. 1961. Studies on the behaviour of the black-tailed godwit (*Limosa limosa* (L.)). *Meddelelse fra Naturfredningsrådets reservatudvalg* 66:1–157.

Lindauer, M. 1961. *Communication among social bees*. Cambridge, Mass.: Harvard University Press.

———. 1967. Recent advances in bee communication and orientation. *Annual Review of Entomology* 12:439–470.

———. 1971. The functional significance of the honey bee waggle dance. *American Naturalist* 105:89–96.

Lissmann, H. W. 1958. On the function and evolution of electric organs in fish. *Journal of Experimental Biology* 35:156–191.

Littlejohn, M. J., and A. A. Martin. 1969. Acoustic interaction between two species of leptodactylid frogs. *Animal Behaviour* 17:785–791.

Lloyd, J. E. 1968. Illumination, another function of firefly flashes? *Entomological News* 79:265–268.

———. 1971. Bioluminescent communication in insects. *Annual Review of Entomology* 16:97–122.

———. 1975. Aggressive mimicry in *Photuris* fireflies: Signal repertoires by femmes fatales. *Science* 187:452–453.

Lockner, F. R., and R. E. Phillips. 1969. A preliminary analysis of the decrescendo call in female mallards (*Anas platyrhynchos* L.). *Behaviour* 35:281–287.

Loeb, F. F., Jr. 1968. The microscopic film analysis of the function of a recurrent behavioral pattern in a psychotherapeutic session. *Journal of Nervous and Mental Diseases* 147:605–618.

Long, Sally Y. 1972. Hair-nibbling and whisker-trimming as indicators of social hierarchy in mice. *Animal Behaviour* 20:10–12.

Lorenz, K. 1937. Über die Bildung des Instinktbegriffes. *Die Naturwissenschaften* 25:289–300, 307–318, 324–331.

————. 1941. Vergleichende Bewegungsstudien an Anatinen. *Journal für Ornithologie* 89, Erganzungsband 19–29, and 194–293. Reprint (1951–1953, in parts). Comparative studies on the behaviour of the Anatinae. Avicultural Magazine 57–59.

————. 1950. The comparative method in studying innate behaviour patterns. *Symposium of the Society for Experimental Biology* 4:221–268.

————. 1958. The evolution of behavior. *Scientific American* 199(6):67–82. Reprinted in J. L. McGaugh, N. M. Weinberger, and R. E. Whalen, eds., *Psychobiology readings from Scientific American*. San Francisco: Freeman. 1966.

————. 1966. Evolution of ritualization in the biological and cultural spheres. *Philosophical Transactions of the Royal Society of Britain* 251: 273–284.

————. 1974. Analogy as a source of knowledge. *Science* 185:229–234.

Lorenz, K., and Tinbergen, N. 1938. Taxis und Instinkthandlung in der Eirollbewegung der Graugans I. *Zeitschrift für Tierpsychologie* 2:1–29.

Lyons, J. 1972. Human language. In R. A. Hinde, ed., *Non-verbal communication*. New York: Cambridge University Press.

McAllister, D. E. 1967. The significance of ventral bioluminescence in fishes. *Journal of the Fisheries Research Board of Canada* 24:537–554.

McAllister, N. M., and R. W. Storer. 1963. Copulation in the pied-billed grebe. *Wilson Bulletin* 75:166–173.

McBride, G., J. James, and R. N. Shoffner. 1963. Social forces determining spacing and head orientation in a flock of domestic hens. *Nature* 197: 1272–1273.

McBride, G., I. P. Parer, and F. Foenander. 1969. The social organization and behavior of the feral domestic fowl. *Animal Behaviour Monographs* 2:127–181.

McClintock, M. K. 1971. Menstrual synchrony and suppression. *Nature* 229:244–245.

McCullough, D. R. 1969. The tule elk: Its history, behavior, and ecology. *University of California Publications in Zoology* 88:1–191.

MacDonald, S. D. 1968. The courtship and territorial behavior of Franklin's race of the spruce grouse. *Living Bird* 7:5–25.

McFarland, D. J. 1974. Time-sharing as a behavioral phenomenon. In D. S. Lehrman, J. S. Rosenblatt, R. A. Hinde, and E. Shaw, eds., *Advances in Behavior*, vol. 5. New York: Academic Press, pp. 201–255.

McFarland, D., and R. Sibly. 1972. "Unitary drives" revisited. *Animal Behaviour* 20:548–563.

MacKay, D. M. 1972. Formal analysis of communicative processes. In R. A. Hinde, ed., *Non-verbal communication*. Cambridge, England: Cambridge University Press.

McKinney, F. 1961. An analysis of the displays of the European Eider *Somateria mollissima mollissima* (Linnaeus) and the Pacific Eider *Somateria mollissima v. Nigra* Bonaparte. *Behaviour Supplement* 7:1–24.

————. 1965a. The displays of the American green-winged teal. *Wilson Bulletin* 77:112–121.

————. 1965b. The comfort movements of Anatidae. *Behaviour* 25:120–220.

————. 1965c. The spring behavior of wild Steller's eiders. *Condor* 67:273–290.

————. 1973. Ecoethological aspects of reproduction. In D. S. Farner, ed., *Breeding biology of birds*. Washington: National Academy of Sciences.

————. 1975. The evolution of duck displays. In G. Baerends, C. Beer, and A. Manning, eds., *Function and evolution in behaviour*. Oxford: Clarendon Press.

MacKintosh, N. J. 1965. Selective attention in animal discrimination learning. *Psychological Bulletin* 64:124–150.

Maier, R. A. 1964. The role of the dominance-submission in social recognition of hens. *Animal Behaviour* 12:59–60.

Manley, G. H. 1960a. Some aspects of social communication in black-headed gulls. *Animal Behaviour* 8:234.

————. 1960b. Agonistic and pair formation behaviour of the black-headed gull. D. Phil. dissertation, Oxford.

Marler, P. 1955a. Characteristics of some animal calls. *Nature* 176:6–8.

————. 1955b. Studies of fighting in chaffinches. (2) The effect on dominance relations of disguising females as males. *British Journal of Animal Behaviour* 3:137–146.

————. 1956a. The voice of the chaffinch and its function as a language. *Ibis* 98:231–261.

————. 1956b. Behaviour of the chaffinch *Fringilla coelebs*. *Behaviour Supplement* 5:1–184.

————. 1959. Developments in the study of animal communication. In P. R. Bell, ed., *Darwin's biological work. Some aspects reconsidered*. New York: Wiley.

————. 1960. Bird songs and mate selection. In W. E. Lanyon and W. N. Tavolga, eds., *Animal sounds and communication*. American Institute of Biological Sciences Publication 7.

————. 1961a. The logical analysis of animal communication. *Journal of Theoretical Biology* 1:295–317.

————. 1961b. The filtering of external stimuli during instinctive behaviour. In W. H. Thorpe and O. L. Zangwill, eds., *Current problems in animal behaviour*. Cambridge, England: Cambridge University Press.

————. 1965. Communication in monkeys and apes. In I. DeVore, ed., *Primate behavior*. New York: Holt, Rinehart and Winston.

————. 1967a. Animal communication signals. *Science* 157:769–774.

————. 1967b. Comparative study of song development in sparrows. *Proceedings of the 14th International Ornithological Congress, 1966*, pp. 231–244.

————. 1968. Aggregation and dispersal: Two functions in primate communication. In P. L. Jay, ed., *Primates. Studies in adaptation and variability*.

New York: Holt, Rinehart and Winston.

———. 1969. Vocalizations of wild chimpanzees. *Recent Advances in Primatology* 1:94–100.

———. 1970a. A comparative approach to vocal learning: Song development in white-crowned sparrows. *Journal of Comparative Physiology and Psychology* 71, Monograph 2(2):1–25.

———. 1970b. Vocalizations of East African monkeys. I. Red colobus. *Folia Primatalogica* 13:81–91.

———. 1972. Vocalizations of East African monkeys. II. Black and white colobus. *Behaviour* 42:175–197.

———. 1975. On the origin of speech from animal sounds. In J. F. Kavanagh and J. Cutting, eds., *The role of speech in language*. Cambridge, Mass.: MIT Press.

———. in press. An ethological theory of the origin of vocal learning. *Annals of the New York Academy of Science*.

Marler, P., and W. J. Hamilton III. 1966. *Mechanisms of animal behavior*. New York: Wiley.

Marler, P., and M. Tamura. 1962. Song "dialects" in three populations of white-crowned sparrows. *Condor* 64:368–377.

———. 1964. Culturally transmitted patterns of vocal behavior in sparrows. *Science* 146:1483–1486.

Marschak, J. 1965. Economics of language. Behavioral Science 10:135–140.

Marsden, H. M., and F. H. Bronson. 1964. Estrous synchrony in mice: Alteration by exposure to male urine. *Science* 144:1469.

Martin, S. E. 1964. Universals of language. *Harvard Educational Review* 34:353–355.

Mason, W. A., and J. H. Hollis. 1962. Communication between young rhesus monkeys. *Animal Behaviour* 10:211–221.

Mattingly, I., A. Liberman, A. Syrdal, and T. Halwes. 1971. Discrimination in speech and nonspeech modes. *Cognitive Psychology* 2:131–157.

Maurus, M., and H. Pruscha. 1973. Classification of social signals in squirrel monkeys by means of cluster analysis. *Behaviour* 47:106–128.

Maynard Smith, J. 1958. *The theory of evolution*. Baltimore: Penguin.

———. 1965. The evolution of alarm calls. *American Naturalist* 99:59–63.

———. 1974. The theory of games and the evolution of animal conflicts. *Journal of Theoretical Biology* 47:209–221.

Maynard Smith, J., and G. A. Parker. 1976. The logic of asymmetric contests. *Animal Behaviour* 24:159–175.

Maynard Smith, J., and G. R. Price. 1973. The logic of animal conflict. *Nature* 246:15–18.

Mayr, E. 1946. The number of species of birds. *Auk* 63:64–69.

———. 1958. Behavior and systematics. In A. Roe and G. G. Simpson, eds., *Behavior and evolution*. New Haven: Yale University Press.

———. 1960. The emergence of evolutionary novelties. In S. Tax, ed., *The evolution of life*. Chicago: University of Chicago Press.

———. 1963. *Animal species and evolution*. Cambridge, Mass.: Harvard University Press.

————. 1974a. Teleological and teleonomic, a new analysis. *Boston Studies in the Philosophy of Science* 14:91–117.

————. 1974b. Behavior programs and evolutionary strategies. *American Scientist* 62:650–659.

Mech, L. D. 1970. *The wolf: The ecology and behavior of an endangered species.* New York: Natural History Press.

Melchior, H. R. 1971. Characteristics of arctic ground squirrel alarm calls. *Oecologia* 7:184–190.

Menzel, E. W., Jr. 1969. Naturalistic and experimental approaches to primate behavior. In E. P. Willems and H. L. Raush, eds., *Naturalistic viewpoints in psychological research.* New York: Holt, Rinehart and Winston, pp. 78–121.

————. 1971. Communication about the environment in a group of young chimpanzees. *Folia primatologica* 15:220–232.

Messmer, E., and I. Messmer. 1957. Die Entwicklung der Lautäusserungen und einiger Verhaltensweisen der Amsel (*Turdus merula merula* L.) unter natürlichen Bedingungen und nach Einzelalufzucht in schalldichten Raümen. *Zeitschrift für Tierpsychologie* 13:341–441.

Meyerriecks, A. J. 1960. Comparative breeding behavior of four species of North American herons. *Nuttall Ornithological Club Publication* 2:1–158.

Michael, R. P., and E. B. Keverne. 1968. Pheromones in the communication of sexual status in primates. *Nature* 218:746–749.

Michael, R. P., and D. Zumpe. 1970. Sexual initiating behaviour by female rhesus monkeys (*Macaca mulatta*) under laboratory conditions. *Behaviour* 36:168–186.

Michael, R. P., E. B. Keverne, and R. W. Bonsall. 1971. Pheromones: Isolation of male sex attractants from a female primate. *Science* 172:964–966.

Michael, R. P., R. W. Bonsall, and P. Warner. 1974. Human vaginal secretions: Volatile fatty acid content. *Science* 186:1217–1219.

Miller, George A. 1956. The magical number seven, plus or minus two. *Psychological Review* 63:81–97.

————. Language and psychology. In E. H. Lenneberg, ed., *New directions in the study of language.* Cambridge, Mass.: MIT Press.

Miller, J. R., T. C. Baker, R. T. Cowde, and W. L. Roelofs. 1976. Reinvestigation of oak leaf roller sex pheromone components and the hypothesis that they vary with diet. *Science* 192:140–143.

Miller, R. C. 1922. The significance of the gregarious habit. *Ecology* 3:122–126.

Miller, R. E. 1967. Experimental approaches to the physiological and behavioral concomitants of affective communication in rhesus monkeys. In S. A. Altmann, ed., *Social communication among primates.* Chicago: University of Chicago Press. 125–134.

Miller, R. E., W. F. Caul, and I. A. Mirsky. 1967. Communication of affects between feral and socially isolated monkeys. *Journal of Personality and Social Psychology* 7:231–239.

Milligan, M. M., and J. Verner. 1971. Inter-populational song dialect discrimination in the white-crowned sparrow. *Condor* 73:208–213.

Minks, A. K., W. L. Roelofs, F. J. Ritter, and C. J. Persoons. 1973. Reproductive isolation of two torticid moth species by different ratios of a two-compound sex attractant. *Science* 180:1073–1074.

Mock, D. W. 1976. Pair-formation displays of the great blue heron. *Wilson Bulletin* 88:185–230.

Möglich, M., U. Maschwitz, and B. Hölldobler. 1974. Tandem calling: A new kind of signal in ant communication. *Science* 186:1046–1047.

Morin, J. G., A. Harrington, K. Nealson, N. Krieger, T. O. Baldwin, and J. W. Hastings. 1975. Light for all reasons: Versatility in the behavioral repertoire of the flashlight fish. *Science* 190:74–76.

Morris, C. W. 1946. Signs, language, and behavior. New York: Prentice-Hall. Reprint (1955). New York: Braziller.

Morris, D. 1952. Homosexuality in the ten-spined stickleback. *Behaviour* 4:233–261.

———. 1955. The causation of pseudofemale and pseudomale behaviour: A further comment. *Behaviour* 8:46–56.

———. 1956a. The feather postures of birds, and the problem of the origin of social signals. *Behaviour* 9:75–113.

———. 1956b. The function and causation of courtship ceremonies. In M. Autori, ed., *L'instinct dans le comportement des animaux et de l'homme*. Paris: Masson.

———. 1957. "Typical intensity" and its relation to the problem of ritualisation. *Behaviour* 11:1–12.

———. 1966. Abnormal rituals in stress situations. *Philosophical Transactions of the Royal Society of Britain* 251:327–330.

———. 1971. *Intimate behaviour*. New York: Random House.

Morris, R. L., and C. J. Erickson. 1971. Pair bond maintenance in the ring dove (*Streptopelia risoria*). *Animal Behaviour* 19:398–406.

Morse, D. H. 1970. Territorial and courtship songs of birds. *Nature* 226:659–661.

Morton, E. S. 1970. Ecological sources of selection on avian sounds. Ph.D. dissertation, Yale University.

Morton, E. S. 1975. Ecological sources of selection on avian sounds. *American Naturalist* 109:17–34.

Morton, E. S., and M. D. Shalter. ms. Vocal response to predators in pair-bonded Carolina wrens.

Moss, R. 1972. Social organization of will ptarmigan in Alaska. *Condor* 74:144–151.

Moynihan, M. 1955a. Some aspects of reproductive behaviour in the black-headed gull (*Larus ridibundus ridibundus* L.) and related species. *Behaviour*, Supplement IV.

———. 1955b. Types of hostile display. *Auk* 72:247–259.

———. 1955c. Remarks on the original sources of display. *Auk* 72:240–246.

————. 1956. Notes on the behaviour of some North American gulls. I. Aerial hostile behaviour. *Behaviour* 10:126–178.

————. 1958a. Notes on the behaviour of some North American gulls. II. Non-aerial hostile behaviour of adults. *Behaviour* 12:95–182.

————. 1958b. Notes on the behaviour of the flying steamer duck. *Auk* 75:183–202.

————. 1958c. Notes on the behaviour of some North American gulls. III. Pairing behaviour. *Behaviour* 13:112–130.

————. 1959a. A revision of the family Laridae (Aves). *American Museum Novitates* 1928:1–42.

————. 1959b. Notes on the behaviour of some North American gulls. IV. The ontogeny of hostile behaviour and display patterns. *Behaviour* 14:214–239.

————. 1960. Some adaptations which help to promote gregariousness. *Proceedings of the 12th International Ornithological Congress (Helsinki 1958)*, pp. 523–541.

————. 1962a. Hostile and sexual behaviour patterns of South American and Pacific Laridae. *Behaviour*, Supplement VIII:1–365.

————. 1962b. Display patterns of tropical American "nine-primaried" songbirds. I. *Chlorospingus. Auk* 79:310–344.

————. 1962c. The organization and probable evolution of some mixed species flocks of neotropical birds. *Smithsonian Miscellaneous Collection* 143(7):1–140.

————. 1962d. Display patterns of tropical American "nine-primaried" songbirds. II. Some species of *Ramphocelus. Auk* 79:655–686.

————. 1963a. Display patterns of tropical American "nine-primaried" songbirds. III. The green-backed sparrow. *Auk* 80:116–144.

————. 1963b. Inter-specific relations between some Andean birds. *Ibis* 105:327–339.

————. 1964. Some behavior patterns of platyrrhine monkeys. I. The night monkey (*Aotus trivirgatus*). *Smithsonian Miscellaneous Collection* 146(5):1–84.

————. 1966a. Communication in the titi monkey *Callicebus. Journal of Zoology of London* 150:77–127.

————. 1966b. Display patterns of tropical American "nine-primaried" songbirds. IV. The yellow-rumped tanager. *Smithsonian Miscellaneous Collection* 149(5):1–34.

————. 1967. Comparative aspects of communication in New World primates. In D. J. Morris, ed., *Primate ethology*. London: Weidenfeld and Nicolson.

————. 1968a. The "Coerebini": A group of marginal areas, habitats, and habits. *American Naturalist* 102:573–581.

————. 1968b. Social mimicry; character convergence versus character displacement. *Evolution* 22:315–331.

————. 1970a. Some behavior patterns of platyrrhine monkeys II. *Saguinus geoffroyi* and some other tamarins. *Smithsonian Contributions in Zoology* 28:1–77.

————. 1970b. The control, suppression, decay, disappearance and replacement of displays. *Journal of Theoretical Biology* 29:85–112.

————. 1975. Conservatism of displays and comparable stereotyped patterns among cephalopods. In G. Baerends, C. Beer, and A. Manning, eds., *Function and evolution in behaviour*. Oxford: Clarendon Press.

————. 1976. *The New World primates*. Princeton: Princeton University Press.

Müller-Schwarze, D. 1971. Pheromones in black-tailed deer (*Odocoileus hemionus columbianus*). *Animal Behaviour* 19:141–152.

Mulligan, James A. 1966. Singing behavior and its development in the Song Sparrow *Melospiza melodia*. *University of California Publications in Zoology* 81:1–76.

Mundinger, P. C. 1970. Vocal imitation and individual recognition of finch calls. *Science* 168:480–482.

————. 1972. Annual testicular cycle and bill color change in the eastern American goldfinch. *Auk* 89:403–419.

Myers, M. T. 1959. Display behavior of bufflehead, scoters and goldeneyes at copulation. *Wilson Bulletin* 71:159–168.

Mykytowycz, R. 1965. Further observations on the territorial function and histology of the submandibular cutaneous (chin) glands in the rabbit, *Oryctolagus cuniculus* (L.). *Animal Behaviour* 13:400–412.

————. 1974. Odor in the spacing behavior of mammals. In M. C. Birch, ed., *Pheromones*. New York: Elsevier.

Mykytowycz, R., and S. Gambale. 1969. The distribution of dung-hills and the behaviour of free-living wild rabbits, *Oryctolagus cuniculus* (L.), on them. *Forma et Functio* 1:333–349.

Mykytowycz, R., and M. M. Ward. 1971. Some reactions of nestlings of the wild rabbit, *Oryctolagus cuniculus* (L.), when exposed to natural rabbit odours. *Forma et Functio* 4:137–148.

Narin, P. M., and R. R. Capranica. 1976. Sexual differences in the auditory system of the tree frog *Eleutherodactylus coqui*. *Science* 192:378–380.

Nelson, J. B. 1965. The behaviour of the gannet. *British Birds* 58:233–288, 313–336.

————. 1967. Colonial and cliff nesting in the gannet compared with other Sulidae and the kittiwake. *Ardea* 55:60–90.

Nelson, K. 1973. Does the holistic study of behavior have a future? In P. P. G. Bateson and P. H. Klopfer, eds., *Perspectives in ethology*. New York: Plenum Press.

Nethersole-Thompson, D. 1951. *The greenshank*. London: Collins.

Nice, M. M. 1943. Studies in the life history of the song sparrow. *Transactions of the Linnean Society of New York* 6:1–328.

Nicolai, J. 1959. Familientradition in der Gesangsentwicklung des Gimpels (*Pyrrhula pyrrhula* L.). *Journal für Ornithologie* 100:39–46.

Noble, G. K. 1936. Courtship and sexual selection of the flicker (*Colaptes auratus luteus*). *Auk* 52:269–282.

Noirot, E., and D. Pye. 1969. Sound analysis of ultrasonic distress calls of mouse pups as a function of their age. *Animal Behaviour* 17:340–349.

Norton-Griffiths, M. 1969. The organization, control and development of parental feedings in the oystercatcher (*Haematopus ostralegus*). *Behaviour* 34:55–114.

Nottebohm, F. 1969. The song of the Chingolo, *Zonotrichia capensis* in Argentina: Description and evaluation of a system of dialects. *Condor* 71: 299–315.

———. 1970. Ontogeny of bird song. *Science* 167:950–956.

———. 1972. The origins of vocal learning. *American Naturalist* 106:116–140.

———. 1975. Continental patterns of song variability in *Zonotrichia capensis:* Some possible ecological correlates. *American Naturalist* 109: 605–624.

Ogden, C. K., and I. A. Richards. 1936. *The meaning of meaning,* 4th ed. New York: Harcourt Brace.

Ogle, K. N. 1962. The visual space sense. *Science* 135:763–771.

Oppenheimer, J. R. 1968. Behavior and ecology of the white-faced monkey, *Cebus capucinus,* on Barro Colorado Island. Ph.D. dissertation, University of Illinois, Urbana.

Oppenheimer, J. R., and E. C. Oppenheimer. 1973. Preliminary observations of *Cebus nigrivittatus* (Primates: Cebidae) on the Venezuelan llanos. *Folia primatologica* 19:409–436.

Orcutt, A. B. 1974. Sounds produced by hatching Japanese quail (*Coturnix coturnix japonica*) as potential aids to synchronous hatching. *Behaviour* 50:173–184.

Orians, G. H. 1969. On the evolution of mating systems in birds and mammals. *American Naturalist* 103:589–603.

Orians, G. H., and G. M. Christman. 1968. A comparative study of the behavior of red-winged, tricolored, and yellow-headed blackbirds. *University of California Publications in Zoology* 84:1–81.

Oring, L. W. 1968. Vocalizations of the green and solitary sandpipers. *Wilson Bulletin* 80:395–420.

Otte, D. 1972. Simple *versus* elaborate behavior in grasshoppers. An analysis of communication in the genus *Syrbula. Behaviour* 42:291–322.

Otte, D., and J. Loftus-Hills. ms. Acoustic interactions in *Syrbula:* Cooperation, competition, interference or concealment?

Otte, D., and J. Smiley. ms. Synchronous flashing in Texas fireflies (*Photinus:* Lampyridae). A consideration of interaction models.

Parker, G. A. 1974a. Courtship persistence and female-guarding as male time investment strategies. *Behaviour* 48:157–184.

———. 1974b. Assessment strategy and the evolution of fighting behaviour. *Journal of Theoretical Biology* 47:223–243.

Parkes, A. S., and H. M. Bruce. 1961. Olfactory stimuli in mammalian reproduction. *Science* 134:1049–1054.

Parkin, D. T., A. W. Ewing, and H. A. Ford. 1970. Group diving in the blue-footed booby *Sula nebouxii. Ibis* 112:111–112.

Patterson, I. J. 1965. Timing and spacing of broods in the black-headed gull *Larus ridibundus. Ibis* 107:433–459.

Payne, R. B. 1967. Interspecific communication signals in parasitic birds. *American Naturalist* 101:363–375.

———. 1971. Duetting and chorus singing in African birds. *Ostrich,* Supplement 9:125–146.

———. 1973a. Duetting and antiphonal singing in birds, its extent and significance. (Review) *Auk* 90:451–453.

———. 1973b. Vocal mimicry of the paradise whydahs (*Vidua*) and response of female whydahs to the songs of their hosts (*Pytilia*) and their mimics. *Animal Behaviour* 21:762–771.

Payne, R. S., and D. Webb. 1971. Orientation by means of long range acoustic signaling in baleen whales. *Annals of the New York Academy of Sciences* 188:110–142.

Peek, F. W. 1972. An experimental study of the territorial function of vocal and visual display in the male red-winged blackbird (*Agelaius phoeniceus*). *Animal Behaviour* 20:112–118.

Pengelly, W. J., and J. Kear. 1970. The hand rearing of young blue duck. *Wildfowl* 21:115–121.

Perdeck, A. C. 1958. The isolating value of specific song patterns in two sibling species of grasshoppers (*Corthippus brunneus* Thumb. and *C. biguttulus* L.). *Behaviour* 12:1–75.

Perrins, C. M. 1970. The timing of birds' breeding seasons. *Ibis* 112:242–255.

Petrinovich, L. 1973. Darwin and the representative expression of reality. In P. Ekman, ed., *Darwin and facial expression: A century of research in review.* New York: Academic Press.

———. 1974. Individual recognition of pup vocalization by northern elephant seal mothers. *Zeitschrift für Tierpsychologie* 34:308–312.

Petrinovich, L., T. Patterson, and H. V. S. Peeke. 1976. Reproductive condition and the response of white-crowned sparrows (*Zonotrichia leucophrys nuttali*) to song. *Science* 191:206–207.

Petrinovich, L., and H. V. S. Peeke. 1973. Habituation to territorial song in the white-crowned sparrow (*Zonotrichia leucophrys*). *Behavioral Biology* 8:743–748.

Phillips, R. E. 1972. Sexual and agonistic behaviour in the killdeer (*Charadrius vociferus*). *Animal Behaviour* 20:1–9.

Portmann, A. 1961. *Animals as social beings.* New York: Harper.

Potash, L. M. 1972. A signal detection problem and possible solution in Japanese quail (*Coturnix coturnix japonica*). *Animal Behaviour* 20:192–196.

Potter, P. E. 1972. Territorial behavior in savannah sparrows in southeastern Michigan. *Wilson Bulletin* 84:48–59.

Powers, J. B., and S. S. Winans. 1975. Vomeronasal organ: Critical role in mediating sexual behavior of the male hamster. *Science* 187:961–963.

Premack, D. 1971. Language in chimpanzee? *Science* 172:808–822.

Price, P. W. 1970. Trail odors: Recognition by insects parasitic on cocoons. *Science* 170:546–547.

Pumphrey, R. J. 1962. Introduction (to a symposium on Biological Acoustics). *Symposium of the Zoological Society of London* 7:1–5.

Quastler, H. 1958. A primer on information theory. In H. P. Yockey, ed., *Symposium on information theory in biology.* New York: Pergamon Press.

Radesäter, T. 1974. On the ontogeny of orienting movements in the triumph ceremony in two species of geese (*Anser anser* L. and *Branta canadensis* L.). *Behaviour* 50:1–15.

————. 1975. Biting in the triumph ceremony of the Canada goose. *Wilson Bulletin* 87:554–555.

Ralls, Katherine. 1971. Mammalian scent marking. *Science* 171:443–449.

Rand, A. S. 1961. A suggested function of the ornamentation of East African forest chameleons. *Copeia* 1961:411–414.

————. 1967. Predator prey interactions and the evolution of aspect diversity. *Atlas do Simposio sobre a Biota Amazonica* 5 (*Zoologia*):78–83.

————. ms. The displays of *Engystomops pustulosus.*

Rand, A. S., and G. E. Drewry. ms. Communication strategies in Puerto Rican frogs of the genus *Eleutherodactylus.*

Rand, A. S., and E. E. Williams. 1970. An estimation of redundancy and information content of anole dewlaps. *American Naturalist* 104:99–103.

Rand, W. M., and A. S. Rand. in press. Agonistic behavior in nesting iguanas: A stochastic analysis of dispute settlement dominated by the minimization of energy cost. *Zeitschrift für Tierpsychologie.*

Ransom, T. W., and B. S. Ranson. 1971. Adult male-infant relations among baboons (*Papio anubis*). *Folia primatologica* 16:179–195.

Rasa, O. A. E. 1973. Marking behaviour and its social significance in the African dwarf mongoose, *Helogale undulata rufula. Zeitschrift für Tierpsychologie* 32:293–324.

Raveling, D. G. 1969. Preflight and flight behavior of Canada geese. *Auk* 86:671–681.

————. 1970. Dominance relationships and agonistic behavior of Canada geese in winter. *Behaviour* 37:291–319.

Regnier, F. E., and E. O. Wilson. 1971. Chemical communication and "propaganda" in slave-maker ants. *Science* 172:267–269.

Rhoads, J. G., and J. S. Friedlaender. 1975. Language boundaries and biological differentiation on Bougainville: Multivariate analysis of variance. *Proceedings of the National Academy of Sciences* 72:2247–2250.

Ribbands, C. R. 1954. Communication between honeybees. I: The response of crop-attached bees to the scent of their crop. *Proceedings of the Royal Entomological Society of London* (A) 29:141–144.

Ricklefs, R. E., and K. O'Rourke. 1975. Aspect diversity in moths: A temperate-tropical comparison. *Evolution* 29:313–324.

Ridpath, M. G. 1972. The Tasmanian native hen, *Tribonyx mortierii.* I. Patterns of behaviour. *Commonwealth Scientific and Industrial Research Organization Wildlife Research* 17:1–51.

Rimland, B. 1964. *Infantile autism. The syndrome and its implications for a neural theory of behavior.* New York: Appleton-Century-Crofts.

————. 1975. Autism, stress, and ethology. *Science* 188:401–402.

Ristau, Carolyn A. 1974. Infant vocal communication: A comparison of normal and mongoloid humans, other primates and carnivores. Ph.D. dissertation, University of Pennsylvania, Philadelphia.

Robinson, F. N. 1974. The function of vocal mimicry in some avian displays. *Emu* 74:9–10.

Robinson, M. H. 1969. Defense against visually hunting predators. In T. Dobzhansky, M. K. Hecht and W. C. Steere, eds., *Evolutionary biology* 3:225–259.

Roelofs, W. L., and A. Comeau. 1969. Sex pheromone specificity: Taxonomic and evolutionary aspects in Lepidoptera. *Science* 165:398–400.

Rohwer, S. A. 1972. The multivariate assessment of interbreeding between the meadowlarks, *Sturnella. Systematic Zoology* 21:313–338.

————. 1973. Significance of sympatry to behavior and evolution of Great Plains meadowlarks. *Evolution* 27:44–57.

Rommel, S. A., Jr., and J. D. McCleave. 1972. Oceanic electric fields: Perception by American eels? *Science* 176:1233–1235.

Rood, J. P. 1970. Ecology and social behavior of the desert cavy (*Microcavia australis*). *American Midland Naturalist* 83:415–454.

————. 1972. Ecological and behavioural comparisons of three genera of Argentine cavies. *Animal Behaviour Monographs* 5:1–83.

Rosenblum, L. A., and I. C. Kaufman. 1967. Laboratory observations of early mother-infant relations in pigtail and bonet macaques. In S. A. Altmann, ed., *Social communication among primates.* Chicago: University of Chicago Press.

Rothstein, S. I. 1971. A reanalysis of the interspecific invitation to preening display as performed by the brown-headed cowbird (*Molothrus ater*). *American Zoologist* 11(4):89.

————. ms. New data on a unique display performed by cowbirds.

Rottman, S. J., and C. T. Snowdon. 1972. Demonstration and analysis of an alarm pheromone in mice. *Journal of Comparative Physiological Psychology* 81:483–490.

Rovner, J. S. 1975. Sound production by nearctic wolf spiders: A substratum-coupled stridulatory mechanism. *Science* 190:1309–1310.

Rowell, T. E. 1962. Agonistic noises of the rhesus monkey (*Macaca mulatta*). *Symposium of the Zoological Society of London* 8:91–96.

————. 1966. Hierarchy in the organization of a captive baboon group. *Animal Behaviour* 14:430–443.

————. 1972. *Social behaviour of monkeys.* Baltimore: Penguin.

————. 1974. Contrasting adult male roles in different species of nonhuman primates. *Archives of Sexual Behavior* 3:143–149.

Rowell, T. E., and R. A. Hinde. 1962. Vocal communication by the rhesus monkey (*Macaca mulatta*). *Proceedings of the Zoological Society of London* 138:279–294.

Rowley, I. 1965. The life history of the superb blue wren *Malurus cyaneus*. *Emu* 64:251–297.

———. 1974. Display situations in two Australian ravens. *Emu* 74:47–52.

Royama, T. 1966. A re-interpretation of courtship feeding. *Bird Study* 13:116–129.

Rozin, P. 1968. Specific aversions and neophobia as a consequence of vitamin deficiency and/or poisoning in half-wild and domestic rats. *Journal of Comparative and Physiological Psychology* 66:82–88.

Ryan, E. P. 1966. Pheromone: Evidence in a decapod crustacean. *Science* 151:340–341.

Sade, D. S. 1972. Sociometrics of *Macaca mulatta* I. Linkages and cliques in grooming matrices. *Folia primatologica* 18:196–223.

Sales, G. D. 1972. Ultrasound and aggressive behaviour in rats and other small mammals. *Animal Behaviour* 20:88–100.

Sales, G., and D. Pye. 1974. *Ultrasonic communication by animals*. New York: Wiley.

Sapir, E. 1921. *Language*. New York: Harcourt, Brace, World.

Sapir, J. D. 1970. *Kujaama: Symbolic separation among the Diola-Fogny*. *American Anthropologist* 72:1330–1348.

Sauer, E. G. F. 1962. Ethology and ecology of golden plovers on St. Lawrence Island, Bering Sea. *Psychologische Forschung* 26:399–470.

Schaller, G. B. 1963. *The mountain gorilla. Ecology and behavior*. Chicago: University of Chicago Press.

———. 1965. The behavior of the mountain gorilla. In I. DeVore, ed., *Primate behavior*. New York: Holt, Rinehart and Winston.

———. 1972. *The Serengeti lion*. Chicago: University of Chicago Press.

Scheflen, A. E. 1964. The significance of posture in communication systems. *Psychiatry* 27:316–331.

———. 1967. On the structuring of human communication. *American Behavioral Scientist* 10(8):8-12.

Schegloff, E. A. 1968. Sequencing in conversational openings. *American Anthropologist* 70:1075–1095.

Schegloff, E. A., and H. Sacks. 1973. Opening up closings. *Semiotica* 8:289–327.

Scherer, K. R. 1972. Judging personality from voice: A cross-cultural approach to an old issue in interpersonal perception. *Journal of Personality* 40:191–210.

———. 1974. Acoustic concomitants of emotional dimensions: Judging affect from synthesized tone sequences. In S. Weitz, ed., *Nonverbal communication*. New York: Oxford University Press.

Schleidt, W. M. 1973. Tonic communication: continual effects of discrete signs in animal communication systems. *Journal of Theoretical Biology* 42:359–386.

Schleidt, W. M., and M. D. Shalter. 1972. Cloacal foam gland in the quail *Coturnix coturnix*. *Ibis* 114:558.

Schloeth, R. 1956. Quelques moyens d'intercommunication des taureaux de Camargue. *La Terre et la Vie* 2:83–93.

Schneider, D. 1974. The sex-attractant receptor of moths. *Scientific American* 231(1):28–35.

Schreiber, R. W. in press. Maintenance behavior and communication in the brown pelican *Pelecanus occidentalis*. *American Ornithologists' Union Ornithological Monographs*.

Schubert, G. 1971. Experimentelle Untersuchungen Uber die Artkennzeichnenden Parameter in Gesang des Zizalps, *Phylloscopus c. collybita* (Vieillot). *Behaviour* 38:289–314.

Scott, W. C. M. 1955. A note on blathering. *International Journal of Psychoanalysis* 36:3–4.

Sebeok, T. A. 1962. Coding in the evolution of signaling behavior. *Behavioral Science* 7:430–442.

———. 1965. Animal communication. *Science* 147:1006–1014.

———. 1968a. Goals and limitations of the study of animal communication. In T. A. Sebeok, ed., *Animal communication*. Bloomington: University of Indiana Press.

———. 1968b. *Animal communication*. Bloomington: University of Indiana Press.

———. 1969. Semiotics and ethology. In T. A. Sebeok and A. Ramsay, eds., *Approaches to communication*. Paris: Mouton.

———. 1970. The word "zoosemiotics." *Language Sciences* 10:36–37.

———. in press. How animals communicate. Bloomington: University of Indiana Press.

Sebeok, T. A., and A. Ramsay. 1969. *Approaches to animal communication*. Paris: Mouton.

Selander, R. K. 1966. Sexual dimorphism and differential niche utilization in birds. *Condor* 68:113–151.

———. 1972. Sexual selection and dimorphism in birds. In B. G. Campbell, ed., *Sexual selection and the descent of man 1871–1971*. Chicago: Aldine.

Selander, R. K., and C. J. LaRue, Jr. 1961. Interspecific preening invitation display of parasitic cowbirds. *Auk* 78:473–504.

Selous, E. 1933. *The evolution of habit in birds*. London: Constable.

Sewell, G. D. 1967. Ultrasound in adult rodents. *Nature* 215:512.

———. 1968. Ultrasound in rodents. *Nature* 217:682–683.

———. 1970. Ultrasonic communication in rodents. *Nature* 227:410.

Shalter, M. D.. ms. Differential responses of wild and domestic fowl to graded parameters of an acoustic alarm signal.

Shannon, C. E., and W. Weaver. 1949. *The mathematical theory of communication*. Urbana: University of Illinois Press.

Sherzer, J. 1973. Verbal and nonverbal deixis: the pointed lip gesture among the San Blas Cuna. *Language and Society* 2:117–131.

Short, L. L. 1971. The evolution of terrestrial woodpeckers. *American Museum Novitiates* 2467:1–23.

Sick, H. 1939. Über die Dialektbildung beim Regenruf des Buckfinken. *Journal für Ornithologie* 87:568–592.

———. 1959. Die Balz der Schmuckvögel (Pipridae). *Journal für Ornithologie* 100:269–302.

————. 1967. Courtship behavior in the manakins (Pipridae). A review. *Living Bird* 6:5–22.

Simmons, J. A., E. G. Weaver, and J. M. Pylka. 1971. Periodical cicada: Sound production and hearing. *Science* 171:212–213.

Simmons, K. E. L. 1951. The nature of the predator-reactions of breeding birds. *Behaviour* 4:161–171.

————. 1953. Some aspects of aggressive behaviour in three closely related plovers (*Charadrius*). *Ibis* 95:115–127.

————. 1955a. The nature of the predator-reactions of waders towards humans; with special reference to the role of the aggressive-, escape- and brooding-drives. *Behaviour* 8:130–173.

————. 1955b. Studies on Great Crested Grebes. *Avicultural Magazine* 6:3–13, 93–102, 131–146, 181–201, 235–253, and 294–316.

Simpson, G. G. 1949. *The meaning of evolution.* New Haven: Yale University Press.

Simpson, M. J. A. 1968. The display of the Siamese fighting fish, *Betta splendens. Animal Behaviour Monographs* 1:1–73.

————. 1973a. Social displays and the recognition of individuals. In P. P. G. Bateson and P. H. Klopfer, eds., *Perspectives in ethology.* New York: Plenum.

————. 1973b. The social grooming of male chimpanzees. In R. P. Michael and J. H. Crook, eds., *Comparative ecology and behaviour of primates.* New York: Academic Press.

Slater, P. J. B. 1973. Describing sequences of behavior. In P. P. G. Bateson and P. H. Klopfer, eds., *Perspectives in ethology.* New York: Plenum Press.

Smith, N. G. 1966a. Evolution of some arctic gulls (*Larus*): an experimental study of isolating mechanisms. *American Ornithologists' Union Ornithological Monographs* 4:1–99.

————. 1966b. Adaptations to cliff-nesting in some arctic gulls (*Larus*). *Ibis* 108:68–83.

————. 1969. Provoked release of mobbing—a hunting technique of *Micrastur* falcons. *Ibis* 111:241–243.

————. in press. Oropendolas, caciques, and cowbirds: Their lives among friends, associates, and enemies. Portland, Oregon: Chiron Press.

Smith, S. T. 1972. Communication and other social behavior in *Parus carolinensis. Nuttall Ornithological Club Publication* 11:1–125 (Museum of Comparative Zoology, Cambridge, Mass.).

Smith, W. John. 1963. Vocal communication of information in birds. *American Naturalist* 97:117–125.

————. 1965. Message, meaning, and context in ethology. *American Naturalist* 99:405–409.

————. 1966. Communication and relationships in the genus *Tyrannus. Nuttall Ornithological Club Publication.* 6:1–250 (Museum of Comparative Zoology, Cambridge, Mass.).

————. 1967. Displays of the vermilion flycatcher (*Pyrocephalus rubinus*). *Condor* 69:601–605.

————. 1968. Message-meaning analyses. In T. A. Sebeok, ed., *Animal communication*. Bloomington: University of Indiana Press.

————. 1969a. Displays and messages in intraspecific communication. *Semiotica* 1:357–369.

————. 1969b. Messages of vertebrate communication. *Science* 165:145–150.

————. 1969c. Displays of *Sayornic phoebe* (Aves Tyrannidae). *Behaviour* 33:283–322.

————. 1970a. Song-like displays in the genus *Sayornis*. *Behaviour* 37:64–84.

————. 1970b. Displays and message assortment in *Sayornis* species. *Behaviour* 37:85–112.

————. 1970c. Courtship and territorial displaying in the vermilion flycatcher, *Pyrocephalus rubinus*. *Condor* 72:488–491.

————. 1971a. Behavior of *Muscisaxicola* and related genera. *Bulletin of the Museum of Comparative Zoology* 141:233–268.

————. 1971b. Behavioral characteristics of serpophaginine tyrannids. *Condor* 73:259–286.

————. 1974. Zoosemiotics: ethology and the theory of signs. In T. A. Sebeok, ed., *Current trends in linguistics*. 12:561–626.

————. in press. Communication in birds. In T. A. Sebeok, ed., *How animals communicate*. Bloomington: University of Indiana Press.

Smith, W. John, J. Chase, and A. K. Lieblich. 1974. Tongue showing: A facial display of humans and other primate species. *Semiotica* 11:201–246.

Smith, W. J., J. Pawlukiewicz, and S. T. Smith. in press. Kinds of activities correlated with singing patterns of the yellow-throated vireo. *Animal Behaviour*.

Smith, W. John, S. L. Smith, E. C. Oppenheimer, J. G. deVilla, and F. A. Ulmer. 1973. Behavior of a captive population of black-tailed prairie dogs. Annual cycle of social behavior. *Behaviour* 46:189–220.

Smith, W. John, S. L. Smith, J. G. deVilla, and E. C. Oppenheimer. in press 1. The jump-yip display of the black-tailed prairie dog, *Cynomys ludovicianus*. *Animal Behaviour*.

Smith, W. John, S. L. Smith, E. C. Oppenheimer, and J. G. deVilla. in press 2. Vocalizations of the black-tailed prairie dog, *Cynomys ludovicianus*. *Animal Behaviour*.

Smythe, N. 1970. On the existence of "pursuit invitation" signals in mammals. *American Naturalist* 104:491–494.

Snow, B. K. 1972. A field study of the calfbird *Perissocephalus tricolor*. *Ibis* 114:139–162.

Snow, B. K., and D. W. Snow. 1968. Behavior of the swallow-tailed gull of the Galapagos. *Condor* 70:252–264.

Snow, D. W. 1958. *A study of blackbirds*. London: Allen and Unwin.

————. 1963. The evolution of manakin displays. *Proceedings of the XIIIth International Ornithological Congress*, pp. 553–561.

Snow, D. W., and D. Goodwin. 1974. The black-and-gold cotinga. *Auk* 91: 360–369.

Snowdon, C. T., and Y. V. Pola. ms. Intraspecific and interspecific responses to synthesized pygmy marmoset vocalizations.

Solomon, R. L., and L. Postman. 1952. Frequency of usage as a determinant of recognition thresholds for words. *Journal of Experimental Psychology* 43:195–201.

Somers, Preston. 1973. Dialects in southern Rocky Mountain pikas, *Ochotona princeps* (Lagomorpha). *Animal Behaviour* 21:124–137.

Sommer, R. 1967. Small group ecology. *Psychological Bulletin* 67:145–152.

Southern, W. E. 1974. Copulatory wing-flagging: A synchronizing stimulus for nesting ring-billed gulls. *Bird-Banding* 45:210–216.

Sparks, J. 1967. Allogrooming in primates: A review. In D. Morris, ed., *Primate ethology*. London: Weidenfield and Nicolson.

Spence, K. W. 1951. Theoretical interpretations of learning. In S. S. Stevens, ed., *Handbook of experimental psychology*. New York: Wiley.

Spieth, H. T. 1952. Mating behavior within the genus *Drosophila* (Diptera). *Bulletin of the American Museum of Natural History* 99:399–474.

Spurway, H., and J. B. S. Haldane. 1953. The comparative ethology of vertebrate breathing. I. Breathing in newts, with a general survey. *Behaviour* 6:8–34.

Stefanski, R. A., and J. B. Falls. 1972. A study of distress calls of song, swamp, and white-throated sparrows (Aves: Fringillidae). I. Intraspecific responses and functions. *Canadian Journal of Zoology* 50:1501–1512.

Stehn, R. A., and M. E. Richmond. 1975. Male-induced pregnancy termination in the prairie vole, *Microtus ochrogaster*. *Science* 187:1211–1213.

Steiner, A. L. 1974. Body-rubbing, marking, and other scent-related behavior in some ground squirrels (Sciuridae), a descriptive study. *Canadian Journal of Zoology* 52:889–906.

Stevens, K. 1973. Potential role of property detectors in the perception of consonants. *MIT Research Laboratory of Electronics Quarterly Progress Report* 110:155–168.

Stokes, A. W. 1960. Nest-site selection and courtship behaviour of the blue tit *Parus caeruleus*. *Ibis* 102:507–519.

————. 1961. Voice and social behavior of the chukar partridge. *Condor* 63:111–127.

————. 1962. Agonistic behaviour among blue tits at a winter feeding station. *Behaviour* 19:118–138.

————. 1963. Agonistic and sexual behaviour of the chukar partridge (*Alectoris graeca*). *Animal Behaviour* 11:121–134.

————. 1967. Behaviour of the bobwhite, *Colinus virginianus*. *Auk* 84:1–33.

————. 1971. Parental and courtship feeding in red jungle fowl. *Auk* 88: 21–29.

Stokes, A. W., and H. W. Williams. 1971. Courtship feeding in gallinaceous birds. *Auk* 88:543–559.

———. 1972. Courtship feeding calls in gallinaceous birds. *Auk* 89:177–180.

Storer, R. W. 1967. Observations on Rolland's grebe. *El Hornero* 10:339–350.

———. 1969. The behavior of the horned grebe in spring. *Condor* 71:180–205.

———. 1971. The behavior of the New Zealand dabchick. *Notornis* 18:175–186.

Storer, R. W., W. R. Siegfried, and J. Kinahan. 1976. Sunbathing in grebes. *Living Bird* 14 (1975):45–57.

Stout, John F., C. R. Wilcox, and L. E. Creitz. 1969. Aggressive communication in *Larus glaucescens*. I. Sound communication. *Behaviour* 34:29–41.

Stout, J. F., and M. E. Brass. 1969. Aggressive communication by *Larus glaucescens*. II. Visual communication. *Behaviour* 34:42–54.

Strongman, K. T., and B. G. Champness. 1968. Dominance hierarchies and conflict in eye contact. *Acta Psychologica* 28:376–386.

Struhsaker, T. T. 1967. Auditory communication among vervet monkeys *Cercopithecus aethiops*). In S. A. Altmann, ed., *Social communication among primates*. Chicago: University of Chicago Press.

Symons, D. 1974. Aggressive play and communication in rhesus monkeys (*Macaca mulatta*). *American Zoologist* 14:317–322.

Thielcke, G. 1962. Versuche mit Klangattrapen zur Klärung der Verwandtschaft der Baumläufer *Certhia familiaris* L., *C. brachydactyla* Brehm und *C. americana* Bonaparte. *Journal für Ornithologie* 103:266–271.

———. 1966. Ritualized distinctiveness of song in closely related sympatric species. *Philosophical Transactions of the Royal Society of London* 251:493–497.

———. 1969. Geographic variation in bird vocalizations. In R. A. Hinde, ed., *Bird vocalizations*. Cambridge, England: Cambridge University Press.

———. 1970. Die sozialen Funktionen der Vogelstimme. *Vogelwarte* 25:204–229.

———. 1971. Versuche zur Kommunikation und Evolution der Angst-, Alarm- und Rivalenlaute des Waldbaumläufers (*Certhia familiaris*). *Zeitschrift für Tierpsychologie* 28:505–516.

Thompson, T. I. 1966. Operant and classically conditioned aggressive behaviour in Siamese fighting fish. *American Zoologist* 6:629–641.

Thompson, W. R., J. Meinwald, D. Aneshansley, and T. Eisner. 1972. Flavonols: pigments responsible for ultraviolet absorption in nectar guide of flower. *Science* 177:528–530.

Thönen, W. 1962. Stimmgeographische, ökologische und verbreitunsgeschlictliche Studien über die Mönchmeise (*Parus montanus* Conrad). *Ornithologische Beobachter* 59:101–172.

Thorpe, W. H. 1958. The learning of song patterns by birds, with special reference to the song of the chaffinch, *Fringilla coelebs*. *Ibis* 100:535–570.

————. 1961. *Bird song: The biology of vocal communication and expression in birds.* Cambridge, England: Cambridge University Press.

————. 1967a. Animal vocalization and communication. In C. H. Milliken and F. L. Darley, eds., *Brain mechanisms underlying speech and language.* New York: Grune and Stratton.

————. 1967b. Vocal imitation and antiphonal song and its implications. *Proceedings of the XIVth International Ornithological Congress, 1966,* pp. 245–262.

————. 1972. Duetting and antiphonal singing in birds. Its extent and significance. *Behaviour Supplement* 18:1–197.

Thorpe, W. H., and M. E. W. North. 1966. Vocal imitation in the tropical Bou-bou shrike *Laniarius aethiopicus major* as a means of establishing and maintaining social bonds. *Ibis* 108:432–435.

Tinbergen, E. A., and N. Tinbergen. 1972. Early childhood autism—an ethological approach. Advances in ethology (*Zeitschrift für Tierpsychologie Supplement*). 10:1–53.

Tinbergen, N. 1935. Field observations of east Greenland birds. I. The behaviour of the red-necked phalarope (*Phalaropus lobatus* L.) in spring. *Ardea* 24:1–42.

————. 1939a. Field observations of east Greenland birds. II. The behaviour of the snow bunting (*Plectrophenax nivalis subnivalis* (Brehm)) in spring. *Transactions of the Linnaean Society of New York* 5:1–94.

————. 1939b. On the analysis of social organization among vertebrates, with special reference to birds. *American Midland Naturalist* 21:210–234.

————. 1952. "Derived" activities; their causation, biological significance, origin and emancipation during evolution. *Quarterly Review of Biology* 27:1–32.

————. 1953a. *Social behaviour in animals.* London: Methuen.

————. 1953b. *The herring gull's world.* London: Collins.

————. 1958. *Curious naturalists.* New York: Basic Books.

————. 1959a. Behaviour, systematics, and natural selection. *Ibis* 101:318–330.

————. 1959b. Comparative studies of the behaviour of gulls (Laridae): A progress report. *Behaviour* 15:1–70.

————. 1959c. Einige Gedanken über Beschwichtigungs-Gebaerden. *Zeitschrift für Tierpsychologie* 16:651–665. Translated and reprinted in N. Tinbergen. 1972. *The animal in its world.* Cambridge: Harvard University Press.

————. 1962. The evolution of animal communication—a critical examination of methods. *Symposium of the Zoological Society of London* 8:1–6.

————. 1963. On the aims and methods of ethology. *Zeitschrift für Tierpsychologie* 20:410–433.

————. 1964. Aggression and fear in the normal sexual behaviour of some animals. In I. Rosen, ed., *The pathology and treatment of sexual deviation.* Oxford: Oxford University Press.

————. 1965. Some recent studies of the evolution of sexual behavior. In F. A. Beach, ed., *Sex and behavior*. New York: Wiley.

————. 1972. *The animal in its world*. Cambridge, Mass.: Harvard University Press.

————. 1975. Autism, stress, and ethology. *Science* 188:405–406.

Tinbergen, N., and D. J. Kuenen. 1939. Über die auslösenden und die richtunggebenden Reizsituationen der Sperrbewegung von jungen Drosseln (*Turdus m. merula* L. und *T. e. ericetorum* Turton). *Zeitschrift für Tierpsychologie* 3:37–60.

Tinbergen, N., and M. Moynihan. 1952. Head flagging in the black-headed gull; its function and origin. *British Birds* 45:19–22.

Tinbergen, N., and A. C. Perdeck. 1950. On the stimulus situation releasing the begging response in the newly hatched herring gull chick (*Larus argentatus argentatus* Pont). *Behaviour* 3:1–39.

Trager, G. L. 1958. Paralanguage: A first approximation. *Studies in Linguistics* 13:1–12.

Trivers, R. L. 1971. The evolution of reciprocal altruism. *Quarterly Review of Biology* 46:35–57.

Tschanz, B. 1968. Trottellummen. *Zeitschrift für Tierpsychologie*, Supplement 4.

Tuck, L. M. 1960. The murres. Their distribution, populations and biology. A study of the genus *Uria. Canadian Wildlife Series* 1:1–260.

Tumlinson, J. H., C. E. Yonce, R. E. Doolittle, R. R. Heath, C. R. Gentry, and E. R. Mitchell. 1974. Sex pheromones and reproductive isolation of the lesser peachtree borer and the peachtree borer. *Science* 185:614–616.

Turner, L. W. 1973. Vocal and escape responses of *Spermophilus beldingi* to predators. *Journal of Mammalogy* 54:990–993.

————. 1966. The syntax of symbolism in an African religion. *Philosophical Transactions of the Royal Society of Britain* 251:295–304.

————. 1967. *The forest of symbols*. Ithaca, N.Y.: Cornell University Press.

Tyler, S. J. 1972. The behaviour and social organization of the New Forest ponies. *Animal Behaviour Monographs* 5(2):87–196.

van Bergeijk, W. A. 1967. Reductionism and real biology. *Science* 158:857–859.

van Hooff, J. A. R. A. M. 1962. Facial expressions in higher primates. *Symposium of the Zoological Society of London* 8:97–125.

————. 1967. The facial displays of the catarrhine monkeys and apes. In D. Morris, ed., *Primate ethology*. London: Weidenfeld and Nicolson.

————. 1972. A comparative approach to the phylogeny of laughter and smiling. In R. A. Hinde, ed., *Non-verbal communication*. Cambridge, England: Cambridge University Press.

van Lawick–Goodall, J. 1968. The behaviour of free-living chimpanzees in the Gombe stream reserve. *Animal Behaviour Monographs* 1:161–311.

van Tets, G. F. 1965. A comparative study of some social communication patterns in the Pelecaniformes. *American Ornithologists' Union Ornithological Monograph* 2:1–88.

Verner, J., and M. M. Milligan. 1971. Responses of male white-crowned sparrows to playback of recorded songs. *Condor* 73:56–64.

Vince, Margaret A. 1969. Embryonic communication, respiration and the synchronization of hatching. In R. A. Hinde, ed., *Bird vocalizations*. Cambridge, England: Cambridge University Press.

Vine, I. 1971. Risk of visual detection and pursuit by a predator and the selective advantage of flocking behaviour. *Journal of Theoretical Biology* 30:405–422.

Voipio, P. 1962. Significance of interspecific and intraspecific bird calls in the predator-prey relation. *Ornis Fennica* 39:96–102.

von Frisch, K. 1941. Über einer Schreckstoff der Fischhaut und seine biologische Bedeutung. *Zeitschrift Vergleichende Physiologie* 29:46–145.

———. 1950. *Bees: Their vision, chemical senses and language.* Ithaca, N.Y.: Cornell University Press.

———. 1964. *Aus dem Leben der Bienen.* Berlin: Springer-Verlag.

———. 1967. Honeybees: Do they use direction and distance information provided by their dancers? *Science* 158:1073–1076.

———. 1974. Decoding the language of the bee. *Science* 185:663–668.

Waldron, Ingrid. 1964. Courtship sound production in two sympatric sibling *Drosophila* species. *Science* 144:191–193.

Ward, P. 1965. Feeding ecology of the black-faced Dioch *Quelea quelea* in Nigeria. *Ibis* 107:173–214.

Ward, P., and A. Zahavi. 1973. The importance of certain assemblages of birds as "information centres" for food-finding. *Ibis*:517–534.

Ward, R. 1966. Regional variation in the song of the Carolina chickadee. *Living Bird* 5:127–150.

Wardwell, E. 1960. Children's reactions to being watched during success and failure. Ph.D. dissertation, Cornell University, Ithaca, N.Y.

Warner, G. F. 1970. Behaviour of two species of grapsid crab during intraspecific encounters. *Behaviour* 3:9–19.

Waters, R. S., and W. A. Wilson, Jr. 1976. Speech perception by rhesus monkeys: The voicing distinction in synthesized labial and velar stop consonants. *Perception and Psychophysics* 19:285–289.

Watton, D. G., and M. H. A. Keenleyside. 1974. Social behaviour of the artic ground squirrel, *Spermophilus unulatus. Behaviour* 51:77–99.

Weaver, W. 1949. Recent contributions to the mathematical theory of communication. In C. E. Shannon and W. Weaver, eds., *The mathematical theory of communication*. Urbana: University of Illinois Press.

Weber, I. 1973. Tactile communication among free-ranging langurs. *American Journal of Anthropology* 38:481–486.

Weeden, J. S. and J. B. Falls. 1959. Differential response of male ovenbirds to recorded songs of neighboring and more distant individuals. *Auk* 76: 343–351.

Weidmann, U., and J. Darley. 1971. The role of the female in the social display of mallards. *Animal Behaviour* 19:287–298.

Weitz, Shirley. 1974. *Nonverbal communication*. New York: Oxford University Press.

Wells, M. J., and S. K. L. Buckley. 1972. Snails and trails. *Animal Behaviour* 20:345–355.

Wells, M. J., and J. Wells. 1972. Sexual displays and mating of *Octopus vulgaris* Cuvier and *O. Cyanea* Gray and attempts to alter performance by manipulating the glandular condition of the animals. *Animal Behaviour* 20:293–308.

Wenner, A. M. 1959. The relationship of sound production during the waggle dance of the honey bee to the distance of the food source. *Bulletin of the Entomological Society of America* 5:142.

————. 1967. Honey bees: Do they use the distance information contained in their dance maneuver? *Science* 155:847–849.

Wenner, A. M., P. H. Wells, and D. L. Johnson. 1969. Honey bee recruitment to food sources: Olfaction or language? *Science* 164:84–86.

White, S. J. 1971. Selective responsiveness by the gannet (*Sula bassana*) to played-back calls. *Animal Behaviour* 19:125–131.

Whitten, W. K. 1966. Pheromones and mammalian reproduction. In A. McLaren, ed., *Advances in reproductive physiology, 1*. New York: Academic Press.

Whitten, W. K., F. H. Bronson, and J. A. Greenstein. 1968. Estrus-inducing pheromone of male mice: Transport by movement of air. *Science* 161:584–585.

Wickler, W. 1968. *Mimicry in plants and animals*. New York: McGraw-Hill.

————. 1972a. *The sexual code*. New York: Doubleday.

————. 1972b. Aufbau und Paarspezifität des Gesangduettes von *Laniarius funebris* (Aves, Passeriformes, Laniidae). *Zeitschrift für Tierpsychologie* 30:464–476.

Wiener, M., S. Devoe, S. Rubinow, and J. Geller. 1972. Nonverbal behavior and nonverbal communication. *Psychological Review* 79:185–214.

Wiener, N. 1967. *The human use of human beings*. New York: Avon Books.

Wieser, W. 1967. Reductionism and real biology. *Science* 158:859–861.

Wilcox, R. S. 1972. Communication by surface waves: Mating behavior of a water strider (Gerridae). *Journal of Comparative Physiology* 80:255–266.

Wiley, R. H. 1971. Song groups in a singing assembly of little hermits. *Condor* 73:28–35.

————. 1973. The strut display of male sage grouse: a "fixed" action pattern. *Behaviour* 47:129–152.

————. 1975. Multidimensional variation in an avian display: Implications for social communication. *Science* 190:482–483.

————. 1976a. Affiliation between sexes in common grackles. I. Specificity seasonal progression. *Zeitschrift für Tierpsychologie* 40:59–79.

————. 1976b. Affiliation between sexes in common grackles, II. Spatial and vocal coordination. *Zeitschrift für Tierpsychologie* 40:244–264.

————. in press. Communication and spatial relationships in a colony of common grackles. *Animal Behaviour*.

Williams, G. C. 1964. Measurement of consociation among fishes. *Michigan State University Museum Publications* 2:349–384.

———. 1975. Sex and evolution. Princeton: Princeton University Press.

Williams, H. W., A. W. Stokes, and J. C. Wallen. 1968. The food call and display of the bobwhite quail (*Colinus virginianus*). *Auk* 85:464–476.

Willis, E. O. 1963. Is the zone-tailed hawk a mimic of the turkey vulture? *Condor* 65:313–317.

———. 1967. The behavior of bicolored antbirds. *University of California Publications in Zoology* 79:1–127.

Willis, E. O. 1972. The behavior of plain-brown woodcreepers, *Dendrocincla fuliginosa*. *Wilson Bulletin* 84:377–420.

Wilson, E. O. 1959. Communication by tandem running in the ant genus *Cardiocondyla*. *Psyche* 66:29–34.

———. 1962a. Chemical communication among workers of the fire ant *Solenopsis saevissima* (Fr. Smith). 2. An information analysis of the odour trail. *Animal Behaviour* 10:148–158.

———. 1962b. Chemical communication among workers of the fire ant *Solenopsis saevissima* (Fr. Smith). 3. The experimental induction of social responses. *Animal Behaviour* 10:159–164.

———. 1962c. Chemical communication among workers of the fire ant *Solenopsis saevissima* (Fr. Smith). 1. The organization of mass-foraging. *Animal Behaviour* 10:134–147.

———. 1965. Chemical communication in the social insects. *Science* 149:1064–1071.

———. 1968. Chemical Systems. In T. A. Sebeok, ed., *Animal communication*. Bloomington: University of Indiana Press.

———. 1971a. *The insect societies*. Cambridge, Mass.: Harvard University Press.

———. 1971b. The prospects for a unified sociobiology. *American Scientist* 59:400–403.

Wilson, E. O., and W. H. Bossert. 1963. Chemical communication among animals. *Recent Progress in Hormone Research* 19:673–716.

Wilson, J. R., Norman Adler, and B. LeBoeuf. 1965. The effects of intromission frequency on successful pregnancy in the female rat. *Proceedings of the National Academy of Sciences* 53:1392–1395.

Wilson, M. 1972. Assimilation and contrast effects in visual discrimination by rhesus monkey. *Journal of Experimental Psychology* 93:279–282.

Wilson, S. C., and D. G. Kleiman. 1974. Eliciting play: A comparative study. *American Zoologist* 14:341–370.

Wilz, K. J. 1970. Causal and functional analysis of dorsal pricking and nest activity in the courtship of the three-spined stickleback *Gasterosteus aculatus*. *Animal Behaviour* 18:115–124.

Winter, P., D. Ploog, and J. Latta. 1966. Vocal repertoire of the squirrel monkey (*Saimiri sciureus*), its analysis and significance. *Experimental Brain Research* 1:359–384.

Wolf, L. L., and F. R. Hainsworth. 1971. Time and energy budgets of territorial hummingbirds. *Ecology* 52:980–988.

Wolfheim, J. H., and T. E. Rowell. 1972. Communication among captive talapoin monkeys (*Miopithecus talapoin*). *Folia primatologica* 18:224–255.

Wooton, R. J. 1971. Measures of the aggression of parental male three-spined sticklebacks. *Behaviour* 40:228–262.

————. 1972. The behaviour of the male three-spined stickleback in a natural situation: A quantitative description. *Behaviour* 41:232–241.

Young, R. E., and C. F. E. Roper. 1976. Bioluminescent countershading in midwater animals: Evidence from living squid. *Science* 191:1046–1048.

Zajonc, R. B. 1966. *Social psychology: An experimental approach.* Belmont, Calif.: Brooks/Cole.

Zippelius, H. M., and W. Schleidt. 1956. Ultraschall-Laute bei jungen Mäuse. *Naturwissenschaften* 43:502.

Zoloth, S. 1975. Arousal from sleep in the female rat: Effects of stimulus relevance and motivational state. Ph.D. dissertation, University of Pennsylvania.

Zucker, Naida. 1974. Shelter building as a means of reducing territory size in the fiddler crab, *Uca terpsichores* (Crustacea: Ocypodidae). *American Midland Naturalist* 91:224–236.

Index

Abortion, 257
Access, state of, 446
Accommodation, mutual, of participants, 445
Acetate, tetradecenyl, 325
Acids
  aliphatic, 253
  carboxylic, 38
Adaptive behavior, 3, 6, 9, 47, 70
Adaptiveness, of message assortment, 179
Adler, N. T., 35, 190
Adornment, of self, 250
Advantages, adaptive, 271
Advertising behavior, 164, 165, 166, 173, 430
Affect, 220–222
Affix, 188, 419
Age, 161, 163, 178, 230–231, 275, 298, 361
Agility, 345
Agonistic behavior, 52, 61, 68, 94, 102, 104, 125, 148, 430
Agonistic events
  functions of displays, 295–299
  meanings of displays, 295–299
Ainley, D., 26, 83, 89, 108, 109, 115, 155, 400
Ainsworth, M. D. S., 153
Alarm calls, 71, 72, 157, 167, 181, 232, 268–270, 283, 377
Albatrosses (Diomedeidae), 35
Albert, S., 455

Alexander, B. K., 161
Alexander, R. D., 33, 37, 285, 292, 301, 359, 367, 369
Allogrooming, 35, 123, 124, 146, 147, 150, 178, 254, 270, 294, 296, 334
Allokine, 337
Allomarking, 254
Allopreening, 35, 36, 123, 124, 146, 178, 294, 334
Alternative acts, incompatible, 75, 113
  range, 114
Altmann, S. A., 188, 217, 227, 231, 236–238, 261, 274
Ambiguity, 173, 184, 328, 353, 421
Ambrose, J. A., 398
Amphibians, 326, 363
Amplexus behavior, 97, 150, 183
Amplification of speech, 148
Amplitude, of sound, 167
Analyses
  costs-benefits, by honeybees, 152
  of display behavior, 192–223
  electrophoretic, 160
  of function, 148
  of information, 148
  levels, 194–195
  message oriented, 193, 199–207
  motivational, 177, 194–195, 199–207
  superordinate units, 195
Anatomy, comparative, 196
Anderson, P. W., 195
Andersson, M., 129, 272–273, 346
Andrew, R. J., 90, 91, 111, 170, 197,

THE GRANGER ANTHOLOGY
Series I

The World's Best Poetry

The World's Best Poetry

THE GRANGER ANTHOLOGY
SERIES I VOLUME IX

TRAGEDY AND HUMOR